MOLECULAR EVOLUTION
Prebiological and Biological

SIDNEY W. FOX

A Volume Commemorating the Sixtieth Birthday of
SIDNEY W. FOX

MOLECULAR EVOLUTION
Prebiological and Biological

Edited by
Duane L. Rohlfing
Department of Biology
University of South Carolina
Columbia, South Carolina

and

A. I. Oparin
A. N. Bakh Institute of Biochemistry
Academy of Sciences of the USSR
Moscow, USSR

ℙ PLENUM PRESS • NEW YORK - LONDON • 1972

Library of Congress Catalog Card Number 72-93444

ISBN 978-1-4684-2021-0   ISBN 978-1-4684-2019-7 (eBook)
DOI 10.1007/978-1-4684-2019-7

© 1972 Plenum Press
Softcover reprint of the hardcover 1st edition 1972

A Division of Plenum Publishing Corporation
227 West 17th Street, New York, N.Y. 10011

United Kingdom edition published by Plenum Press, London
A Division of Plenum Publishing Company, Ltd.
Davis House (4th Floor), 8 Scrubs Lane, Harlesden, NW10 6SE,
London, England

LIST OF CONTRIBUTORS

(Numbers in parentheses indicate the pages
on which the authors' contributions begin.)

AKABORI, S., Protein Research Foundation, Ina, Minoo, Osaka, Japan (189).

BARKER, W., National Biomedical Research Foundation, Georgetown University Medical Center, Washington, D. C. 20007 (297).

BERNSTEIN, W., Laboratory of Chemical Biodynamics and Lawrence Radiation Laboratory, University of California, Berkeley, California 94720 (151).

BOGATYREVA, S., A. N. Bakh Institute of Biochemistry, Academy of Sciences of the USSR, Moscow, USSR (291).

BRICTEUX-GRÉGOIRE, S., Laboratoire de Biochimie, Institut Léon Fredericq, Université de Liège, Belgique (319).

CALVIN, M., Laboratory of Chemical Biodynamics and Lawrence Radiation Laboratory, University of California, Berkeley, California 94720 (151).

DAYHOFF, M., National Biomedical Research Foundation, Georgetown University Medical Center, Washington, D. C. 20007 (297).

DEBORIN, G., A. N. Bakh Institute of Biochemistry, Academy of Sciences of the USSR, Moscow, USSR (343).

DOSE, K., Max-Planck-Institut fur Biophysik, Frankfurt (Main), Germany (199).

EVREINOVA, T., Department of Plant Biochemistry and Biophysical Institute, Academy of Sciences of the USSR, Moscow, USSR (361).

FLORKIN, M., Laboratoire de Biochimie, Institut Léon Fredericq, Université de Liège, Belgique (319).

FOUCHE, C., Department of Biology, University of South Carolina, Columbia, South Carolina 29208 (219).

FOX, J., Division of Biophysics and Physiology, Department of Zoology, University of Texas at Austin, Austin, Texas 78712 (23).

FOX, R., Department of Physics, University of California, Berkeley, California 94720 (79).

FOX, T., Department of Neuropathology, Harvard Medical School, Boston, Massachusetts 02115 (35).

HARADA, K., Institute for Molecular and Cellular Evolution & Department of Chemistry, University of Miami, Coral Gables, Florida 33134 (157).

HARDEBECK, H., Institut fur Anatomie und Physiologie der Haustiere, Universitat Bonn, Germany (233).

HAYAKAWA, T., Institute of High Polymer Research, Faculty of Textile Science and Technology, Shinshu University, Ueda, Japan (247).

HEINRICH, M., Division of Planetary Sciences, Ames Research Center, Moffett Field, California 94035 (331).

HSU, L., Institute for Molecular and Cellular Evolution, University of Miami, Coral Gables, Florida 33134 (371).

HUNT, L., National Biomedical Research Foundation, Georgetown University Medical Center, Washington, D. C. 20007 (297).

JUNGCK, J., Department of Biology, Merrimack College, North Andover, Massachusetts 01845 (101).

KARNAUKHOV, V., Department of Plant Biochemistry and Biophysical Institute, Academy of Sciences of the USSR, Moscow, USSR (361).

KEOSIAN, J., Department of Zoology and Physiology, Rutgers University (Professor Emeritus), The State University of New Jersey (9).

KRASNOVSKY, A., A. N. Bakh Institute of Biochemistry, Academy of Sciences of the USSR, Moscow, USSR (141).

LACEY, J., Laboratory of Molecular Biology, University of Alabama in Birmingham, Alabama 35233 (171).

LEMMON, R., Laboratory of Chemical Biodynamics and Lawrence Radiation Laboratory, University of California, Berkeley, California 94720 (151).

LIPMANN, F., The Rockefeller University, New York, New York 10021 (261).

LOZOVAYA, G., Institute of Botany, Academy of Sciences of the Ukrainian SSR, Kiev, Ukrainian SSR (353).

MAMONTOVA, T., Department of Plant Biochemistry and Biophysical Institute, Academy of Sciences of the USSR, Moscow, USSR (361).

McLAUGHLIN, P., National Biomedical Research Foundation, Georgetown University Medical Center, Washington, D. C. 20007 (297).

MUELLER, G., Institute for Molecular and Cellular Evolution, University of Miami, Coral Gables, Florida 33134 (379).

MULLINS, D., Laboratory of Molecular Biology, University of Alabama, Birmingham, Alabama 35233 (171).

OKAWARA, T., Department of Chemistry, University of Miami, Coral Gables, Florida 33134 (157).

OPARIN, A., A. N. Bakh Institute of Biochemistry, Academy of Sciences of the USSR, Moscow, USSR (1, 343).

OSHIMA, T., Department of Agricultural Chemistry, Faculty of Agriculture, University of Tokyo, Tokyo, Japan (399).

PAULING, L., Department of Chemistry, Stanford University, Stanford, California 94302 (113).

PIRIE, N., Department of Biochemistry, Rothansted Experimental Station, Harpenden, Herts., Britain (67).

PRICE, C., Department of Chemistry, University of Pennsylvania, Philadelphia, Pennsylvania 19104 (461).

RAUCHFUSS, H., Max-Planck-Institut fur Biophysik, Frankfurt (Main), Germany (199).

ROHLFING, D., Department of Biology, University of South Carolina, Columbia, South Carolina 29208 (1, 219).

ROSENTHAL, S., Institute for Experimental Therapeutics, University of Pennsylvania School of Medicine, Philadelphia, Pennsylvania 19104 (271).

SCHULTZ, J., Papanicolaou Cancer Research Institute, 1155 N. W. 14th Street, Miami, Florida 33136 (271).

SCHWARTZ, A., Department of Exobiology, University of Nijmegen, Nijmegen, The Netherlands (129).

SEREBROVSKAYA, K., A. N. Bakh Institute of Biochemistry, Academy of Sciences of the USSR, Moscow, USSR (353).

SPIRIN, A., A. N. Bakh Institute of Biochemistry, Academy of Sciences of the USSR, Moscow, USSR (291).

SZENT-GYORGYI, A., The Institute for Muscle Research, Marine Biological Laboratory, Woods Hole, Massachusetts 02543 (111).

TRIFONOV, E., A. N. Bakh Institute of Biochemistry, Academy of Sciences of the USSR, Moscow, USSR (291).

UMRIKHINA, A., A. N. Bakh Institute of Biochemistry, Academy of Sciences of the USSR, Moscow, USSR (141).

VEGOTSKY, A., Department of Biology, Wells College, Aurora, New York 13026 (449).

WELCH, C., Department of Biology, Macalester College, Saint Paul, Minnesota 55101 (443).

WOOD, A., Waite Agricultural Research Institute, The University of Adelaide, Glen Osmond, South Australia 50614 (233).

YAMAMOTO, H., Institute of High Polymer Research, Faculty of Textile Science and Technology, Shinshu University, Ueda, Japan (247).

YAMAMOTO, M., Protein Research Foundation, Ina, Minoo, Osaka, Japan (189).

YANOPOLSKAYA, N., A. N. Bakh Institute of Biochemistry, Academy of Sciences of the USSR, Moscow, USSR (343).

YOUNG, R., Headquarters, National Aeronautics and Space Administration, Washington, D. C. 20546 (43).

YUYAMA, S., Department of Zoology, University of Toronto, Toronto, Ontario, Canada (425).

ZUCKERKANDL, E., Department de Biochimie Macromoleculaire, Centre de Recherches Biophysiques et Biochimiques, Du C. N. R. S., 34-Montpellier, France (113).

PREFACE

For much of his professional career, Sidney W. Fox has devoted his thought and research to studies of molecular evolution. MOLECULAR EVOLUTION: PREBIOLOGICAL AND BIOLOGICAL is a dedicatory volume of thirty-five contributed papers commemorating, on the occasion of his sixtieth birthday, his many achievements.

The volume had its conception in the USSR (by AIO), had much of its development in the USA (by DLR), and was made possible by the enthusiastic responses and encouragements of fifty-eight contributors from ten nations and many disciplines. These numbers connote not only the esteem in which S. W. Fox is regarded, but also the international and interdisciplinary nature of studies of molecular evolution.

The term "molecular evolution" is often associated with abiotic or prebiotic evolution; it is also used to denote processes of biotic evolution at the molecular level. The point of merger of these two sub-areas, at "life," represents but one stage (albeit a very important one) in the total process of the evolution of matter, from hydrogen to Homo sapiens and beyond. This volume considers aspects of molecular evolution in this broader sense. Accordingly, the contributors include persons experienced in the prebiological and also the biological aspects of molecular evolution; several "outside" viewpoints are provided by persons whose principal interests lie in other disciplines. The contributions are both experimental and theoretical. (Within those major sections of the volume where both prebiological and biological aspects are represented, papers dealing with the prebiological precede those emphasizing the biological.)

In a somewhat similar volume of symposium papers published in 1965 (THE ORIGINS OF PREBIOLOGICAL SYSTEMS, S. W. Fox, editor), the papers concerned with the micromolecular level of organization were more numerous than those of any other single division. In this volume, four papers deal solely with the micromolecular level, whereas twelve are concerned with macromolecules, seven with cells or pro-

tocells, and others with more than one level of organization. This re-
proportioning (which connotes nothing regarding quality of papers) perhaps
epitomizes an evolution of experiments and thought, concerning prebiotic
aspects, towards more complex levels of organization and with emphasis
on integration of structures and functions.   The variegated nature of the
current contributions, when considered in toto, affords much opportunity
for integration of ideas; indeed, a major objective of this volume is to
stimulate new ideas and new research. With this goal especially in mind,
the editors are proud and honored to present this volume as a dedication
to Sidney W. Fox.

                                                              D. L. R.
                                                              A. I. O.

ACKNOWLEDGMENTS

All papers were reviewed by persons who are not contributors to this volume;* the editors express their appreciation to these many persons and regret that they cannot be recognized by name. We also appreciate the many helpful suggestions and well-wishes of numerous persons who were unable to contribute a paper to the volume. Thanks are due to Linda Baccene for preparation of the justified typescript and to Kathleen T. Poole for help with the index.

---

*The editors do not know whether the reviewers of their own papers were also contributors.

TABLE OF CONTENTS

MAN AND EVOLUTION

MOLECULAR EVOLUTION AND SIDNEY W. FOX

A. I. Oparin and D. L. Rohlfing

Bakh Institute of Biochemistry
Academy of Sciences of the USSR, Moscow, USSR
and
Department of Biology
University of South Carolina, Columbia, S.C. 29208

Three indissolubly linked fundamental problems have attracted the human intellect for centuries. These problems are the essence of life, the origin of life, and the dispersal of life through the Universe.

Life is the most beautiful and, after all, most important among all things present on our planet. The scientific elucidation of the mysteries of living nature is of decisive importance for both theoretical generalizations and for all those branches of practical activity where mankind deals with living organisms--medicine, agriculture, etc.

The deeper the human mind penetrates into the mysteries of life, the greater is the potential for healthy, fruitful, and happy human life. The understanding of the essence of life and of the mode of organization of living matter is thus full of obvious and overpowering fascination for human beings. This understanding, however, is impossible without a knowledge of the origin of life. "The major practical reason for studying the origin of life is that we cannot understand current life without this study and, if we cannot understand it, we cannot control it. " (J. Bernal)

At the present time, we only know life by one singular example, i.e., life on Earth. From it, we must formulate our opinions about the other possible forms of biological organization. Hence, the Earth and the events that took place on it must to a considerable degree serve as a model for our paramount ideas about life in the Universe.

The emergence of life on the Earth was at one time considered to be the result of chance; this view could contribute nothing to a theory concerning life beyond our planet.   However, we now have a basis for considering the emergence of life not as "a lucky happenstance," but as a phenomenon completely regulated by natural laws and an integral and inalienable part of the general evolutionary development of our planet (cf. the paper by Pauling and Zuckerkandl in this volume).

The search for life beyond the Earth (cf. the paper by Young in this volume) is only a part of the more fundamental question concerning the origin of life in the Universe.   The study of the origin of life on Earth presents itself as the investigation of just one example of events that must have occurred in the Universe countless numbers of times.   Therefore, the explanation of the manner in which life arose on the Earth must of necessity give strong evidence in favor of its existence elsewhere in the Universe.

For these reasons, today, in the second half of the twentieth century, the origin of life is broadly studied by scientists of many countries and disciplines.   The works of Sidney W. Fox hold an especially honored place in this scientific cooperation.

Sidney Walter Fox was born on March 24, 1912, in Los Angeles, California.   On the occasion of his sixtieth birthday, the several contributors to this volume recognize and honor him and his achievements.

Dr. Fox received his B. A. degree at the University of California at Los Angeles in 1933 and his Ph.D. degree in biochemistry, under H. M. Huffman and T. H. Morgan, from California Institute of Technology in 1940.   His first scientific article, "The Synthesis of Aspartic Acid" (coauthored with M. S. Dunn), appeared in 1933.   Most of his early research, conducted largely at Iowa State College, was concerned with amino acids and proteins, but without emphasis on origins or evolution.   Some accomplishments of this era include: investigations of enzymatic peptide-bond synthesis; studies of thermodynamic properties of amino acids; the incorporation for the first time of a truly nonorganismic amino acid, p-fluorophenylalanine, into organismic protein (subsequently much developed by others); and considerations of methods for elucidating terminal residues and amino acid sequences of proteins (before the development of the techniques, e. g., by Sanger).   The significances of this early work are exemplified by the large current literature on these subjects.   This era and area of interest is epitomized by his book, with Joseph Foster, "Introduction to Protein Chemistry."

An early interest in evolution, although latent through early publications, was awakened in Dr. Fox by working directly with T. H. Morgan. This interest was further developed by discussions with a graduate student, D. E. Atkinson, and led to a second (and current) era of research activity, dating from 1953 and occurring largely at The Florida State University and the University of Miami. This era has emphasized prebiotic molecular evolution. One major area has concerned the abiotic origins of protein. A geologically relevant experimental approach has resulted from the simple heating together of amino acids, under hypohydrous conditions; polyamino acids (termed "proteinoids") that show many properties in common with contemporary protein are formed, and they are regarded as models for protoprotein. A second major area concerns the abiotic origins of protocells. This area also has a geologically relevant experimental approach that is a continuation of the first approach. Upon contact with water under appropriate conditions, proteinoids are transformed into microscopic, organized units termed "microspheres." Microspheres mimic processes of contemporary cells (e.g., selective absorption of material, division, budding, "communication"--see the paper by Hsu in this volume) and are regarded as models for protocells. The model differs from other models in that the only material (proteinoid) needed for the formation of microspheres is itself the product, not of biosyntheses, but of other simulated prebiotic syntheses.

These two major areas, extensively explored by S. W. Fox, are part of an evolutionary continuum (or continuality, as used by T. O. Fox in this volume) of increasing complexity: simple compounds (gases, e.g.,) → amino acids → proteinoids → protocells → associations of protocells and of protocellular functions. S. W. Fox has investigated each stage of the scheme, and his efforts provide plausible and experimentally documented answers for such quandaries as: 1) the origin of proteins in the absence of proteins; 2) the origin of order in proteins when no large molecules and no code existed; 3) the origin of enzymes when no enzymes to make them existed; 4) the origin of metabolism in the absence of metabolizing cells; 5) the origin of membranes when no membrane-containing microsystems existed; 6) the origin of cells when no cells existed to produce them; and 7) the origin of reproduction. Dr. Fox's constructionistic approach is documented in approximately 50 papers, many reviews, and two books (1965 and in press).

The impact of the work of Dr. Fox and others on general science is evidenced by the numerous textbooks that now have sections or even chapters devoted to the origin of life (cf. the paper by Vegotsky in this volume).

Lehninger's book, "Biochemistry," perhaps epitomizes the growth and recognition of this discipline by the statement: "Not too long ago, inquiry into the origin of life was considered to be a matter of pure armchair speculation with little hope of yielding conclusive information. But many scientific advances made in the last decade have given encouragement to the view that valid answers to some of these questions may be deduced and that at least some of the steps in the origin of biomolecules and of living cells may be simulated in the laboratory." Lehninger and others document their treatises with numerous examples stemming from Dr. Fox's laboratory.

S. W. Fox has not restricted his attention to evolution at the prebiological level. In 1953, he published a paper entitled "A Correlation of Observations Suggesting a Familial Mode of Molecular Evolution as a Concomitant of Biological Evolution." As the title implies, this paper focused attention on evolution at the molecular level, at a time when such was largely ignored by biochemists. This type of thinking has been expanded by Dr. Fox in three reviews (1953, 1962, and 1965). The reviews point out, among other things, the importance of closely associating genetic and proteinous evolution with cellular evolution and the importance of evolutionary processes in prebiotic eras.

These and many other studies by Dr. Fox help indicate that the biological, as well as the prebiological, fall within the domain of molecular evolution; the title of this volume echos this concept. Much insight into investigations of the prebiological has been gained from knowledge of the biological; it is increasingly evident that the reverse situation is becoming true. In a sense, the origin of life can be regarded as one focal point of both branches of the discipline. However, as N. W. Pirie points out in this volume, there is no magic dividing line between the living and the nonliving. Accordingly, one is forced to recognize that those evolutionary processes termed for convenience "biological" are simply continuations at higher levels of complexity (and by more refined mechanisms) of processes that occurred or occur at relatively simple levels of complexity, termed for convenience "prebiological" (or, by Pirie, "probiological").

Sidney W. Fox has been the mentor of twenty-nine Ph.D. students and seventeen postdoctoral associates. No attempt is made to number his colleagues. The contributors to this volume, some former students, some current and former postdoctoral associates, some current colleagues (including three sons) hope that the several articles will not only connote the esteem in which he is regarded, but also will contribute to the understanding of molecular evolution.

One of us (DLR) was privileged to have obtained graduate degrees under Dr. Fox, at a time when prebiological molecular evolution was not widely recognized as a discipline. The laboratory environment then was exciting. The discipline yet is exciting; many hurdles--not all of them scientific--have been overcome, and many more remain to be encountered. The vigor and enthusiasm of S. W. Fox has contributed much to past endeavors and no doubt will contribute to future ones. Dr. Fox recently said, "I am sixty, I feel forty, and I think [an enthusiastic] eighteen." The editors will be satisfied if this volume will transmit to others some of this vigor and enthusiasm, to encourage inquiry into the many remaining puzzles and interrelationships of molecular evolution-- prebiological and biological.

HISTORY AND SCOPE

THE ORIGIN OF LIFE PROBLEM--A BRIEF CRITIQUE

John Keosian

Department of Zoology and Physiology
Rutgers University, Professor Emeritus
The State University of New Jersey

The probability that organic compounds could have formed abiotically on the primitive Earth has been documented over the past two decades and needs no elaboration here (1). But historically there was a time when, from the middle of the nineteenth to the middle of the twentieth centuries, it was stoutly proclaimed that organic compounds, especially proteins, could be synthesized only by living things, and consequently before there was life, there could be no organic compounds. Conversely, it was held that without organic compounds life could not originate. Eventually, the question of the origin of life was considered to be a metaphysical, not a scientific problem, one that was beyond human ken. So firmly and generally was this view held, that scientists of that period did not care nor dare to discuss the problem seriously.

The origin of life problem was further beclouded by the alleged disproof of the concept of spontaneous generation. This centuries-old concept is supposed to have been disproved by Redi (2), Harvey, and a few others. In fact, each, while disproving the spontaneous generation of a particular form, believed in the spontaneous generation of some other form or forms. Redi accepted the spontaneous generation of gall flies while Harvey maintained a belief in the spontaneous generation of insects and intestinal worms. The unsung heroes who really demolished the concept of spontaneous generation were the numerous naturalists, anatomists, and the physiologists of the seventeenth and eighteenth centuries, whose cumulative work on the anatomy, taxonomy, habitat, life cycles and reproduction, embryology and physiology of species after species of plants

9

and animals gradually dispelled the belief in the spontaneous generation of higher (than microbes) organisms. Thus, was the concept abandoned, not disproved.

The history of the alleged disproof of the spontaneous generation of microbes has followed a similar course. Pasteur (3) reviewing his experiments on spontaneous generation in his lecture at the Sorbonne in 1864 proclaimed dramatically, "and, therefore, gentlemen, I could point to that liquid and say to you, I have taken my drop of water from the immensity of creation, and I have taken it full of the elements appropriated to the development of microscopic organisms. And I wait, I watch, I question it... begging it to recommense for me the beautiful spectacle of the first creation. But it is dumb, dumb ever since these experiments were begun several years ago; it is dumb because I have kept it from the only thing that man does not know how to produce: from the germs which float in the air, from Life, for Life is a germ and a germ is Life. Never will the doctrine of spontaneous generation recover from the mortal blow of this simple experiment."

Pasteur was blind to the fact that although his media were "full of the elements appropriated to the development (nutrition?) of microscopic organisms" they did not necessarily contain substances essential for the origin of microorganisms. We are just beginning to appreciate what materials and conditions were appropriate for the origin of the first living things, and they are certainly not those that Pasteur employed. Most important, the conclusion drawn from a negative result can go no further than the observation that the method, conditions and materials employed did not give a positive result. From such experiments (and Pasteur's were of surprisingly small number and scope), one is not justified in concluding, as Pasteur (and posterity) did, that no combination of materials, conditions, and methods will give a positive result.

The issue here is not whether spontaneous generation is or was a possibility, but rather, one of experimental design and the validity of conclusions in terms of the design. The lesson to be remembered from this episode is the possible fallibility of even universally held scientific views. This oft-repeated but rarely observed caution is particularly pertinent today when, with the rapid accumulation of "facts," views are unhesitatingly propounded as dogmas.

Two major concepts compete today in the discussion and contemplation of the manner of origin of life on the primordial earth. Muller (4) has referred to these as the "primacy of the gene" versus the "primacy of protoplasm and metabolism. " By the first is meant that the gene is the basis of life and its origin. This is a development of the mechanistic approach to life of the latter part of the nineteenth century and the beginning of the twentieth century (5). The second is an outgrowth of the thinking of the materialist philosophers of the nineteenth century (6). Both the mechanistic and materialist views are naturalistic approaches to the problem. Both seek to explain life and its origin through natural laws. Each attempts also to counter and reject the vitalism that was in its ascendency during that period.

The gene theory of the origin of life has its foundations in a paper written in 1914 by L. T. Troland (7). The ideas expressed in it were extraordinary for those times, for Troland, a Harvard physicist, took a strong stand against his contemporary biologists who had embraced a vitalistic outlook on the subject of life. Troland wrote in part, "The last years of the nineteenth and the first of the twentieth century, which have constituted a period of tremendous progress in physical science, have unfortunately witnessed a recrudescence of that "cult of incompetence" in biology, vitalism. Neo-vitalism under the leadership of Driesch in Germany and of Bergson in France, asserts that the phenomena of life are not determined by law-abiding forces, but by a form of activity the effects of which are unpredictable, and which, consequently, must be regarded as chaotic and beyond the range of science. "

"It is the purpose of the present paper to combat the thesis of the new vitalism by showing how a single physico-chemical conception may be employed in the rational explanation of the very life-phenomena which the neo-vitalists regard as inexplicable on any but mystical grounds. " Troland put this "single physico-chemical conception" in these words: "Let us suppose that at a certain moment in earth-history, when the ocean waters are yet warm, there suddenly appears at a definite point in the oceanic body a small amount of a certain catalyzer or enzyme. " And further, "The original enzyme was the outcome of a chemical reaction, that is to say, it must have depended on the collision and combination of separate atoms or molecules, and it is a fact well known among physicists and chemists that the occurrence and specific nature of such collisions can be predicted only by use of the so-called laws of chance... consequently we are forced to say that the production of the original life enzyme was a chance event. " Then, "The striking fact that the enzymic theory of life's origin, as we have outlined it, necessitates the production of only a single molecule of the original catalyst, renders the objection of improbability almost absurd... and when one of these enzymes first appeared, bare of all body, in the aboriginal seas it followed as a consequence of its characteristic regulative nature that the phenomenon of life came too."

Troland wrote at a time when the few remaining mechanists were searching for evidence in support of their belief in a naturalistic basis for the origin of life. The enzyme theory of life's origin appeared to supply the answer. But the theory has flaws which seem to have escaped the attention of its adherents. Accident and the occurrence of remote possibilities play a key role in the theory. However, large molecules of even the inorganic variety do not form by the simultaneous accidental combination, all at once, of the separate atoms. They are the result of a stepwise increase in complexity. The sudden combination of all the atoms composing the 'life enzyme' is not, as Troland intimated, a remote-possibility event which was sure to take place because the synthesis of only one molecule is needed. It is a near zero-probability event. More-over, this theory, based as it is on accidents, is not susceptible to ex-perimental testing. Further, the hypothetical life-enzyme is a living thing only by Troland's say-so. He assigned the properties of autocata-lysis (self replication) and heterocatalysis (metabolic regulation), his criteria of life and living, to the life-enzyme, and then made the double assumption that a protein molecule would arise suddenly and spontane-ously, and would embody his two criteria. In his words, "...and when one of these enzymes first appeared, bare of all body, in the aboriginal seas it followed as a consequence of its 'regulative nature that the phe-nomenon of life came too.'" (my italics)

But granting, for the sake of argument, the appearance of such an enzyme, what would be the result of its presence? Limiting ourselves to the two "regulative" properties, it would, through autocatalysis, catalyze the specific combination of elements that produced it, in the first place, to the exclusion of all other random permutations. Rapidly (the rate pro-gresses exponentially) as many molecules of the enzyme would form as the amount of the limiting constituent element would permit. As for hetero-catalysis, there was nothing in the "aboriginal seas" of Troland that would require or permit metabolic regulation. In later articles (8, 9) Troland favored nucleic acid as the composition of the life enzyme in place of pro-tein with no substantial change in details.

As long as very little was known about biochemistry and nothing at all about prebiotic chemical evolution, one could fill the void with broad and logical sounding assertions. It is no longer permissible to dismiss the problem of how a single auto- and heterocatalytic molecule can serve as the ancestor of all future life. The theory is an origin of life by edict and is thus akin to the vitalistic view.   In fact, it is a form of vitalism disguised in mechanistic terms.

The first detailed statement of the gene theory of life was made by Muller in an address before the International Congress of Plant Sciences in 1926. It was published in 1929 in the Proceedings of that Congress (10) The gene theory of the origin of life was lengthily defended by Muller in 1966 (4).   Earlier, Muller had substituted the naked gene for Troland's enzyme "bare of all body" and had added mutability to the two properties of autocatalysis and heterocatalysis.   His theory, however, borrows all of the fatal faults of Troland's theory. A naked gene is alive only because Muller so argues. The gene theory of the origin of life is also untestable for the same reason that a meaningful experiment can not be devised to test an event that depends on remote accidents.   And an accidental com- bination of atoms to form a functional gene is also a near zero-probability event. Muller attempted to sidestep this shortcoming by substituting nu- cleotides for elements in the accidental appearance of a gene on the primi- tive Earth. But in the absence of specific enzymes, the accidental simul- taneous coming together of thousands of nucleotides to form a functional molecule of DNA is no more believable. Moreover, the demonstration of the possible abiotic synthesis of nucleotides on the primitive Earth em- ploys methods and conditions that are hardly compatible with what is known concerning primitive Earth conditions.   Even at that, yields are unim- pressive.

Again, granting for the sake of argument, the all-at-once appearance of a functional gene on the primordial Earth, what are the possibilities of its serving as the ancestor of all life? Ignoring all limitations, it would serve as a template for synthesis of more of itself at an exponential rate; it would mutate, and mutant genes would similarly increase in number and themselves mutate; original and each mutant variety would be respon- sible for the synthesis of a particular protein for which it is coded; and, finally, each protein would be specific for the catalysis of some reaction. Trying to stay within the confines of these events, it is difficult to see what else would occur beside the synthesis, in a short time, of an ocean- ful of the original gene, lesser quantities of various mutants, and amounts of the corresponding proteins. All of this overlooks the requirement for specific enzymes in the replication of DNA, and an apparatus, RNA, and enzymes for the synthesis of proteins.

The gene theory is thus also an origin of life theory by edict. There is also an element of mysticism in it.   The gene has been made the re- pository of all of the mystical powers of the "elan vital" of the vitalists.

The gene alone is equated with life. All else--the cytoplasm of the cell, the tissues, organs and organ-systems of higher forms--are relegated to the role of appertenances devised by the gene for its survival and perpetuation, to be discarded when their purpose has been served, while the gene goes on to a new abode of its own designing.

Muller's lengthy defense of the gene theory largely overlooks what has been termed "chemical evolution" stemming from highly productive world-wide experimental activity on the part of numerous competent investigators inspired by the testable alternative hypothesis of Oparin (11). It was therefore, perhaps out of desperation that he characterized that hypothesis, wrongly identified as "the primacy in life of protoplasm and metabolism" and its author in these terms: "It is a curious anachronism, however, that even today some of the most eminent biochemists and biologists, doing very valuable work in their respective fields, still adhere to this view and its corollary concerning life's origin. Unfortunately, it became much publicized and elaborated, beginning in the 1930's by the Lysenkoist Oparin in his book The Origin of Life (1938 et seq.), as part of an attempt to down-rate the significance of genetics. His part of that attempt was most subtly carried out" (reference 4, p. 494). By 1966 Oparin's hypothesis had been put to the test for more than a dozen years with marked success by increasing numbers of workers from various fields, attracted by the soundness and implications of the hypothesis. That hypothesis does not exclude a nucleic acid approach to the origin of life problem, as witness Oparin's own use of nucleic acids as a component of the materials out of which he constructs some of his coacervates. The hypothesis is so broad, indeed, that it has supplied the first clues for an experimental approach to the problem by gene-oriented scientists.

The materialist theory of the origin of life from inanimate beginnings recognizes the role of chance in the interactions of matter in the universe, but views the overall development as in no way accidental; on the contrary it is looked upon as an inevitable, almost inexorable, outcome of the emergence and operation of natural laws. Bastian (12) in 1872 expressed this sketchily in chemical terms. He penetratingly saw the shortcomings of Pasteur's experiments and maintained a belief in the spontaneous generation of microorganisms. That doctrine had been so discredited that Bastian's work (which had errors of its own) as well as his thinking were ignored or rejected. Referring to the problem of origins, he wrote: "We know that the molecules of elementary or mineral substances combine to form acids and bases by virtue of their own 'inherent' tendencies; that these acids and bases unite so as to produce salts, which, in their turn, will often again combine and give rise to 'double-salts'. And at each stage in this series of ascending molecular complexities, we find the products endowed with properties wholly different from those of their constituents. Similarly, amongst the carbon compounds there is abundance of evidence to prove the existence of internal tendencies or molecular properties, which may and do lead to the evolution of more and more complex chemical compounds. And it is such synthetic processes occurring amongst the molecules of colloidal and allied substances, which seems so often to engender or give 'origin' to a kind of matter possessing that subtle combination of properties to which we are accustomed to apply the epithet 'living'."

In a more general vein, Schafer (13) in a presidential address to the British Association for the Advancement of Science in 1912, said, "Setting aside as devoid of scientific foundation the idea of immediate supernatural intervention in the first production of life, we are not only justified in believing, but compelled to believe, that living matter must have owed its origin to causes similar in character to those which have been instrumental in producing all other forms of matter in the universe, in other words, to a process of gradual evolution." Of that address, Moore (14), in the same year, wrote, "The great merit of Schafer's Presidential Address (1912) to the British Association of Dundee lies in this, that it has once more centered the attention of the scientific world on the main inquiry, and marked it out as a problem that may be solved and one demanding experimental enquiry. Life probably arose as a result of the operation of causes which may still be at work today causing life to arise afresh. Although Pasteur has conclusively proven that life did not originate in certain ways, that does not exclude the view that it arose in other ways. The problem is one that demands thought and experimental work, and is not an exploded chimera. Therein lies the value of Schafer's contribution to the question, and it is a most refreshing and valuable one."

These views came in the midst of a long period when most scientists had given up the problem of origin as hopeless. The belief in a naturalistic explanation remained only as a spark in a few quarters for the ensuing four decades. One of those sparks (in addition to Troland and Muller) was a small treatise in Russian published in 1924 by the Russian biochemist, A. I. Oparin, and entitled The Origin of Life. The book, in turn, was an expansion of an address he had delivered a few years earlier. During the ensuing years, Oparin gathered supporting evidence for his hypothesis from various fields of science and, in 1936, published his now familiar hypothesis under the title Origin of Life (11). Briefly, he contended that the atmosphere of the primitive Earth contained the simple gases methane, ammonia, hydrogen, and water vapor, but lacked oxygen. He went on to elaborate how the interaction among these and their products would lead to the formation of more and more complex organic compounds including biochemicals like amino acids and proteins. He believed that the origin of cells had its beginnings in the formation of microdroplets from colloidal substances like protein, and was able to demonstrate the formation of such microscopic particles by the separation of dissolved colloids from the rest of the liquid, in the form of microdroplets called coacervates. He believed that coacervates increased in complexity, eventually to the level of cells.

Since Oparin's hypothesis begins with the interactions among methane, ammonia, hydrogen, and water vapor, leading to the formation of organic compounds including protein, it is tempting to cite that most interesting passage from a letter by Darwin in 1871, brought to light some years ago by Hardin (15). It has been referred to, since, as proof of Darwin's disbelief in spontaneous generation. An even greater significance of Darwin's statement has been largely overlooked. Darwin wrote, "It is often said that all the conditions for the first production of a living organism are now present, which could ever have been present. But if (and oh! what a big if!) we could conceive in some warm little pond, with all sorts of ammonia, phosphoric salts, light, heat, electricity, etc., present, that a proteine compound was chemically formed ready to undergo still more complex changes, at the present day such matter would be instantly devoured or absorbed, which would not have been the case before living creatures were formed." (my italics) Can the great significance of the portion in italics be missed by anyone familiar with present experiments that start out with ammonia, methane, water, and light (U. V.), or electricity (sparking), or heat?

In that passage Darwin questions the relevance, in modern times, of a supposedly commonly held view ('it is often said') that life began when warm water, ammonia and phosphoric salts reacted, in the presence of energy of one sort or another, to form a protein which then went on, through changes, to develop into a living thing. In the absence of evidence to the contrary (our search so far is fruitless), this was not a commonly held view, but, more likely, one of the many ideas which that great mind must have been mulling, but which, unlike the present one, never slipped out.

The passage, however, has been much quoted in support of the thesis that, although life does seem to have originated at one time from organic compounds, this no longer is possible because, today, organic compounds

on their way to generating life would be consumed immediately by existing life, which was not the case in the first place. This and other arguments
against the possibility of the re-origin of life have been considered and
rejected (16, 17). It is doubtful that Darwin, informed on present developments, would have concluded as he did.

A picture of the chemical origin of life that is beginning to emerge is
the following. The atmosphere of the primitive Earth contained methane,
ammonia, hydrogen, water vapor and perhaps hydrogen sulfide as well as
other gases, but not oxygen. It was, thus, a chemically reducing atmosphere. A high flux of short wave ultraviolet rays, now mostly screened
out by the present ozone layer, reached the Earth. Lightning, heat, and
ionizing radiation (cosmic and also decay of radioactive elements of the
primordial crust) were more abundant on the primitive Earth than at present. The oceans were warm. Energy in one form or another drove endergonic reactions among the gases resulting in an abundance of organic
and biochemical compounds of a great variety. Products acting as substrates for further reactions produced an ever increasingly complex mixture of compounds among which were amino acids and amino acid polymers (proteinoids). This is the basis for the supposed condition of the
primitive waters frequently referred to as the 'hot dilute soup,' a term
first used by J. B. S. Haldane. The proteinoids separated from this
'soup' in the form of microscopic droplets referred to as coacervates
(Oparin) or microspheres (Fox) or prebiological systems. Life is supposed to have originated from this chemically complex mixture as a completely heterotrophic microorganism which obtained all of its organic requirements from it.

Thus, the crucial step in the chemical origin of life is the spontaneous
formation of microdroplets. Fox is preeminent in this field. His "microspheres" are acceptable models of prebiological systems. Significantly,
they are produced not from contemporary materials but from proteinoids
that can be formed under believably prebiological conditions (18). He has
shown (19 - 21) that such microspheres can be made to exhibit--depending
on the materials and conditions of formation--growth, division and budding,
catalytic activity, osmotic phenomena, non-random movement, and communication.

Selection to autotrophism began as compound after compound became
depleted, killing off all organisms except those which, by random mutation, had developed a metabolic pathway for the synthesis of the depleted
compound. Eventually, complete autotrophism would replace heterotrophism, after which random mutations would introduce metabolic deficien-

cies in some of the autotrophs establishing gradually, in this way, the variety of heterotrophism present today. Horowitz (22) proposed the heterotroph-autotroph-heterotroph hypothesis in 1945. The autotrophs also supplied an excess of oxygen which eventually contributed to the present oxidizing atmosphere and made possible the evolution of an oxidative metabolism to supplement the anaerobic or fermentative metabolism of the original heterotrophs. In turn, oxidative metabolism is held responsible for the greatly increased rate of evolution which began about 8 or 10 x 10^8 years ago. Another effect of the excess production of oxygen was the formation of the ozone layer of the upper atmosphere which is an effective shield against short wave ultraviolet rays. Until a sufficient diminution occurred in the amount of such ultraviolet rays reaching the Earth, organisms were thought to have been confined to a depth of water sufficient to absorb the harmful rays.

This picture presents some difficulties. The great variety of compounds claimed to have been synthesized under presumed primitive Earth conditions can be accepted only if we ignore the fact that the list is an accumulation of data from experiments representing a variety of reactants under a variety of conditions, some mutually exclusive. For example, an experiment designed to produce nucleic acids produces little else. On the other hand, experiments of the Miller type, while producing many organic compounds including some amino acids, produces only traces of bases and no nucleic acids or proteins. Strictly speaking, none of the conditions employed today in the abiotic synthesis of organic compounds can be accepted as being "simulated primitive Earth conditions." For reasons of expediency, all employ apparatus which too greatly confines the reactants and products, a condition which can greatly alter the course of reactions and the nature and amount of end-products. Pattee (23) has suggested a setting measured in hundreds of cubic meters with sand and tides. The expenditure of time and money implied by the scale of such operations is more than justified by the grandiose plans proposed for testing for evidence of chemical evolution or life on other planets, the results from which will be misinterpreted if our present assumptions are incorrect.

The abiotic formation of organic compounds seems now to be well established not only for the primitive Earth, but also for space. A striking phenomenon in experiments employing continuous sparking as a source of energy is the early formation of colloidal material at a time when only traces of a few amino acids have appeared. Much of the colloid is reported to be inorganic, but some of it is apparently polyamino acid which yields amino acids on hydrolysis. Also present are microdroplets presumably derived from the colloids both inorganic and organic (24). It is mostly taken for granted that amino acids form first and then polymerize rapidly, thus accounting for the low concentration of amino acids early in the experiment. Amino acids do not polymerize spontaneously in water, and even under contrived conditions, the yield is small and of very low molecular weight. These facts give greater weight to alternative hypotheses (25-27) which suggest the direct formation of polyamino acids, the free amino acids resulting from the subsequent hydrolysis of the polyamino acids.

There is the possibility, then, that the first microdroplets were, if not inorganic, nevertheless stable in the presence of U.V. and other normally destructive conditions existing at the time of their formation. Indeed, the first cells and all the systems leading to them must be considered to have been stable to the conditions under which they were formed. Chemical reactants within the microdroplets--microspheres--at first may have been similar to those of the medium, endergonic (synthetic) reactions being driven by physical energy as in the environment. Due to the small volume and the semipermeability of the boundary, reactions within the microspheres would, on many counts, soon take a different course than in the medium. For one, a molecular species can accumulate beyond its solubility product and precipitate out as a solid phase in the droplet, whereas this may never occur in the medium. Similarly, concentrations of potential reactants may early reach a reaction level, whereas in the medium this may take longer or never take place. Coupling of reactions can take place within droplets especially when a microscopic or submicroscopic structure develops. This last is especially important in the coupling of exergonic and endergonic reactions which eventually must replace physical forces in the synthesis of compounds.

Considering the primitive methods of multiplication of microspheres, the retention by successive generations of an ever increasingly complex structure and chemistry must have been precarious. This might explain the long period of non-genetic evolution of microdroplets. Microfossils, said to be two and a half to three billion years old, may represent various stages in that chemical history. How, when, and in what role nucleotides, polynucleotides, and nucleic acids became incorporated in this chemistry is hazardous to guess. As for coded nucleic acids, the sudden great increase in the rate of evolution, about a billion years ago, is better explained by the appearance of coded nucleic acids in already chemically complex microspheres, than by the hypothesis that it was due to the advent of oxidative metabolism.

REFERENCES

1. Keosain, J., "The Origin of Life" (second ed.), pp. 32-38. Reinhold Publ. Corp., New York, 1968.
2. Redi, F., Esperienze intorno alla generazione degl' insetti, Firenze,
   Italy, 1668.
3. Valery-Radot, R., "The Life of Pasteur," p. 109. Dover Publications, New York, 1960.
4. Muller, H. J., Am. Naturalist 100, 493 (1966).
5. Loeb, J., "The Mechanistic Conception of Life," Harvard University Press, Cambridge, 1964. (Reprint of original by University of
   Chicago Press, 1912.)
6. Engels, F., "Dialectics of Nature," International Publishing Company, N. Y., 1940 (translation).
7. Troland, L. T., The Monist 24, 92 (1914).
8. Ibid., Cleveland Med. Jour. 15, 377 (1916).
9. Ibid., Am. Naturalist 51, 321 (1917).
10. Muller, H. J., Proc. Int. Cong. Plant Sciences 1, 897 (1929).
11. Oparin, A. I., "Origin of Life," Dover Publications, New York,
    1953. (Republication of translation by S. Morgulis, MacMillan Co.
    New York, 1938).
12. Bastian, H. C., "The Beginnings of Life," p. VIII. D. Appleton &
    Co., New York, 1872.
13. Schafer, E. A., "Presidential address, Rep. Brit. Assoc. Adv.
    Science (1912), p. 3.
14. Moore, B., "The Origin and Nature of Life," p. 163. Henry Holt
    and Co., New York, 1912.
15. Hardin, G., Sci. Monthly 70, 178.
16. Keosian, J., Science 131, 497 (1960).
17. Ibid., "The Origin of Life," (first ed.), pp. 98-111. Reinhold Publ.
    Corp., New York, 1964.
18. Fox, S. W., in "Biology and the Exploration of Mars" (C. S. Pit-
    tendrigh, W. Vishniac, and J. P. T. Pearman, eds), p. 213. Na-
    tional Academy of Sciences--National Research Council, Washington,
    1966.
19. Ibid., Nature 205, 328 (1965).
20. Ibid., Naturwissenschaften 56, 1 (1969).
21. Ibid., Chem. Eng. News 49, 46 (Dec., 1971).
22. Horrowitz, N. H., Proc. Natl. Acad. Sciences U.S. 31, 153 (1945).
23. Pattee, H. H., in "Molecular Evolution," Vol. 1 (R. Buvet, C.
    Ponnamperuma, eds.), p. 48. North-Holland Publ. Co., Amster-
    dam, 1971.

24. Grossenbacher, R. A., and Knight, C. A., in "The Origins of Pre-
    biological Systems," (S. W. Fox, ed.), p. 173. Academic Press,
    New York, 1965.
25. Akabori, S., in "I. U. B. Symp. Series," Vol. 1, p. 189. Acade-
    mic Press, New York, 1959.
26. Matthews, C. N., in "Molecular Evolution," Vol. 1 (R. Buvet, C.
    Ponnamperuma, eds.), p. 231. North-Holland Publ. Co., Amster-
    dam, 1971.
27. Akabori, S., and Yamamoto, M., "Model Experiments on the Pre-
    biological Formation of Protein," this volume.

Received 13 March, 1972

ORIGINS

J. Lawrence Fox

Division of Biophysics and Physiology, Department of Zoology
University of Texas at Austin 78712

The A. I. Oparin theory (1) presented in 1924 and its subsequent elaborations constitute today's most commonly accepted theory of the origin of life on the Earth. The lucidity and scope of this monumental contribution has withstood many tests of time. It has provided the impetus for many contemporary workers to enter this area of research (2). On the basis of laboratory results and new insights, certain stages of Oparin's theory are now open to conceptual changes.

OPARIN'S THEORY

Oparin begins with a hot, but cooling, Earth surrounded by a scant reducing atmosphere. When the earth cooled to temperatures above $1,000^{\circ}C$ the formation of hydrogen-carbon bonds became stable and unsaturated hydrocarbons appeared. As the earth cooled below $1,000^{\circ}C$, the nitrogen and oxygen bonds to hydrogen became stable and so ammonia and water were formed. These molecules are well known to add across unsaturated carbon bonds at such elevated temperatures creating a host of compounds containing oxygen and nitrogen in addition to carbon and hydrogen. At temperatures around $100^{\circ}C$, the condensation of water vapor produced rains, washing these organic molecules into pools, lakes, and oceans. Haldane coined the expression "hot dilute soup" for these dilute organic seas (3). Macromolecular synthesis was proposed to partially occur here. The organic molecules of this hot dilute soup then formed a second aqueous phase evolving as coacervates. The coacervates possessed levels of structure which permitted selective retention of molecules and conducted additional macromolecular synthesis. Oparin proposes that the coacervates underwent a long period of chemical evolution (perhaps millions of years) in which reactions and molecules were selected for their suitability.   Ultimately the proper set were obtained and the first living cell was formed.

OPARIN'S SELECTIVITY

Oparin (21) has elaborated upon the levels of molecular selection which directed the chemical evolution of coacervates into cells in more recent discussions of his theory. The open system nature of these organized structures is greatly emphasized. Oparin describes these systems as dynamic stationary states which represent a balance between input and output.   Their stability is maintained by the release of free energy obtained from molecules selectively removed from the environment.   This dynamic stationary state is kinetically defined, not thermodynamically, so its description might more simply be stated as a kinetic steady state.

Molecular selection begins with the relative abundances (or absences) of molecules in the milieu from which the coacervates first formed. These compositions also serve to determine the future properties of a coacervate. Having formed, preservation of coacervates is then determined by their relative stabilities.   Coacervates possessing growth capabilities possessed a greater opportunity for preservation. "The systems did not merely become more dynamically stable, they also became more dynamic." (21) Reproduction or proliferation at both the molecular and "cellular" levels gives additional selective advantage to these primitive systems.

Several types of activities may be distinguished in Oparin's discussion of selectivity. Relative levels of enzymic activity confer selective advantages. This may be simply the relative kinetic rates for a given enzymic function, or it may be the diversity of enzymic functions which are present in a given system. In situations of diverse enzymic activities selection between levels of the coordination of enzymes in enzymic networks may exist.   Even within a given level of coordination, selection may be made by the degree of compartmentalization present. This serves to optimize the sequential nature of a reaction network which is well established from contemporary examples as being a criterion for kinetic and energetic efficiency.

ORIGIN OF SIMPLE MOLECULES

Oparin's description of the formation of organic molecules from a primordial atmosphere under thermal conditions is strongly supported by experimental investigation. [See Keosian (4) for an extensive cataloging of experiments and Kenyon and Steinman (5) for more recent experiments]. In addition to thermal energy, simple molecules have been synthesized by a variety of energy sources such as electric discharge or spark, sunlight, UV light, alpha, beta, and gamma rays, electron beams, ultrasonic vibration, and ambient conditions (4). An objective appraisal of these energy sources indicates two considerations: 1) Most of these energies are ionizing radiations which tend to be destructive rather than constructive. An organic chemist is aware of the high molecular weights one obtains from overheating simple molecules. Thus, heat alone of these energys sources is a constructive form of energy. 2) Heat is a byproduct in the experiments using nearly all of these other energy sources so that thermal components of each synthesis may not be conveniently separated, even when caution has been taken to include thermal jacketing of the reaction vessel.

Thus, the relevance of many of these experiments to pre-biological conditions is unclear. Extensive exposure to ionizing radiations will produce considerable decomposition which will alter the determined products and their yields. Allowing for some selective data picking, the generalization can be offered that thermal syntheses permit the production of molecules selectively and quantitatively in a manner reflected by contemporary molecular compositions (6). The previous, somewhat backward sentence expresses the idea that nature has thermodynamically directed which molecules could selectively form from the very beginning. Thus, contemporary organisms reflect only nature's initial preferences. It is further contended here that evolution has not evoked numerous changes in these compositions.

This generalization can be supported, in part, by an examination of biological aromatic molecules. Molecular orbital calculations of these molecules permit an estimation of the additional degrees of stability which result from the electron delocalization of aromatic molecules by the calculation of so called "resonance energies" (7). This is the amount by which the energy of the aromatic system is lowered relative to a corresponding number of individual unsaturated bonds (π electron systems). The biological molecules are among nature's most stable as witnessed by their unusually high resonance energies. In addition, correlation of simple Huckel calculations with spectra requires the evaluation of a term labelled β , the resonance integral.   The value of β necessary to make this correlation for biological molecules is near its maximum value for any organic molecule (8, 9).

If biological molecules were indeed preferred in their synthesis, abio-genesis becomes a moot point.   There would no longer be a requirement for extended periods of chemical evolution.   I shall return to this point shortly.

ORIGIN OF MACROMOLECULES

Many properties of living cells can be attributed to the functions of macromolecules. One such example is the stable, semipermeable mem-brane which isolates cellular cytoplasm from its environs.   The coacer-vates of Bungenberg de Jong (10) and Oparin (11) and the highly versatile microspheres of Fox (12; 13; the majority of references to Fox refer to S. W. Fox and not the present author) require either biogenic or chemi-cally synthesized macromolecules for these properties and biphasic sta-bilities.

The formation of protein requires peptide bonds, the formation of nucleic acids requires carbon-nitrogen, ester, and anhydride bonds, the formation of carbohydrate requires glycosidic bonds, and the formation of fat requires ester bonds. Each of these macromolecular linking bonds is formed with the loss of the elements of water as seen in Fig. 1.   Dia-grammatic representations of these reactions are given in Fig. 2.   The Law of Mass Action clearly indicates that such a reaction would be most unfavorable in the presence of a large excess of water, i.e., the 55.5 M concentration of an aqueous milieu.   These reactions, therefore, will proceed to the left side.   This observation is further supported by equi-librium thermodynamic experiments which yield a positive free energy of 3-4 Kilocalories/mole (14) for dipeptide formation. In other words, only one in ten thousand molecules would exist as a dipeptide.   A number of the experiments performed in the context of prebiological synthesis of macromolecules have neglected this thermodynamically determined finite concentration of low molecular weight products in their discussions of low molecular weights and low yields.   To produce a molecule of insulin half the size of the smallest proteins without an energy source would be an impossible task (1 to 10^{168}) by simple equilibria.

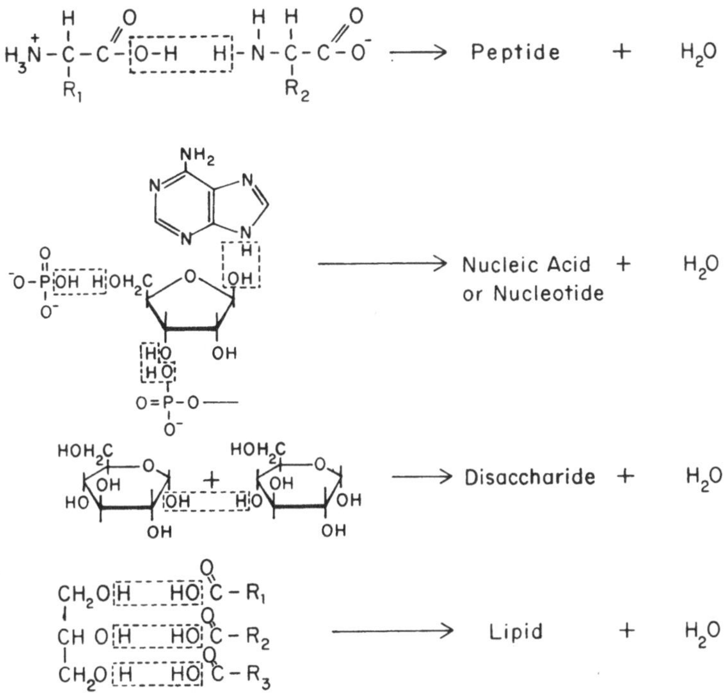

Peptide + H_2O

Nucleic Acid + H_2O
or Nucleotide

Disaccharide + H_2O

Lipid + H_2O

Fig. 1. Formation of dehydro linking bonds of biological macromolecules starting with (top to bottom)
amino acids; nucleic acid bases, ribose, and phosphoric acid; glucoses; and glycerol and three fatty
acids where R=linear hydrocarbon.

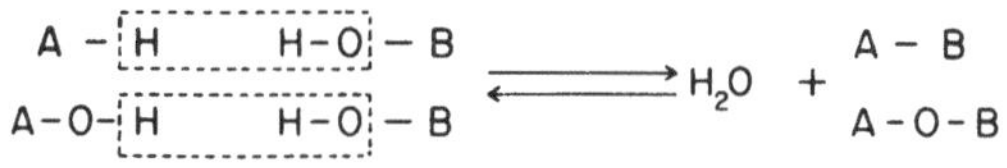

$$A - [H \quad\quad H-O] - B \quad\quad \rightleftharpoons H_2O + \begin{array}{c} A - B \\ A-O-B \end{array}$$
$$A-O-[H \quad\quad H-O] - B$$

Fig. 2. Idealized diagram of dehydro linking bonds of macromolecules.

On the other hand, Le Chatelier's Principle tells us that removal of water will force the equilibrium in Fig. 2 to shift to the right. Hypo- or an-hydrous conditions are essential. These reactions are not conceivable in aqueous milieu. Thus, this stage of prebiological macromolecule formation must be removed from the "hot dilute soup" environment.

The experiments of Harada and Fox (15) show conclusively that temperatures between 50 and 180°C are more than adequate to produce hypohydrous polymerization of amino acids, for example. The myriad of properties and the similarities of this thermal proteinoid to protein are well documented (16). Contemporary proteins reflect the amino acid compositions observed in these materials. A limited, discrete set of sequences are generated in these proteinoids (17). Nature selects by her own principles against the vast majority of potentially unique sequences and presents us with only a handful. The significance of this point is covered in the next section. Fox has also demonstrated (18) the geological relevancy of this model by locating temperatures in excess of 200°C within an inch of the surface of a number of contemporary volcanic cinder cones. Thermal regions above 100°C other than cinder cones would also suffice.

The formation of macromolecules is interpretable as the production of a metastable state. The energetics of this process must be handled by non-equilibrium thermodynamics (19). Unfortunately the formalism to fully handle this does not exist in a satisfactory state. The forces which determine amino acid sequence are largely unknown. These macromolecules can be considered as energy storage states. Since they are obviously stable they must represent an entropically and energetically favored state. Insofar as the interpretation of elevated energy states as informational levels is correct, these macromolecules may, thus, be considered as information storage molecules (20).

As a metastable state of an open system, it is more appropriate to refer to these molecules as a minimized steady state. They represent a balance between formation and hydrolysis reaction rates. Thus, our ultimate description of the energetics of formation of these molecules will require a kinetic component (21).

Questions concerning the stability of macromolecules have been raised in the literatures. The decarboxylation reaction of alanine in dilute aqueous solution, for instance, has been calculated to possess a half life of 10^5 years at 100°C by Abelson (22). At 40°C, the half life exceeds estimates of the Earth's age. In addition, peptides and proteins possess degrees of secondary and tertiary structure which lower their energy levels

to more stable states. X-ray diffraction studies indicate that the hydrophobic groups cluster internally and the hydrophilic groups are exterior,
permitting interaction with water and probably protecting the interior
linkages (23). Thus, it is quite likely that macromolecules are even more
stable as a function of time, indicating nearly any time scale is appropriate, although the discussions of this article consider only short time
scales to be appropriate.

ORIGINS OF ENZYMES

Catalytic activity is noted in thermal proteinoids of a weak, polyfunctional type (24).   In other words, one fairly homogeneous thermal
proteinoid can be active catalytically in several reactions. This is consistent with the first living cells containing a limited number of proteins
of polyfunctional activities.   Under genetic control, gene duplication and
mutation would lead to diversification and specialization of enzyme activities resulting in divergent evolution of enzyme function (25).   The result
of this selective process would be highly specific and highly catalytic proteins.   It would be naive, however, to require that primitive enzymes
would possess contemporary reaction rates and restricted specificities.

The thermal polymerization of amino acids results in a brownish
colored product (26).   Two pigments have been identified from this system which are only poorly characterized (27). Although they bear resemblances to flavins, they are not equivalent to contemporary cofactors.
We have identified the production of these cofactors as depending only
upon glutamate and glycine for one and upon glutamate, glycine, and as-
parate for the other pigment.   This is indicative of the basic chemical
principles from which life ultimately arose since the contemporary biological heterocyclics are biologically synthesized from these same amino
acids (28).

In another set of experiments, we have demonstrated that the final
amino acid content of an adenylate proteinoid (29) obtained from the polymerization of 18 amino acyl adenylates is altered by the presence during
polymerization of flavin, hemin or chlorophyll cofactors (30). Thus, the
apoenzyme produced is at least partially determined by its cofactor. Weak
catalytic activities with 3-50 fold reaction rate enhancements as opposed
to the free coenzyme are obtained with these systems (30, 31).   Thus,
complex holoenzymes are possible at the earliest stages of macromolecule synthesis.

ORIGINS OF CELLS AND METABOLISMS

The association of anhydrously prepared macromolecules and simple molecules into an organized structure requires a rehydration step. Rains or splashing waves offer geologically feasible mechanisms. The resulting "protobiont" then is analogous to the coacervate in the Oparin theory. However, here the spectre of chemical evolution again is raised. What molecules are present? What chemistry could these structures conduct? What selective processes were necessary before they developed all of the attributes of life?

Many experimental answers come from Fox's laboratory (13). One may incorporate into microspheres a great variety of molecules within the limitations of membrane selective permeability as shown by proteinoid microspheres (16, 32). These limitations are analogous to contemporary membranes. I have already pointed out that appropriate molecules were selectively produced. Thus, only the proper set was present to warrant consideration and were therefore included from the beginning.

The development of crude enzymes produced a compartmentalization of reactions in such a way as to better utilize intermediate levels of the energetically downhill path of metabolism. For example, the oxidation of glucose by oxygen in a calorimeter is a simple reaction producing CO_2 and H_2O heat. In a biological system somewhere between 36 and 38 ATP/glucose molecule are generated as metastable energy storage states, so that the potential energy content of glucose may be utilized by living cells in ways other than heat production. However, water and carbon dioxide are still the products of aerobic oxidation. A number of catalytic steps in glucose metabolism have already been demonstrated in proteinoid microspheres (24).

The heterotrophic metabolism of these first cells is generally recognized. It is to be recalled that a reducing atmosphere was present and so electron acceptors other than oxygen were required. With little or no CO_2 present, photosynthetic organisms were unlikely. Extrapolations from contemporary atmospheres support this contention and place the evolution of photosynthetic systems at about 1.2 billion years ago (33).

Thus, microspheres satisfy the criteria that the first primitive cells must have had the ability to selectively retain molecules, metabolize, synthesize macromolecules from monomeric units, and proliferate. The proliferation phase is not a mystical property of living cells, but is determined

by the thermodynamics of surface tension.   Oil droplets can be seen to
proliferate under a variety of conditions (13). Contemporary cells possess
a strict genetically controlled reproduction system which permits the fixation of mutations in a very rigid manner.   If one defines primitive life
in terms of the ability of a system to mutate and fix that mutation, then
nucleic acids are an essential constituent of life. Definitions of primitive
life have seldom been that restrictive.

ORIGIN OF GENETIC CONTROL

Did the first organism require DNA?   The importance of protein to
contemporary living systems is well established.   They possess structural, storage, transport, catalytic, and colloidal roles.   A contemporary cell could not function without these attributes. The nucleic acids, on
the other hand, are largely restricted to functioning as a genetic memory
and as translators in the expression of the genetic memory.   It is clear
that contemporary organisms depend upon their nucleic acid memory for
replication and evolution.   Did the first primitive cells?

Nearly all criteria of primitive living cells can be met by proteinoid
microspheres, an experimentally derived system (13).   The ability of
proteinoid microspheres to catalyze the formation of more proteinoid
raises serious questions as to the essentiality of nucleic acids in a primitive cell. Protein synthesis by a simple system without nucleic acid participation has been discussed by Lipmann (34). He credits these questions
with directing his attention to polypeptide antibiotic biosynthesis as a possible vestige of an earlier, more primitive protein synthesizing system
(35). The extrapolation is tenuous, at best, as these antibiotic synthetase
systems are highly specialized and, hence, now highly evolved.

The complexity of the contemporary genetic mechanism in light of
the simplicity available in microsphere function suggests it is too much
to ask of a simple system.   Since heterotrophic proliferation of microspheres is evident from experiment, the nonessentiality of the nucleic
acids is emphasized. Fox (37) has produced a real time film which indicates that synthesis from simple gases to microspheres may be achieved
easily within minutes. Thus, a primitive cell with most of the attributes
of life can "spontaneously" arise in a brief period of time.   Protracted
periods of chemical evolution are unnecessary.   It is unfortunate that
"spontaneous generation" possesses the connotation commonly given to
that term now. It is the most appropriate way of indicating the time scale
necessary to achieve a living cell. It must be emphasized that spontaneous

generation here refers only to a most primitive cell and not to the highly evolved species to which the term was historically applied.

This argument has not presented us with the essential attribute of contemporary life which ultimately produced man, genetic control. I propose that following the spontaneous generation of primitive, repro- ducing forms of life that an evolution leading to genetic control arose. This period I will label genogenesis. I currently envisage this as the lengthy period of time in life's origins. Indications of nucleic acid inter- actions with amino acids exist (36) but no clear picture has yet emerged. The end of this period marks the beginning of biological evolution.

THE THERMAL THEORY

In summary, the genesis which I have presented extends from simple gases to cellular models. It has required modification of the Oparin theory in several instances. These modifications can be fit into a theory best labelled the Thermal Theory. Its outline consists of these points:

1) Hot earth origin with scant, reducing atmosphere.
2) Thermal synthesis of simple molecules.
3) Condensation of water vapor to produce rains which washed organic molecules into pools.
4) Hypo- or an-hydrous macromolecule synthesis step.
5) Rehydration to form microsphere like cells, progenitors of contem- porary cells.

ACKNOWLEDGEMENTS

The author wishes to thank Dr. S. W. Fox and his wife Raia for dem- onstrating their understanding of the origin of life by creating it.

REFERENCES

1. Oparin, A. I., Proiskhozhdenie Zhizni Izd. Moskovskii Rabochii, Moscow, 1924; "The Origin of Life," 2nd Ed., Dover Publications, New York, 1953.
2. Fox, S. W., private communication.
3. Haldane, J. B. S., "Science and Human Life," Harper and Brothers Publishers, New York, 1933.

4. Keosian, J., "The Origin of Life," 2nd Ed., Reinhold Book Corporation, New York, 1968.

5. Kenyon, D. H., and Steinman, G., "Biochemical Predestination," McGraw-Hill Book Company, New York, 1969.

6. Harada, K., and Fox, S. W., in "Origins of Prebiological Systems," (S. W. Fox, ed.), Academic Press, Inc., New York, 1965.

7. Pullman, B., and Pullman, A., "Quantum Biochemistry," Interscience Publishers, New York, 1963.

8. Streitwieser, A., Jr., "Molecular Orbital Theory for Organic Chemists," John Wiley and Sons, Inc., New York, 1961.

9. Isenberg, I., and Szent-Gyorgyi, A., Proc. Nat. Acad. Sci. 45, 519 (1959).

10. Bungenberg de Jong, H. G., Protoplasma 15, 110 (1932).

11. Oparin, A. I., "Life: Its Nature, Origin, and Development," Oliver and Boyd, Edenburgh, 1961.

12. Fox, S. W., Nature 205, 328 (1965).

13. Fox, S. W., Harada, K., Krampitz, G., and Mueller, G., Chem. Eng. News, p. 80 (June 22, 1970); Fox, S. W., Chem. Eng. News, p. 46 (December 6, 1971).

14. Huffman, H. M., J. Phys. Chem. 46, 885 (1942).

15. Harada, K., and Fox, S. W., J. Am. Chem. Soc. 80, 2694 (1958).

16. Fox, S. W., "Encyclopedia of Polymer Science and Technology," Vol. 9, John Wiley and Sons, Inc., New York, 1968.

17. Fox, S. W., Naturwissen 56, 1 (1969).

18. Fox, S. W., Nature 201, 336 (1964).

19. Prigogine, I., "Introduction to Thermodynamic of Irreversible Processes," Charles C. Thomas, Springfield, Illinois, 1955.

20. Quastler, H., "Information Theory in Biology," University of Illinois Press, Urbana, 1953.

21. Oparin, A. I., "The Origin of Life on the Earth," 3rd Ed., Academic Press, Inc., New York, 1957.

22. Abelson, P. H., in "Organic Geochemistry" (I. A. Breger, ed.), Pergamon Press, Oxford, 1963.

23. Dickerson, R. E., and Geis, I., "The Structure and Action of Proteins," Harper and Row, Publishers, New York, 1969.

24. Rohlfing, D. L., and Fox, S. W., Advances in Catalysis 20, 373 (1969).

25. Bryson, V., and Vogel, H. J., eds., "Evolving Genes and Proteins," Academic Press, Inc., New York, 1965.

26. Fox, D., and Fox, S. W., unpublished results.

27. Hudson, P., and Fox, J. L., unpublished results.

28. Fruton, J. S., and Simmonds, S., "General Biochemistry," 2nd Ed.,
    John Wiley and Sons, Inc., New York, 1958.
29. Nakashima, T., Lacey, J. C., Jungck, J., and Fox, S. W., Na-
    turwissen 57, 67 (1970).
30. Fox, J. L., 8th International Congress of Biochemistry, Interlaken,
    Switzerland, 1970; Fox, J. L., and Muska, C. F., Experientia 28,
    143 (1972).
31. Dose, K., and Zaki, L., Z. Naturforsch. 26, 144 (1971).
32. Experiments performed in freshman biology laboratory, University
    of Texas at Austin.
33. Berkner, L. V., and Marshall, L. C., Proc. Natl. Acad. Sci. 53,
    1215 (1965).
34. Lipmann, F., in "The Origins of Prebiological Systems" (S. W.
    Fox, ed.), Academic Press, Inc., New York, 1965.
35. Lipmann, F., Science 173, 875 (1971).
36. Nakashima, T., and Fox, S. W., Proc. Nat. Acad. Sci. 69, 106
    (1972).
37. Fox, S. W., "The Origin of Life" (Film Loops), Harper and Row,
    New York, 1972.

Received 27 September, 1971

EVOLUTION OF LEVELS OF EVOLUTION

Thomas O. Fox*

Department of Neuropathology
Harvard Medical School, Boston, Massachusetts

Within each field of science evolutionary processes are observed. Their elucidation in such diverse disciplines as cosmology, biology, and sociology suggests that an overall evolutionary process continues through atoms, through cells, and beyond the confines of the human brain. But while most scientific attention has dealt singly with individual levels of evolution--observable by each corresponding scientific discipline--we can also recognize principles that describe the overall evolutionary process and, particularly, the <u>emergence</u> of each level of evolution.

Aimed at experimentally describing the origins of biological systems, the field of molecular evolution is well suited for an examination of the principles governing the emergence of new levels of evolution. Using selected findings of that field, combined with those of other fields of science, we can describe a new class of evolutionary principles.

In this presentation, several levels of evolution are compared, and new systems that assemble from them are described. The requirement by each level for a faster rate of selection and the resulting selective interactions among different levels are discussed.

---

* Postdoctoral Research Fellow of the National Institute of Child Health and Human Development.

LEVELS OF EVOLUTION

Levels of evolution involve systems that are distinguished by their primary constituents, the interactions of these constituents, and the focus of selection by which each system changes in time. One catalog of some levels, the constituents of their systems and their foci of selection is presented in Table I (1). Levels historically have appeared sequentially, with simpler systems and levels having emerged earlier and over wider expanses of time and space (2).

The <u>focus of selection</u> of a system is the major property, or set of properties, that selectively changes in increments. These properties are both the major source of variability in the system and the portion of the system upon which selective pressures are "focused." The selective pressures arise from the interaction of the changing environment with expressed properties of the system. Thus the term "focus of selection" applies to the overall process of selective pressures upon the source of variability.

These focal properties--either the constituents themselves or their interactions--are unique to each level of evolution. Within a level, selection partially is based on the foci of selection of older levels. However, each older factor receives less of the total selective pressure with the emergence of each successive level.

As an example, in the genetic level of evolution, the focus of selection is the set of genes (see Table I). The increments of change, called mutations in this level, involve the sequences of nucleotides that make up genes. Selective pressures act upon the structures and functions of the "products" of genes--such as proteins, intracellular particles, and membrane phases. These determine both quantitative and qualitative aspects of evolution. An overall rate of selection, characteristic of each evolutionary level, reflects both the source of variability and the pressures of selection.

SYNTHESES OF NEW SYSTEMS

When levels of evolution develop to particular stages, emergent systems can assemble from the components present. These systems will, if selected, yield new levels of evolution. The synthesis of proteinoids and the assembly of microspheres are examples of new systems (3). When the organic level of evolution contains sufficient amounts of an ap-

TABLE I

LEVEL OF EVOLUTION	CONSTITUENTS OF CORRESPONDING SYSTEM	FOCUS OF SELECTION
atomic	neutrons, electrons and protons	stability of atomic nuclei
molecular	atoms	covalent bonds in small molecules
macromolecular	monomeric organic molecules	weak interactions of polymers
cellular	macromolecular phases	membranes
genetic	cells	genes
multicellular	genetically communicating cells	development
neural (organismic)	cellular circuits (multicellular compartments)	dendritically associated cells (multicellular interactions)
mental	neural-environmental processes of humans	thought patterns
social	humans	communications among humans
supra-social	----	----

propriate mixture of amino acids, an energy-requiring synthesis of proteinoids can occur. The proteinoids are polymers of the amino acids and
display many biological-like properties that are not present in the monomeric amino acid mixture. Likewise, the appropriate proteinoids assemble in water to yield microspheres. These cell-like systems display
several membrane and propagative characteristics that resemble properties of contemporary living cells (4).

It is, in fact, this feature of assembly with the resultant "whole...
greater than the sum of the parts" (3) that provides us with the notion of
discrete levels of evolution. This description applies similarly to the
origins of the other levels of evolution.

SELECTION OF, AND EVOLUTION OF, NEW LEVELS

A new system can be selected and can evolve only if it provides a
new focus of selection that allows a faster rate of selection than does the
selective focus of the older system at that time. More rapid selection is
essential to enable the new system to adapt to a changing environment
more quickly than do the components of the older system. If the older
level evolves too rapidly, those of its components that are needed as inputs to, or as entities of, the new system may be depleted. Since other
assembled systems from the older level could cause these depletions, we
see that there is competitive selection of levels of evolution.

The competitive advantage of a new system is a direct consequence
of the faster adaptive rate provided by the new focus of selection. The
faster rate causes the new focus of selection to become more significant
relative to selection based on the foci of older systems. Within the new
system, then, selection based on older foci is restricted to changes that
do not preclude existence of the new system.

Discussion of the mental level of evolution illuminates these factors.
This system has evolved from the neural and earlier levels (see Table I).
Selection involves thought processes, which include the associations of
experiences. Further evolution of the neural and earlier levels could
lead to neural circuits that no longer have the necessary components or
organization for thought processes. But if thought patterns were evolving
rapidly, there would not be sufficient selective pressure for this type of
evolution of the neural and earlier levels; primary pressure would select
individuals with the best thought patterns, and these would have to be
individuals with operative neural systems. Thus the faster rate of selec-

tion in the mental level would restrict selection in the slower neural (and earlier) systems.   On the other hand, if a mental system were slow and of little advantage to humans, then further neural, multicellular, (et cetera) evolution could lead to alternate systems.

In addition to potential evolution, some of the properties that were manifest for a given component in the older system are incompatible with the newer system.   These properties are not observed within the newer level. Thus properties of older components are restricted within a newer level of evolution to the extent that they can limit expression of new properties and components.

Acetic acid, for example, is a chemical from the organic level of evolution and is also a metabolite of cellular genetic systems.   Within living systems, which include cells and their immediate environments, acetic acid is restricted to aqueous acetate solutions.   Although geologically it can exist as solid or gaseous acetic acid, these states restrict or destroy the activities of the living cell. In other words, the temperature, concentration, ionic form, and state of acetic acid must exist within narrow limits in the region of a living system.   Thus the form of acetic acid is restricted by the requirements of the living cell, and these restrictions, in turn, limit the possible evolution of acetic acid.

It must be emphasized that the competition among levels of evolution differs from Darwinian competition within a single level.   Darwinian evolution can be conceptualized as occurring within a single plane, and the selection of one individual may lead to the extinction of competitors. Evolution of levels involves an added vertical dimension. Preceding levels necessarily do not disappear, although they continue to evolve with the restrictions described above.

CONTINUALITY OF EVOLUTION

In addition to the levels discussed above, other levels of evolution can be described.   In the neural system, the different components of the brain can be described as subsequent evolutionary levels.   With the evolution of reflexive, sensory, and higher associative functions, the brain has progressed through segmental, regional, and increasingly less local systems of neurons and intercellular connections.   Likewise the social institutions have evolved through many levels.   Even within the industrial realm of society, we recognize at least three generations--early industrial

England, mass production in the United States, and modern industrializa-
tion in rebuilt Japan and West Germany. We recognize in these hierarchal
levels the dual principles of new systems with new foci of selection and
more rapid evolution.   In fact, we all recognize these principles daily;
new and improved methods of communications result in faster and more
voluminous accumulation and dissemination of information and ideas.

As we continue to narrow our attention and define more levels, or
sub-levels, we finally reach the level of the individual.   Thus the application of principles for the evolution of levels gradually converges to the
application of Darwinian evolution upon individuals within, and among,
species.   This convergence of evolutionary principles applied to levels
and to individuals is a direct consequence of the continual and incremental
nature of the world.   The two sets of principles both describe continual
processes and both apply to discrete components.   At all levels of evolution, changes occur by discrete steps.

The above continual characteristics yield the concept of a continuality
of evolution, in marked contrast to a smooth continuum.   The word, continuality, emphasizes the discrete, incremental nature of the universe
and of the respective evolutionary processes within different realms of
the universe (discrete because of different foci and rates of selection).
The term is chosen in preference to the term, continuum of evolution,
which has been used often, but which suggests processes that are tem-
porally homogeneous and that occur smoothly in uninterrupted sequence.

SYNTHETIC EXPERIMENTS

Modern species may appear to be fundamentally different--and therefore discontinuous--from the oldest species of a given level.   This occurs
because new properties and the shifting focus of selection that accompany
a new level pull the newer system away from the fundamental focus of
function and selection that prevails in the older level.   Thus origins become obscured and can not be deduced accurately from the modern.

The study of the origin of a level of evolution is in practice the discovery of the new system and new focus of selection that define that level.
Since these origins are synthetic and evolutionary processes which shift
as described above, the relevant experiments must themselves be synthetic
and evolutionary in design.

Processes that we might have predicted from modern biological species must have required extensive genetic evolution--as enzymatic activity and particular membrane and proliferative properties--appear to be basic properties of newly assembled proteinaceous systems. The simultaneous appearance of surprisingly diverse biological-like systems has been obtained with experimentally simple procedures using chemically simple compounds (3). These results alter our interpretations of what are the fundamental characteristics of biological systems and what is the fundamental sequence by which new properties emerged.

This perspective of the relationship between "origins" and the "modern," based on the premise that new levels of evolution are selected as suggested above, should offer alternate approaches to the study of higher levels of evolution. We see that studies of "origins" and of the "modern" yield different information. More extensive combination of these experimental approaches can be foreseen to be fruitful as we continue to pursue the nature of the evolutionary continuality.

NOTES AND REFERENCES

1. Several words in this presentation deliberately have been used with narrow meanings. These include genetic, biological, mental and social. In general usage these have broad and overlapping connotations. Here they have been used narrowly to connote levels of evolution with real and discrete foci of selection. This presentation does not suggest, for example, that the higher levels of evolution are independent of genetic determinates, but rather that the broad "genetic" mechanism itself evolves and that important new foci of selection appear as evolution progresses through and beyond the "biological" realm.

2. Huxley, J., "Evolution in Action," Perennial Library, Harper & Row, Publ., New York, 1966.

3. Fox, S. W., Harada, K., Krampitz, G., and Mueller, G., Chemical & Engineering News 48, pp. 80-94 (22 June 1970).

4. The particular levels listed in Table I, or in alternate lists of evolutionary levels, are partially speculative. Since some levels, systems, and foci of selection are more obvious than others, the listing will evolve with discussions as further experimental data is gathered. We may learn, for instance, that the synthesis of proteinoids can occur from amino acid precursors, not requiring the evolution of an amino acid level, per se. Likewise, genetic cell systems might be

obtained directly from amino acid and nucleic acid derivatives.   Then
these hypothetical systems, along with the amino acid synthesis of proteinoids and the formation of proteinoid microspheres, would all be experimental models and potential historical intermediates.   Changes of
this type, as are anticipated, should not alter the basis of this presentation, only its specific form.

Received 9 September, 1971

SPACE EXPLORATION AND THE ORIGIN OF LIFE

Richard S. Young

NASA Headquarters, Washington, D. C.

The relationship of space exploration to the question of the origin and early evolution of life is not obvious to those not actively engaged in such research. It is my purpose here to indicate how the study of extraterrestrial bodies and even interstellar space can and is contributing to man's understanding of the processes which may well have been involved in the beginning of life (however one wishes to define life) and thus are related to his understanding of his place and role in the universe.

Basic to this discussion is the acceptance of the concept that the origin of life is an integral part of the origin and early evolution of a planet with suitable environmental conditions. The processes by which a planet condenses from a cloud of gas and differentiates into core, mantle, crust, and atmosphere are also responsible for the sequence of events we now call Chemical Evolution. From gaseous precursors, increasingly complex organic molecules are formed, and, on planets where conditions are suitable, this process of Chemical Evolution may lead eventually to a self replicating, metabolizing, mutating system which we would call living. This concept, first put forth by Oparin (1) and Haldane (2), has served as the basis for much of the research on the origin of life question.

There are several kinds of research possible, which contribute directly to our understanding of the origin of life. First, terrestrial research involving the laboratory synthesis of organic compounds, and even living entities, in which we attempt to reconstruct primordial events pre-

sumed to have led to the origin of life; and second, laboratory analyses
of ancient rocks, from which we may extract knowledge of the early
Earth and direct evidence of prebiological and early biological history.
The laboratory syntheses have been fruitful in that they have indicated
pathways through which atoms and molecules could have been converted
to the fairly limited array of organic molecules which probably made up
the first living entities.   The difficulty is that we can perhaps never be
certain that this is what actually happened 4 billion years ago, however
suggestive such experiments may be. The fact that electrical discharges,
ultraviolet light, heat and other sources of energy, when applied in the
laboratory to gas mixtures presumed to represent primitive planetary
atmosphere, produce biologically significant molecules (amino acids,
purines, pyrimidines, ATP, etc.) does not prove that such events ever
took place on the early planet or that this sequence was indeed a step in
the origin of life.   Even if such laboratory experiments produce a self
replicating, metabolizing, mutating system which most biologists are
willing to define as living, we have not necessarily demonstrated how
life began on Earth. Thus, laboratory-synthesis experiments may never
provide the ultimate answers to life's origin, but they certainly will contribute greatly to our understanding of how it could have happened.

A record of the Earth's very early history is presumed to be available for study in precambrian rocks which date back to 3.4 billion years
or older, that is, around a billion years after the Earth's formation.
Laboratory analyses of these rocks indicate that life was already present
on Earth 3.4 billion years ago. These precambrian rocks contain organic molecules and even microfossils which strongly suggest an early
origin for life on Earth. The difficulty lies in the fact that the Earth today is teeming with life, from microbes to man; and residues of that
life are virtually everywhere, very likely masking whatever evidence
may exist of events preceding life. Although the precambrian rocks can
be accurately dated, it is not certain that the organic molecules and microfossils found in them are as old as the rocks themselves or even that
the "microfossils" are evidence of prebiological assembly (3). It is quite
conceivable that these "bio markers" are much younger than the rocks
and are evidence of more recent biological contamination.   Therefore,
the evidence in the precambrian rocks for early life or for a prebiological era is still ambiguous and hotly debated.

A third area of research is concerned with bridging the gap between
Chemical Evolution and Biological Evolution.   Through the combination
and assembly of the products of Chemical Evolution, under geologically
plausible conditions, we can hope to uncover pathway(s) explored in nature

which led to the first cell(s) or self replicating, metabolizing, mutating system which we could call life.

Experiments (4) have been done in several laboratories in which microstructures, with many of the properties of living cells, have been synthesized. Much work has been done to suggest that life could have begun in such a way and that it is possible that a "living system" could today be started in this way.

A fourth area concerns the disassembly and reassembly of contemporary cells and tissues (5) through which we can hope to reconstruct living systems with properties which have not evolved naturally on this planet but could be expected to have evolved under conditions quite different from that of the Earth--on another planet.

Although the Earth as a laboratory and repository is an invaluable aid to our understanding of life and its beginning, it is clearly inadequate. Evidence must be sought elsewhere in the solar system and indeed the universe; thus, there is a need for a fifth kind of research made possible by space exploration: the search for extraterrestrial life.

The most important objects to be studied for this purpose include:
a) Planets
b) Moons or satellites of planets
c) Meteorites, asteroids, and comets.

The fundamental questions to be asked are:
a) Does life exist elsewhere in the universe, or is terrestrial life unique?
b) If life does exist elsewhere, is it identical to, similar to, or unlike terrestrial life in terms of chemical composition, metabolic pathways, general morphology, and distribution?
c) Is its evolutionary history like that on Earth--this is, did it arise at about the same relative time in the planets' history as on Earth and were the same evolutionary pathways followed?
d) If there is no life on a given planet, did life ever exist? How far did it evolve? Why did it fail? At what period in planetary evolution did life arise then decline?
e) If there never was a life form on a given planet, is there a record of Chemical Evolution? When in the history of that planet did Chemical Evolution occur, or is it now occurring? Why did it not produce a living system? What was the nature and distribution of the organic compounds produced? Do they look like the products of abiogenesis in the terres-

trial laboratory experiment, or are they quite different? Do they tend to confirm the apparent precambrian record of the Earth?

f) How does a planet, similar to the Earth, evolve in the absence of a biota? It is clear that the Earth is being profoundly influenced by its biota. The interaction between an evolving planet and its biota needs study if for no other reason than to enable man to come to grips with environmental problems. It will surely be of enormous value to be able to compare with the Earth, the physical environment of a planet which has no biota, but has an origin similar to that of Earth.

None of these questions can be answered on the Earth. The history of life is irrevocably interwoven with the physical history of the planet itself, so that the physical environment is as relevant to the study of the origin of life as is the detection of an existing form of life.

I have already indicated the most important objects for study in space for relevant data. There are many factors which can and do influence the rate and order of exploration of these objectives, including:

1) availability of funds
2) state of development of suitable spacecraft
3) competing scientific objectives
4) state of development of scientific instruments
5) state of knowledge about a given planet, etc.

In spite of that, we can discuss objectives with some attempt at ordering in terms of priority, at least from the point of view of the knowledge to be gained about the origin and evolution of the planets and life. It should be pointed out that planetologists and physical scientists do not necessarily agree on the basic philosophy of planetary exploration--that is, whether it is more important scientifically to explore at least one planet in great detail, including several landed spacecraft, or to survey all the planets before landing for detailed study. However, planetary exploration for the purpose of detecting and studying life and life-related molecules requires landed spacecraft and eventually a surface-roving capability; therefore, a detailed study of at least one planet is ultimately more important than a multi-planet survey.

With these ideas in mind, let us then review the solar system and beyond to see where we stand with regard to suitable targets for exploration, what have we learned and where shall we go from here.

THE MOON

We are exploring the Moon first because of its relative accessability. At this writing, samples from Apollo 11, 12, 14, and 15 are being analyzed by scientists from a variety of disciplines. Lunar studies are far from complete so that conclusions are difficult to draw at this time; however, we can outline the potential of lunar analyses for the life scientist, and the status of our knowledge to date.

The origin of the Moon: Was the Moon formed independently of the Earth and captured by the Earth or somehow torn from the Earth? If it was formed independently, is its chemistry similar to or very different from the Earth? If it came from the Earth, what does it tell us about the early history of the Earth and the origin of life on Earth? Is there a record of life on the Moon? Is there a record of organic chemistry on the Moon? Is the Moon differentiated as is the Earth? It is felt that physical and chemical differentiation on the Earth was related to the origin of life--what can the Moon tell us about these processes? Is the Moon a celestial repository of material in space? Is evidence of life or organic matter from elsewhere in the solar system found on the Moon?

Although the answers to these and other questions are not yet completely resolved, the following summarizing statements can be made:

1. The available evidence argues against the view that the Moon was torn from the Earth; thus the Moon does not provide a direct record of the early Earth (6).

2. Because the Moon either had no atmosphere at its time of formation or lost it very early, its surface has been exposed to meteorite, cosmic ray and solar wind bombardment, resulting in physical and chemical processes quite different than those operating on the Earth's surface (6).

3. Although there is some indication that lunar material can affect (stimulate) the growth of some terrestrial plants, there is no evidence that these effects are due to some unique property of lunar surface material (7). Similar effects can be induced by a variety of terrestrial soils or basalts--that is, growth response to the addition of certain trace elements.

4. There is no evidence of present life on the Moon (8, 9). If life had existed in the past, little or no evidence of its remains were found.

[Carbon isotope ratio studies indicate that lunar carbon is probably not of biological origin (10)].

   5. The record of organic molecules on the Moon is very limited. There is good evidence for indigenous CH_4 and several other light hydrocarbons on the Moon in trace amounts (11), (micrograms per gram of Moon soil).   Although CO and CO_2 have also been found, the available evidence suggests that they are almost entirely chemically generated from other materials during the analysis procedures (10, 12). The presence of about half of the carbon on the Moon can be explained on the basis of solar wind action on the surface; the origin of the remainder is open to speculation at this time (10, 13).   Porphyrins (14) and amino acids (15, 16) (nanograms per gram of Moon soil) have been reported, although other laboratories (17, 18) have failed to confirm the results to date. If amino acids are present in such low amounts, the question still remains as to their origin: They could be contaminants (human or rocket exhaust products), meteorite remnants, indigenous, or artifacts synthesized from other materials in the lunar samples during analysis.   The importance of determining the origin of the organic compounds cannot be overstated. The question of whether these molecules represent contamination of some kind, remnants of biological activity on some extraterrestrial body, or results of non-biological chemical processes should be pursued until it is resolved. In addition, other locales on the Moon and samples from deeper cores, where organic compounds have a better chance of survival, must be studied for the same purposes.   Although there is no detectable evidence of past or present life on the Moon and only tenuous evidence for past organic chemical evolution of the kind presumed to have occurred on Earth, the Moon stands as a valuable yardstick against which to compare the results of surface phenomena and evolution with similar processes that have affected the Earth or other planets. Continued lunar study is important, either in the manned or unmanned mode. Unmanned vehicles have some advantage in that the contamination burden can be much reduced without the presence of man. Some organic chemical analyses can be done in situ, but the advantage of having highly sophisticated laboratory facilities with capabilities of performing extremely sensitive analyses (e.g., for amino acids) favor the return of samples for investigation on Earth.   Such facilities would be extremely expensive and time consuming to automate for flight to the Moon and operation on the lunar surface. Unmanned sample return could be the optimum mode for this kind of work.

METEORITES

The one source of extraterrestrial material (in addition to the Moon) available to scientists for many years are the meteorites, particularly the carbonaceous chondrites. These meteorites, containing up to 5% carbon, have been analyzed over the years in various laboratories and have provided some evidence for the presence of extraterrestrial organic carbon. However, there has been difficulty in establishing that the organic compounds were indigenous to the meteorites and not contaminants picked up by these porous rocks during atmospheric entry, while lying on the ground, or during subsequent handling and storage. If the organics are indigenous, the further problem remains of determining whether they are of biological origin or evidence of non-biological extraterrestrial chemical evolution.

Recent analyses (19, 20) of the freshly fallen Murchison and the Murray meteorites suggest quite strongly that organic matter is produced non-biologically elsewhere in the solar system. Although some of the amino acids found in these meteorites are normally in protein, many are not found in living systems. This is considered evidence that the amino acids are not due to terrestrial (biological) contamination. In addition these amino acids occur in the $\underline{d}$ and $\underline{l}$ form in nearly equal amounts, which is also unexpected if they are recent terrestrial contaminants. These data, therefore, indicate that these amino acids are indigenous to the meteorite and thus of extraterrestrial origin. It is possible, although unlikely because of the presence of both $\underline{d}$ and $\underline{l}$ isomers, that these amino acids are evidence of extraterrestrial life. The above determination plus the finding of heavy ^{13}C isotope values for the extractable organic carbon in this meteorite point to abiogenesis as the source of these molecules, and, therefore, provide evidence that Chemical Evolution has occurred or is occurring elsewhere in the universe.

These meteorite analyses should, of course, continue--especially on meteorites obtained from other planets or the asteroids so that contamination can be rigorously ruled out. The source of these meteorites should be determined--are they from the asteroids as is commonly believed? What can we learn about the source of the asteroids? Are they remnants of unconsolidated bodies formed during condensation of the solar nebula, or are they fragments of a planet on which Chemical Evolution occurred before the planet was destroyed? In either case analysis of asteroidal material should shed light on a period of chemical evolution or planetary evolution which is otherwise inaccessible to study.

INTERSTELLAR SPACE AND COMETS

In recent years radio astronomers have discovered that interstellar space does not consist only of hydrogen atoms at an average density of less than 1 atom/cc. There are also clouds in the interstellar medium where the densities are more like 10 hydrogen atoms/cc. and in some cases may reach 10^4/cc. These clouds apparently contain dust nuclei as well. The dust particles are about 0.1 micron in diameter and about 10^{-12} as abundant as hydrogen atoms. Temperatures in the dust clouds are around $10^{\circ}K$. It appears that these dust particles may serve as catalytic surfaces for the formation of molecules, since it is now known (21) that fairly complex molecules, including organics such as CN, H_2CO, NH_3, HCN, HC_3N, CH_3OH, CH_3CN, and CH_3C_2H, exist in space. Of these species, formaldehyde appears to be one of the most abundant. The similarity of the interstellar molecules to those synthesized in the laboratory under presumed prebiological conditions is striking, suggesting that Chemical Evolution is indeed a universal phenomenon that occurs even under the unlikely conditions of interstellar space. Whether or not this phenomenon is directly related to planet formation and could contribute to the primitive atmosphere of a condensing planet is very doubtful, but understanding what is actually happening is clearly important to understanding Chemical Evolution. It is quite possible that even more complex molecules such as amino acids exist in these clouds; however, at the present time the detection of such large molecules is beyond the sensitivity limit of present telescopes and receivers.

The origin and role of the interstellar dust grains is not understood, and the nature of these grains may be related to the nature of the molecules observed in their presence. The relationship of these organic molecules to those found in comets and in meteorites may be of profound importance to understanding the origin of planets and life, in that they may represent remnants of primitive solar nebulae. If comets are frozen samples of the interstellar medium (22), they could provide a mechanism for dumping large amounts of organic matter from space into the early atmosphere of a planet. Thus, in a sense, the interstellar medium may also be related to the origin of life. Improvements in radio astronomy techniques and spacecraft missions to probe the interplanetary medium and to obtain or analyze cometary material are also needed.

THE PLANETS

Among the giant planets of the outer solar system, Jupiter is exceptionally interesting from the point of view of Planetary Biology.

Jupiter is an extremely massive planet of very low density. The atmosphere contains mostly hydrogen, helium, methane, and ammonia, and probably water as liquid and ice along with traces of ethylene, ethane, silicon hydride, and hydrogen sulfide. This composition resembles in many respects what is thought to have comprised the primitive atmospheres of all planets. Therefore, Jupiter's atmosphere is of great interest because therein may be preserved an example, one that can be examined by deep space probes, of a planetary atmosphere undergoing the processes of chemical evolution which occurred in the very early history of the atmospheres of the terrestrial planets. The occurrence of a high energy flux in the Jovian atmosphere should result in the synthesis of organic molecules; the results of laboratory experiments simulating the Jovian atmospheric environment indicate a high probability that organic molecules have been and are being synthesized on Jupiter (23). Recent observations of Jupiter (24) show an absorption feature at 4.73 microns which could be identical to an absorption feature at 4.72 microns due to an organic polymer synthesized in the laboratory (23).

The compositions of the colored bands and the huge red spot in Jupiter's atmosphere are of great interest to the organic chemists because these features may be regions where highly colored organic compounds are concentrated. Knowledge of the actual chemical and physical processes taking place in the Jovian atmosphere would provide direct evidence to confirm or deny our present notions concerning the early period of a planet's chemical evolution which heretofore were based on very indirect evidence and speculation.

Since the physics and chemistry of the Jovian atmosphere may be of prime importance to our understanding of Chemical Evolution, it becomes important to study that atmosphere by means such as infrared or mass spectrometry. The former can be done from an orbiting or fly-by spacecraft, while an atmospheric probe can employ both means. Such missions should be high on our list of priorities. In the eyes of some scientists, a narrow region of the Jovian atmosphere may be a suitable habitat for life, and that possibility cannot be ruled out.

Titan, one of the satellites of Saturn, is of great interest since it is large enough to hold an atmosphere, which in this case appears to be composed of methane (CH_4) (25). In addition, Titan is a cold body which has a surface on which organic material would collect as it was synthesized in the atmosphere (26). Since there is some question as to whether the organic matter synthesized in the atmosphere of Jupiter would survive or be thermally destroyed as it was swept down toward the interior of the planet, Titan may well be a better object of study for this purpose. Indeed, a fly-by or even a lander on Titan would be an extremely interesting mission.

The remaining three massive planets are Saturn, Uranus, Neptune. We can also consider Pluto in the same category. Although little is known of these planets, it is assumed that with the exception of Pluto, they are at least superficially similar to Jupiter. Some of the satellites of Saturn are probably sufficiently large to have an atmosphere of their own and can therefore be considered of some interest to Planetary Biology. Very little is known of these outer planets and it will probably be some time before a detailed exploration is made of them; so for the time being, our primary interest should lie in the planet Jupiter (27). A closer look at the outer planets can be made in a mission in which a single spacecraft can be launched on a trajectory that makes use of the gravitational field of each successive large planet to assist in accelerating the spacecraft on its flight to and past the remaining planets. Such a trip is of special interest to the planetologists and physicists in that the appropriate alignment of the planets occurs only once in over one hundred years.

The inner planets of the solar system, the so-called terrestrial planets, provide two potentially interesting foci in the search for extraterrestrial life. One, Venus, is shrouded in mystery. The surface of the planet is not visible, being covered by a very dense cloud layer. It is approximately the same size and mass as the Earth, but its atmosphere is massive so that the pressure at the surface is probably between 60 and 100 atmospheres. The surface temperature is extremely high, probably approaching 900° to 1,000°F. The atmosphere is primarily composed of carbon dioxide. In probing the atmosphere of Venus, Soviet spacecraft, Venera 4 and 5, detected the presence of small amounts of water, carbon monoxide, and oxygen, in addition to carbon dioxide (28). Ground-based studies have also shown the presence of traces of hydrofluoric and hydrocholoric acids. Venus is clearly an interesting planet, although the high surface temperatures virtually preclude the possibility of water or organic molecules. Some authors have speculated about the possibility of large amounts of ice in the polar caps of Venus (29), but the evidence for this is practically non-existent. Other speculations have been made concerning the possibility of a bio-zone in the upper atmosphere, perhaps in a region of the cloud layer where the temperatures are in the range of 60-90°F at 1 atmosphere pressure (30), and where carbon dioxide, water vapor, and perhaps oxygen are present. This, however, would require a completely airborne ecology for which there is no terrestrial counterpart. Also, it seems likely that the atmosphere is thoroughly mixed from the surface up to considerable altitudes, so that organic material (or living organisms) in the atmosphere would be continually swept to the surface and destroyed by the high temperatures. Nevertheless, the chemistry of the atmosphere and surface may be of some interest to Planetary Biology. Analyses in the atmosphere at varying altitudes should be considered, perhaps with a buoyant probe with a mass spectrometer in order to determine the presence or absence of organic molecules. Surface temperatures and the distribution of water should also be determined.

Mercury is somewhat larger than the Moon and has a very high density; however, it has little if any atmosphere. The absence of an atmosphere and the very high temperature on the Sun side of Mercury pretty much preclude the possibility of water and of organic molecules. Thus, Mercury does not seem a high priority objective for study, at least as far as Planetary Biology objectives are concerned.

Of all the extraterrestrial bodies in the solar system, the planet Mars is the most interesting from the viewpoint of the search for extraterrestrial life. It is a close neighbor, and we have accumulated more information about it than any other planet except for the Moon (31, 32). Thus our search focuses on Mars. Seen through the telescope the red planet exhibits the gross features that have fascinated astronomers ever since the invention of the telescope, including the orange-reddish color, the polar caps, and the light and dark regions. These gross features have stimulated considerable speculation about the possibilities of life on that planet. The polar cap recedes during the course of the summer and at the same time the dark areas on the surface of the planet become progressively darker. The so called "wave of darkening" has been viewed for many years and has led many astronomers to conclude that this was the response of Martian vegetation to the availability of water from the pole cap. It appeared from ground-based observations and from the 1969 Mariner fly-bys that the pole caps were probably not composed of water, but were primarily composed of frozen carbon dioxide, and that water comprised but a trace of the pole cap material. The day/night cycle on Mars is quite similar to that of the Earth, being almost 24 hours although the seasons are about twice as long as on the Earth. Mars has a tenuous atmosphere, around 10 millibars in pressure at the mean surface, primarily composed of carbon dioxide. The only other molecules that have been detected are water, carbon monoxide, a trace of atomic oxygen in the upper atmosphere, and most recently, O_2 and ozone in small amounts. Water appears to be present in amounts varying from a few to about 50 microns of precipitable water, which is about 1/1000th of that in the Earth's atmosphere. The amount of water in the atmosphere of Mars appears to vary with the season. The temperature at the surface of the planet ranges from as high as 25°C. at the equator during the day to about -100°C. at night so that there is a tremendous diurnal freeze-thaw cycle, even at the equator. The mean temperature is probably about 40°C. below that of the Earth; however, this does not preclude the possibility of biological activity. There appears to be little or no magnetic field, so that the ionizing radiation flux at the surface is quite high. The virtual absence of UV absorbing materials in the atmosphere means that the ultraviolet flux at the surface is virtually unattenuated. This presents a problem for biological activity, but since ultraviolet is relatively easy to shield, it is not considered to preclude biological activity. Nitrogen was not detected in the Martian atmosphere by the Mariner spacecraft; however, atmospheric nitrogen is not a requirement for biological activity. Nitrogen may well be present in the surface material in the form of nitrate or ammonia salts. In addition, the sensitivity of the ultraviolet photometer in Mariner 6 and 7 was such that there could still be as much as 1% nitrogen in the Martian atmosphere and not be detected.

Mariner 6 and 7 flew by Mars in 1969 with a series of instruments on board, including wide and narrow angle television cameras, an infrared radiometer, an infrared spectrometer, and an ultraviolet spectrometer. These flights were extremely successful and produced a great deal of photographic and spectrophotometric data from which the previously cited Martian parameters were derived. These photos included some from the far encounter series showing some of the surface markings on Mars, which turned out to be primarily the result of alignments of craters, and close up photographs of Mars showing the edge of the Martian pole cap and some of the frozen CO_2 filled craters at the edge. Photographs such as this led to the conclusion that the depth of the CO_2 layer at the pole caps may be several meters. Some showed regions on Mars with a heavily cratered surface typical of large areas of the planet, while others illustrate a region of several thousand square kilometers on Mars that is virtually featureless at this resolution. The total absence of craters in this area was rather mystifying. Although the data from Mariner spacecraft have pointed up the hostility of the Mars environment for terrestrial biological activity, they have not scratched the surface of the life question let alone resolved it.   Clearly, landed spacecraft are required. (Table I)

Mariner 9, which orbited Mars in November, 1971, has provided us with a still more provocative view of the planet, based on essentially the same kinds of instruments as on preceding Mariners; television cameras (wide and narrow angle), an ultraviolet spectrometer, an infrared interferometer spectrometer, and an infrared radiometer. The resolution of the Mariner 9 pictures was improved by about a factor of two, and of course, the coverage of the planet by the cameras and other instruments was far greater since this was an orbiting mission rather than a fly-by.

The surface of the planet was obscured by a planet-wide dust storm when Mariner 9 encountered Mars. This, of course, seriously impeded the collection of data for six or more weeks, when the storm began to subside. Eventually the surface cleared so that all the experiments were able to accumulate data over a period of six more weeks before solar conjunction.

In brief, a number of points can be made about Mars in light of Mariner 9 data (still incomplete at this writing) which alters our view of Mars considerably:

1.   The featureless regions of Mars seen in fly-by photos are suggested to be regions where dust is perpetually raised in the atmosphere by continuous winds, thus obscuring the surface from above (33).   This may mean that regions like Hellas which were assumed to be potentially safe landing sites may not be safe after all.   It also suggests that areas on Mars may be periodically or even permanently shielded from solar ultraviolet, which has obvious implications for life.

2.   Some craters appear to be definitely of volcanic origin, indicating a much more geologically active planet than previously thought from earlier Mariner photos.   Some of these volcanoes are of enormous size (by terrestrial standards), and although there is still no evidence of presently active volcanoes, that possibility cannot be ruled out yet (34).

3.   There are numerous areas to be seen on the surface which suggest rather strongly that water erosion has occurred at some time in the past. There are features which are strongly suggestive of river valleys, which are very difficult to explain on the basis of flowing lava or wind erosion. Although only traces of water are detected in the atmosphere, the possibility that water may once have been much more abundant is indicated (35).

VIKING MARS OBSERVATIONS

TABLE I

Sci. Instruments \ Investigators	Upper Atmos: Composition	Upper Atmos: Ion-Species/Conc./Dis	Temp. Profile	Pressure Profile	Lower Atmos: Composition	Temperature	Pressure	Humidity	Wind Direction/Vel.	Clouds	Surface: Topography	Mapping	Phys/Mag. Prop.	Water	Organic Chem.	Inorg. Chem.	Mineralogy	Geology	Biology	Planetology: Variable Features	Ephemeris	Gravity	Natural Satellites	Magnetosphere	Albedo	Rotation Axis	Figure	Other: Relativity	Solar Corona	Interplanetary Medium
Radio		x									x		x								x	x	o	x		x	x	x	x	x
Orbiter Imaging										x	x	x								x			o		x	△	△			
IR Radiometer			x						△	△								o												
IR Spectrometer								x	△	△			△					o					o		x	△	△			
Entry Temperature				x		x																								
Entry Pressure				x			x																							
Upper Atms. MS	x				x																									
Ret. Pot. Anal.		x																												
Lander Active Biol.					△									x	x	x			x	△										
Mol. Anal.					x										x	x	o		x	△										
Meteorology						x	x	x	x										o											
Seismometry																		x												
Surface Sampler													x		x	x	x	x	x											
Magnets												x		o	o	△	o			x										
Imaging									△	△	o			o	o	o	x	o	o	x			△							△

The symbol "x" denotes primary objectives, "o" secondary or supporting functions and "△" areas of potential contribution. In the area of potential contribution, the orbiter IR instruments could support meteorology by changes in location of atmospheric water vapor concentration or, warm or cool air masses; the lander camera contribute to planetology by star fixes or observing Mars natural satellites and; a positive active biology observation could relate to the study of variable features.

4. The polar regions, once believed to be covered only with a CO_2 frost deposit, are now thought to contain massive amounts of water (ice) beneath the surface. There is even the visual suggestion of glacial markings in the polar regions, left after the sublimation of the CO_2 polar caps. It is thought that the large amounts of water tied up beneath the CO_2 polar cap could become available periodically, although at this time it must be considered speculative.

All of the Mariner 9 observations confirm the concept of a much more active and scientifically exciting planet than previously thought, and further study of the enormous amount of data obtained in this mission will provide even more clues into the nature of Mars as well as aid us greatly in the preparations for the Viking lander mission.

The Viking spacecraft, to be launched in 1975, will be an orbiter-lander combination. The orbiter will serve as the relay station for lander data and will perform visual and spectrophotometric experiments in conjunction with the lander. The lander's primary mission has bio-related objectives including the direct search for biological activity, the analysis of soil for organic compounds, the search for water, and meteorological and atmospheric measurements both on the surface and during atmospheric entry. All of the measurements have bearing on the questions of life, organic molecules, and those environmental parameters that are most directly relevant to the existence of life. One instrument is actually three biology experiments in one: it is designed to detect evidence of carbon dioxide and/or carbon monoxide fixation into organic compounds; it will look for the evolution of C_{14} labeled carbon substrates and will measure gas exchange between the atmosphere and a sample of Martian soil as a function of time, with and without the addition of water and/or substrate material to the soil. Such a complex biological experiment is necessary in order to cover a spectrum of possibilities in the metabolic and reproductive characteristics of Martian microorganisms. There will also be a gas chromatograph--mass spectrometer designed for Viking to be used in the search for organic molecules in Martian soil. This is a pyrolysis experiment in which the soil is pyrolyzed and the end products of the pyrolysis are separated chromatographically and identified with the mass spectrometer. In addition to these two experiments there will be cameras, a seismometer, and other instruments designed to monitor pressure, temperature, wind velocities and direction, etc., on the surface.

All of these experiments must, of course, be considered preliminary. It is hoped that they will produce sufficient data to provide us with initial information about the presence of organic matter and the presence or absence of life, so that more sophisticated and detailed experiments can be designed for subsequent missions, including the exploration of Mars by automated roving laboratories, or eventually even by manned expeditions.

The chronology of life-related Mars studies in the 1960's and 1970's can be summarized as follows: (Table II)

Even though there is high probability that organic molecules will be detected on Mars, it cannot be assumed that the presence or absence of Martian life will be established by the Viking Missions. The landings will result in sampling at several spots which may not be highly representative of the total Martian surface. Even if they are representative, it is possible that life may be restricted to some limited Martian environments where liquid water is accessible and protection from ultraviolet light is provided. It is desirable, therefore, to have a mobile laboratory on Mars which will extend the discrete area findings, whether positive or negative, into generalizations for the entire Martian surface.

The bioscience objective of a roving mission, if we obtain positive results on Viking, would be to follow up on such observations in order to determine the nature and distribution of the life detected by Viking. Are they microorganisms? Are they made of the same kinds of molecules as terrestrial forms? Is their metabolism similar or not? Is there more than one form of life on Mars? To answer these questions, uncontaminated and unaltered samples from various sites on Mars must be obtained for chemical analysis with the same general constraints as on Viking-- that is, without prolonged or extreme heating during acquisition, and without being contaminated biologically or chemically by the vehicle.

The following investigations can be anticipated:

I. Life detection--the problem of detecting microbial forms of life with automated mobile equipment is unique to planetary exploration. There are no such pieces of equipment in use on Earth so that each mission requires development of new instruments. Life detection on "Viking" has required the design of an "Active Biology" instrument. A different concept is being developed for post-Viking missions: instead of bringing a surface sample into the instrument for testing, thereby disturbing its natural configuration considerably, we can take the monitoring apparatus to the sample, thus preserving the natural ecology of the surface to be studied.

TABLE II

	Life Related Objectives	Mariner 6 & 7 Results (1969)	Mariner 9 (1971) Investigations	Viking 1975-76 Orbiter	Viking Lander Impact (1976)
Cameras	Visual coverage	Craters, Rough and smooth terrain, little erosion evident, limited coverage H_2O in W-Cloud	Like 6 & 7 plus much more surface coverage, day-night plus seasonal changes Higher resolution	Like 6 & 7 plus higher resolution, better coverage	Landing site selection, Topography
I.R. Radiometer	Temperature of Surface, Pole cap	$-100°$ to $25°C$. $-120°C$.	"	"	Search for warm region
I.R. Spectrometer	Atmosphere Temperature, Atmosphere Comp. H_2O CO_2 Path Length	CO_2 - Atm. CO_2 -Polar Ice CO O - Upper Atm. Topography	Plus water	Plus H_2O	Search for wet region
S-Band Radio (Occultation)	Atmospheric Pressure, Topography	6-7 mb - 3 sites 3.8 mb in Hellespontus	"	"	Search for low regions with higher pressure
U.V. Spectrometer	Clouds, Atmosphere Comp., Radiation flux, Pressure O_3 conc.	N_2 5% U.V. unattenuated at surface No O_3 (Pole?) Topography	"	N.A.	Search for low regions with higher pressure

The instrument concept requires a boom capable of extending beyond the vehicle and its sphere of potential contamination and reaching the surface.   An encapsulating device is required, capable of enclosing (from above) about 6-12 cubic inches of surface material and air space.   This space is monitored for changes in gas composition (O_2, H_2O, N_2, NH_3, CH_4, CO_2, CO, SO_2, H_2S) as a function of time using a gas chromatographic monitoring system.   The gas chromatograph can be similar to the one in the Viking Biology instrument. It is also possible that a small probe containing a gas chromatographic monitoring system, as in Viking, can be inserted into the ground to monitor gas exchange in situ.

The above described experiments will be suitable for providing information on the distribution of organisms, and some information on their metabolism; however, in anticipation of returning samples to Earth for additional study, an assessment of the more detailed characteristics of life forms must be made in order to determine to some degree the potential danger such organisms pose for life on Earth. Such experiments cannot be designed at this time since they will depend on earlier results, but may well include the capability for:
    A. Automated microscopy--resolution comparable to oil immersion
          (1000 x magnification) on Earth.
    B. Culture capability on a variety of media with video-monitoring.
    C. Immunological testing.
All three will require considerable weight and power, as well as sophisticated instrumentation.

II.   Organic Chemistry--the detection of organic matter on Mars by Viking and/or subsequent missions will impose a requirement for further and more detailed analyses.   The detection of life forms will also require additional analytical capabilities.   The basic Viking Gas Chromatograph--Mass Spectrometer (GC-MS) should be retained, but modified as follows: the detection of organic matter by Viking will leave unanswered the questions of a) the distribution of organics, b) the exact molecules detected, 3) whether or not the organics detected are of biological origin and,   d) the composition of life forms, if present.   Modification of the Viking instrument will allow such tests.

The GC-MS could be modified so that: a) it can be used to monitor the gas exchange in acquired soil samples, b) it can monitor the atmosphere, c) it can accept samples from the biology experiment.

In addition, the capability for "wet chemistry" should be added. This includes: a) a soil extraction system. This hardware now exists in breadboard form, and provides the capability of extracting the soil with suit-able solvents, thus separating the organic compounds from the soil par-ticles before further analysis and concentrating the organic compounds to reduce background noise. This will allow better separation of specific molecules, and thus better identification than possible in Viking. For example, we will not only know that amino acids are present, but what amino acids and in what proportion. b) A derivatizer system. This de-vice can be used in making optically active derivatives of the amino acids. In this way we will be able to determine the optical activity of the original amino acids and thus probably determine whether they were of biological or nonbiological origin. The derivatizer system will be used in conjunc-tion with the gas chromatograph or the gas chromatograph-mass spectro-meter.

All of these experiments should be designed to be repeatable each time the rover stops at a new site. Sampling sites should be selected and monitored visually.

III. Detection and measurement of water--the capability of deter-mining the amount and nature (bound and free) of water in and near the Mars surface. This is the most important variable with regard to the possible existence and distribution of life on Mars. A very sensitive technique is needed because of the apparent low amount of water present. Water may well be the key to Martian Biology, and if life is not detected in the Viking mission, subsequent missions should seek out above-average water concentrations on the planet as being high priority sites for de-tailed study.

The exploration of Mars will be a sequential study, in which one series of investigations will be designed to follow up on the results of earlier investigations. Much will depend on the indications from Viking as to the presence or absence of life and organic molecules. Certainly, the detection of life or a rich organic fraction will engender a great deal of interest in continued exploration of the planet. The complete absence of such indicators will not lessen the value of Mars as an object of great scientific curiosity, although such a discovery will very likely lessen the popular interest in the planet.

SUMMARY

In "Priorities for Space Research, 1971-1980" (1971, p. 12) the Space Science Board of the National Research Council concluded that "the highest life-sciences priority in space is the search for extraterrestrial life and for the fundamental understanding of life's origins (Exbiology). This search, coupled with the investigation of the origin and evolution of the solar system, constitute the essential elements of a quest for a 'cosmic perspective' for mankind, a quest that may illuminate our long evolutionary path from elemental origins in the primal fireball, through interstellar dust clouds, for formation of galaxies, solar systems, and planets (inorganic chemical evolution), to the first stirrings of life in the Earth's primitive oceans (organic chemical evolution) and the beginnings of biological evolution. We seek to understand the generality of such a chain of events in the universe so that we may attain a new perspective of life in the cosmos that is both enhanced by knowledge of its universality and unclouded by our geocentricity."

The origin and evolution of life is inextricably bound to the conditions of primitive planetary environments. Planetary evolution may not be conducive to the origin of life; but if it is, and life begins, organisms will interact with, alter, and be altered by the planet's environment. NASA recognizes the intimate connection between the evolutions of life and planets and has made Exobiology an integral element in its Planetary Programs. In a proper search for knowledge concerning extraterrestrial life and the origin of life, NASA must explore as many extraterrestrial bodies as possible in our solar system for the relevant information they can supply. In particular the search should involve exploration of a continuum of planetary possibilities including bodies totally devoid of organic chemicals, those conceivably undergoing (or having undergone) organic chemical evolution and those possibly harboring life.

Because all present biology is based on the evolutionary sequence that occurred on Earth, the discovery of life on another planet could revolutionize the field of biology. The impact on other areas of intellectual and scientific endeavor could be equally important. Indeed, "America's Next Decade in Space: A Report for the Space Task Group" (1969, p. 72) states boldly: "The discovery of extraterrestrial life would likely rank as the greatest scientific discovery of the century. If extraterrestrial life is found to resemble terrestrial life in biochemistry and cellular organization, the implication that life in general evolves along rather similar lines would confirm our notions about chemical and biological evolution and lead to new insights into the nature of living organisms. If extraterrestrial life turns out to be fundamentally different from terrestrial life, then our concepts about comparative biology would require a broadening, reshaping and new synthesis."

The discovery of lifeless planets would provide examples of where chemical or biological evolution ended. On them it may be possible to study the remnants of organic chemical evolution or of past life and to learn how planetary evolution may have broken the thread of chemical or biological evolution. Examination of planets devoid of organic chemicals can provide us either with a calibration point in the process of chemical evolution which corresponds to a time prior to the formation of organic chemicals or with insight into how the remnants of organic chemical evolution may have been obliterated by processes of planetary evolution. Clearly, even the discovery that there is no life on a planet is of high interest and importance because the extant planetary conditions and what we can learn of its past history constitute basic data important to the general theory of the origin of life.

Before the decade of the 1960's, the search for direct evidence of life in extraterrestrial bodies was limited to Earth-bound laboratory studies of meteorites. No evidence of extraterrestrial life was detected, and terrestrial contamination confounded any adequate assessment of the real significance of organic compounds isolated from meteorites. Only recently, in the new fallen Murchison meteorite and in the interstellar medium, were organic compounds of unambiguous extraterrestrial origin discovered, thereby providing clear evidence that organic chemical evolution was taking place or took place elsewhere in the solar system.

The success of the Apollo flights (1969–1971) had made lunar soils and rocks available for close scrutiny. These were the first extraterrestrial samples whose contamination history was closely documented. It was hoped that analysis of the organic substances in lunar material could confirm or shed new light on extraterrestrial chemical evolution and the mechanisms for synthesis of carbon compounds in the solar system. In fact, very low concentration (parts per billion) of organic compounds were found in the samples. This is consistent with geological and mineralogical evidence that high temperature and high energy processes were involved in formation and alteration of the soils and rocks returned from the Moon, making survival of organic compounds unlikely. Apparently, if organic chemical evolution or life occurred on the Moon, little evidence of it remains in the surface material. However, the significance to Planetary Biology of lunar carbon and organic matter remains to be fully evaluated in the light of increasing knowledge of the surface chemistry and the evolutionary history of the Moon's surface.

The exploration of Mars, beginning with the telescope, fly-by and orbiting spacecraft, followed by landed laboratories, and perhaps culminating eventually in manned expeditions, may well provide a significant portion of the data we seek about the origin of life and introduce a new era of understanding of man's place in the universe.

REFERENCES

1. Oparin, A. I., "Proischogdenie Ahizni," Moscovsky Robotchu, Moscow, 1924.
2. Haldane, J. B. S., Rationalist Annual 148, 3 (1928).
3. Fox, S. W., McCauley, R., Joseph, D., Windsor, C. R., and Yuyama, S., "Life Sciences and Space Res. IV" (A. H. Brown and M. Florkin, eds.), pp. 111-120. Spartan, Washington, D. C., 1965.
4. Fox, S. W., editor), "The Originals of Prebiological Systems," Academic Press, New York, 1965.
5. Jeon, K. W., Lorch, J. J., and Danielli, J. F., Science 167, 1626, (1970).
6. Hinners, N. W., Reviews of Geophysics and Space Physics 9, 447 (1971).
7. Walkinshaw, C. H., Sweet, H. C., Venketeswaran, S., and Horne, W. H., Bioscience 20, 1297 (1970).
8. Oyama, V. I., Mercer, E. L., Silverman, M. P., and Boylen, C. W., Proc. Second Lunar Sci. Conf. Geochim. Cosmochim. Acta, Suppl. 2, Vol. 2, p. 1931. M. I. T. Press, Cambridge, 1971.
9. Taylor, G. R., Ellis, W., Johnson, P. H., Kropp, K., and Groves, T., ibid., p. 1939.
10. Kaplan, I. R. and Petrowski, C., ibid., p. 1397.
11. Abell, P. I., Cadogan, P. H., Eglinton, C., Maxwell, J. R., and Pillinger, C. T., ibid., p. 1843.
12. Gibson, E. K., Jr. and Johnson, S. M., ibid., p. 1351.
13. Hayes, J., Space Life Sciences, (1972), in press.
14. Hodgson, G. W., Bunnenberg, E., Halpern, B., Peterson, E., Kvenvolden, K. A., Ponnaperuma, A., Proc. Second Lunar Sci. Conf. Geochim. Cosmochim. Acta., Suppl. 2, Vol. 2, p. 1865.
15. Harada, K., Hare, P. E., Windsor, C. R., and Fox, S. W., Science 173, 433 (1971).
16. Nagy, B., Modzeleski, J. E., Modzeleski, V. E., Mohammad, M. A. J., Nagy, L. A., Scott, W. M., Drew, C. M., Thomas, J. E., Ward, R., and Urey, H. C., Nature 232, 94 (1971).
17. Gehrke, C. W., Aue, W. A., Stalling, D. L., Duffield, A., Kvenvolden, K. A., Ponnamperuma, C., and Zumwalt, R. W., Proc. First Lunar Sci. Conf., Geochim. Cosmochim. Acta. Suppl. 1,

Vol. 2, p. 1845. Pergamon, New York, 1970.

18. Oro, J., Flory, D. A., Gilbert, J. M., McReynolds, J., Lichtenstein, H. A., and Wikstrom, S., Proc. Second Lunar Sci. Conf., Geochim. Cosmochim. Acta. Suppl. 2, Vol. 2, p. 1913. M. I. T. Press, Cambridge, 1971.

19. Kvenvolden, K. A., Lawless, J. G., and Ponnamperuma, C., Proc. Natl. Acad. Sci. 68, 486 (1971).

20. Lawless, J. G., Kvenvolden, K. A., Peterson, E., and Ponnamperuma, C., Science 173, 626 (1971).

21. Snyder, L. E., and Buhl, D., Sky and Telescope 40, 267, 345 (1970).

22. Lyttleton, R. A., "The Comets and Their Origins," Cambridge University Press, Cambridge, England, 1953.

23. Woeller, F., and Ponnamperuma, C., Icarus 10, 386 (1969).

24. Munch, G., and Neugebauer, G., Science 174, 940 (1971).

25. Lewis, J. S., Science 172, 1127 (1971); Icarus 15, 174 (1971)

26. Trafton, L. M., Astrophys. J., (1972), in press.

27. Sagan, C., Space Science Reviews 11, 827 (1971).

28. Avduevskii, S., Marov, I. A., Rozhdestvenskii, M. K., Borodin, N. F., and Kerzhanovich, V. V., J. Atmos. Sci. 27, 561 (1970). ibid., 28, 263 (1971).

29. Libby, W. F., Science 159, 1097 (1968).

30. Morowitz, H., and Sagan, C., Nature 215, 1259 (1967).

31. Ponnamperuma, C., and Klein, H. P., Quart. Rev. of Biology 45, 235 (1970).

32. Special Issue, Journal of Geophysical Research 76, No. 2 (1971).

33. Sagan, C., Veverka, J., and Gierasch, P., Icarus 15, 253 (1971).

34. Masursky, H., Icarus (1972), in press.

35. Sagan, C., Icarus 15, 279 (1971).

Received 15 September, 1971

ON RECOGNISING LIFE

N. W. Pirie

Rothamsted Experimental Station
Harpenden, Herts., Britain

The three essential, one might in this context say <u>vital</u>, components of science are observation, logic and assumption. It is conventional to condemn, or reluctantly condone, the third component, but it is assumption that integrates the scientific structure. It controls the directions of observation, and it is the fashionable body of assumption at any given time that supplies the canon by which the acceptability of logical processes is judged. Assertions such as these are anathema to most of those who teach science and to many who practice it. They like to think (or assume) that science is a steady progression in which old uncertainties are gradually replaced by certainty based on increasing knowledge. They forget, or repress, the fact that now-discarded assumptions were not held tentatively but with positive conviction as strong as that with which contemporary orthodoxies are held. An assumption that is later discarded, or even found in retrospect to be ridiculous, need not at its inception be harmful. Thus the assumption that something was coming out during combustion was formulated into the "Phlogiston Hypothesis" and that guided the activities of those who constructed the main framework of inorganic chemistry. They would have been rudderless without it. As someone remarked "Truth is more likely to emerge from error than from confusion."

To be useful an assumption need not be recognized. Thus the laws of motion were formulated and used more than a century before the principle of "Conservation of Energy" was enunciated. But if, in every-day life, energy had appeared and disappeared casually, the laws of motion

would have had little validity. It is hard to believe that Newton was un-
aware of this. Nevertheless, assumptions become more useful when they
are recognized because experimentation is stimulated by the process of
clarifying the exact nature of the assumptions we like and, for the time
being at any rate, mean to retain; those that are rejected leave a gap
that is uncomfortable once we are aware of it. Discomfort is a stimulus
to experimentation.

The techniques chosen by different groups of scientists planning the
search for extraterrestrial life, will depend partly on the techniques with
which they are familiar in their more conventional lines of research, but
more on the assumptions they, consciously or unconsciously, make about
the nature of life and the essential characteristics required before something is classified as living. The lists of qualities that used to appear at
the beginning of biology textbooks are now usually omitted: too many
students are aware of systems that they wish to include in the living category although they do not manifest some of these characteristics, and of
non-living systems that do. The only characteristic that is absolutely
essential in a still-living system is that it should sometimes do something. Decisions about the status of a system--whether it is to be called
living or not--amount to decisions about what we expect it to do.

The necessary but not sufficient requirement that the system should
do something, entails of necessity the conclusion that an organism must
be able to control energy. The preeminent ways of doing this are to
capture radiant energy, e.g. light, or to establish equilibrium in an environment that some agency other than the organism has disequilibrated.
The familiar photosynthetic organisms use the first mechanism and the
saprophytes the second. Articles dealing with the origins of life now rely
heavily on the primal importance of the second mechanism. They point
out, correctly, that in an illuminated probiotic environment containing
simple compounds of carbon, nitrogen etc., complex molecules would be
synthesised; many examples of non-biological syntheses have been described during the past century. Evidence is now accumulating that molecules containing 6 or 7 atoms can even be formed in space. It is therefore
legitimate to postulate a layer of complex molecules on the surface of a
probiotic planet or asteroid. So far so good. What is not so clear is the
nature of the reaction that an eobiont would catalyse to get the energy it
must control.

Contemporary saprophytes on Earth re-establish the equilibrium
that green plants upset when they produce oxygen and reducing material.
They also "ferment" partly oxidised material, e.g. glucose, into more

and less oxidised products, e.g. alcohol and carbon dioxide. Both mechanisms depend on the continuing existence of living plants, and these would probably be as obvious in any search for xenobionts, or alien life, as the saprophytes that depended on them. Alternatively, for as long as a mixture of simple molecules is illuminated, some level of disequilibrium will be maintained; water will be dissociated and unstable molecules will be formed. In the absence of light, equilibrium would soon be restored, but by participating in the restoration an organism could gain the energy it needs. It is very unlikely that a probiotic environment would be so free from inorganic catalysts that any large concentrations of molecules not in equilibrium would accumulate. There may be a great deal of organic matter, as there seems to be on Jupiter, but it would be largely in equilibrium. The vision held out by some scientists of a vast ocean of organic matter waiting to be seeded, and thereafter teeming with life, is an illusion for, if there were anything for the organic matter to react with, it would have reacted.

There is no unanimity about the amount and composition of the "probiotic soup." Although possibly abundant it would probably have been too inert to sustain anaerobic fermentative reactions unless reactive molecules in it were being constantly replenished by an external energy-source such as light. Other sources, e.g. lightning and wave action, depend, in part at least, on light. I suggest, as a consequence of this line of argument, that eobionts could depend on the metastability of the chemical environment for tens rather than millions of years and thereafter would depend, as life does now, on illumination. Illuminated bodies are therefore the ones most worth searching for signs of life and the level of vital activity to be expected will depend on the intensity of the illumination.

Words such as "living" and "alive" have two antitheses. "Dead" means "no longer alive"; the antithesis with which we are here concerned is "never having been alive." Evidence for past life is therefore just as significant as evidence for persisting life, but evidence for persisting life could be so much more definite that it may conveniently be considered first. Tesla (1901) claimed the reception of messages but Tsiolkovsky (Tsvetikov 1960) thought other intelligences would have disdained communication with us. If it were established that some universal, such as the first few members of the sequence of prime numbers in binary, were being received, we would be forced to believe in vital activity somewhere. If there were a community at our technical level on Mars, it would be able to detect terrestrial TV and would, presumably, respond. Transmissions between more widely separated bodies would involve an intimidating expenditure of power. It may therefore be that in the cosmic situation, unlike the familiar domestic one, everyone is listening and no one talking. Even attempts to listen have not been very assiduous; the absence of unequivocal success may therefore mean little.

The large-scale, seasonal colour changes on Mars can be interpreted either as changes in the texture or hydration of a mineral, or as the growth of organisms. The postulated organisms are usually called plants because one colour phase is greenish and we are more familiar with green than purple photosynthetic organisms. The point is not likely to be cleared up by a fly-by; an instrumented landing could, however, give definite information. Something that looks like a bush or tree will be very much more convincing than any amount of spectroscopic or biochemical evidence. The surface of Mars has probably been available for the growth of organisms for as long as, or longer than, Earth's surface. There is no evidence that the trend of evolution is making plants smaller here. There is no reason therefore to expect a photosynthetic system on Mars to be either small or primitive; as on Earth, the greater the area exposed to light the more successful a photosynthetic organisms is likely to be. Geometrical necessity may lead to a multi-layered structure to facilitate this exposure, but, however grotesque the object seen may be, if only one is seen it will be hard to classify. A grove will be disproportionately more convincing than a tree. Multiplicity of similar objects will also help to distinguish organisms from top-heavy, or in other ways improbable and metastable, products of erosion. Visible growth or spread will be still more convincing. One phrase in a satisfactory definition of life would state that the organism gets its shape by development and not by the differential removal of material. Conviction will come more readily if development is by growth from within rather than accretion from without. That is why crystal growth does not puzzle the would-be framers of a definition of life whereas the structures formed osmotically in sodium silicate solutions do. It would, nevertheless, be difficult to categorise structures such as "desert roses" (sand grains cemented together by barytes or gypsum) by observations made from a distance.

A TV transmitter, landing at random on Earth, would have about a 1 in 50 chance of viewing a macroscopic biological object. If it were aimed so as to avoid the larger areas of water, ice and desert, its chances would increase to about 1 in 5. It is not therefore reasonable to expect a definite answer from some such body as Mars until there have been several landings. If macroscopic organisms are mobile, it will be easier to categorise them, but the opposed effects of fear and curiosity will have unpredictable effects on the chance of seeing them. From this line of argument it follows that ease of categorisation will depend on the rate at which an organism changes, grows, or moves.   If low temperature, feeble illumination, or scarcity of essential metabolites make metabolism very sluggish, categorisation will be difficult.   So too if action is very rapid.   The assertion that some substances are too unstable to play a part in the metabolism of xenobionts, and the assumptions underlying some proposed life-detecting contrivances, are valid only if the terrestrial rate-of-living is universal.

Apart from their preconceptions about time, observers working along the lines discussed in the last three paragraphs will not be seriously inhibited by the assumptions they make.   They will observe as much as they can and interpret the observations as best they can.   Assumptions become more important as the landing of instruments designed to detect organisms too primitve, or small, to send messages or show up on a TV screen becomes more imminent.   The consensus of opinion was that there would be no life on the moon though several scientists argued that there once was life there. Experience seems to support the first opinion, the second may be regarded as still open.   Nevertheless the reasons that led to an apparently correct conclusion are worth examining because they depend on a set of assumptions that, when stated clearly, would be difficult to justify.

The surface of the moon, we are told, is too dry, too exposed to ultraviolet radiation, and too variable in temperature for life.   This amounts to assuming that the only possible form of life is life with the characteristics with which we are familiar.   Those who make this assumption, if they are thoughtful, are then astounded at the perfection with which the earthly environment is adapted for life.   The astounding thing is that the scientists have become so well adapted to inversion.   Organisms have developed in the way they have because the environment is like that: had it been different, organisms would have been appropriately different.   Thus, extreme changes in environmental temperature could be useful as a source of energy, at least for movement. It may be difficult to imagine an organism without a liquid phase, but an involatile hydrocarbon could, with a different chemistry, serve instead of water.   And organisms that relied more on carbohydrates and fats, and less on proteins and nucleic acids, would be less vulnerable to ultraviolet light.   The immediate objection raised to suggestions such as these is that we know of no examples of such organisms. This is not unexpected; they would not be well adapted for living in this environment. Even the incomplete fossil record that we have shows that many more species are extinct than extant.   Incipient organisms, or eobionts, would be especially liable to predation by those that are more highly evolved and better adapted.   The essential point of

my set of assumptions about biology is that all we see here is the successful end-product of a vast amount of elimination, and the end-product
need tell us nothing whatever about the course of biochemical evolution
nor about the alternative courses that would be successful in another environment.

There are, on Earth, unusual environments and it may be argued
that anomalous forms of life, if they can exist, should be found there.
The abyss is the most extreme extensive environment.   So far as is
known, the biochemical processes of the organisms dwelling there are
similar to those at the surface: that is not surprising; because of the
absence of light, these organisms are wholly dependent on detritus contributed by surface organisms.   There is light during the Antarctic summer and it can be argued that, if anhydrous low-temperature life were
possible, it should exist on Antarctica.   The only probable sites would
be permanently exposed rock faces because snow and ice surfaces are
constantly being covered by fresh falls and the universal principles of
chemistry suggest that cold-adapted life would operate slowly.   Those
who have studied Antarctic rock have been so surprised by finding a
sparse cover of more-or-less conventional organisms that they have devoted little attention to the search for anomalous ones.   Nevertheless a
few are already known that use antifreeze.

At the other end of the temperature scale a point arises that is similar to the one raised by the brief existence of an illuminated snow surface before it is covered over.   No organism is known that can multiply
at 100°; although many can survive that temperature in cryptobiotic states.
The total area of illuminated terrestrial surface that approaches 100° is
extremely small and, at our atmospheric pressure, only a concentrated
solution of low molecular weight material could have a greater temperature than this.   A biological generalisation of universal application is
"organisms are opportunistic." If an extensive environment exists for a
reasonable time, they are likely to adapt so as to colonise it, but there
is no advantage in adapting to a trivial environment.   The quantitative
values that should be put to the words "extensive" and "reasonable" are
unknown, though the 2G years that seem to have elapsed on Earth before
biology advanced to the stage of producing something able to form an unequivocal fossil may offer a suggestion.   My thesis is that hot springs at
nearly 100° do not remain active at the same site for the time needed for
selection to produce a set of proteins or other catalysts able to function
at 100°. If an even hotter environment existed somewhere for a prolonged
period, in circumstances where some form of metabolism is possible,
that is to say, where there is a source of energy (e.g. light) or chemical

metastability, there seems to be no reason to assume that organisms would not adapt to it.   The negative evidence that no such system is known here is irrelevant, and the assumption that no such system could exist is unreasonably limiting.

All the organisms that have been thoroughly studied contain protein and nucleic acid; they also contain fat, carbohydrate and other components.   It is widely assumed that the proteins and nucleic acids are the fundamental components and it is unquestionable that they take part in a very wide range of vital processes.   There is no reason for surprise at that considering the immense amount of biochemical effort devoted to them.   This is another example of inverted assumption.   Just as one of a pair of rivers, rising in the same region, will gradually capture some of the tributaries of the other and thus become larger and flow through a more impressive valley, so a successful line of research, with well established techniques, will capture scientific attention.   There is no more reason for surprise at scientists following fashion than at large rivers flowing through large valleys.   Each process is self enhancing.   When someone devises methods for studying the actual conformation and disposition of, let us say, fats in vivo, a very productive fashion will probably start.

Although it may be premature to assume that proteins are essential components of living systems, and prudent to keep the existence of non-protein catalysts in mind, the observation that amino acids are formed non-biologically in various gas mixtures, and that they polymerise to form proteinoids, makes it reasonable to expect that eobionts anywhere will use them. The question is "To what extent?" The facts that protein-like material could have accumulated in probiotic conditions, and that they play a fundamental role in life on Earth, are very interesting; however, they raise another possibility of inverted thinking.   The association of the two facts could be interpreted as a colossal coincidence; it could be used as evidence that unless proteins had appeared spontaneously there could have been no life; or it could mean that eobionts made opportunistic use of the material that was available.   Had there been other polymers, eobionts might well have started by making catalysts out of them.

We do not yet know what it is about protein structure that gives a few proteins such outstanding catalytic power; the first result of gaining that knowledge may be the synthesis of non-protein enzyme analogues. The particular system of catalysts that attains dominance may depend as much on accident as on the initial composition of the surface layers of a

planet.   My preconceptions would attach more importance to the tars
and resins that are formed during <u>in vitro</u> non-biological syntheses than
to the chemically characterised small molecules.   The tar-covered
surface of a mineral, containing one of the more reactive metals, would
seem to have interesting catalytic possibilities from which evolution
might start.

Once organisms have achieved a measure of catalytic coordination,
there will be competition between them, that is to say, some will sequest-
rate more of the useful elements, and occupy better sites, than others.
With the beginnings of predation, biochemical uniformity would be established because of the elimination of the less successful experiments in
living.   That was realised by most of those who wrote on the subject 60
to 100 years ago.   The most successful predator is not necessarily the
most efficient biochemical unit.   Thus we dominate Earth but, unlike
most mammals, do not synthesise ascorbic acid and, unlike plants, do
not synthesise the essential amino acids.   The relevant point is that an
organism that has coordinated one set of interrelated biochemical reactions will be in a position to eliminate an eobiont which, by definition,
is in a preliminary and less coordinated state even although that eobiont
depended on an equally good, or possibly better, set of biochemical
mechanisms. In the early stages, accident rather than merit may decide
which mechanism becomes dominant.

The consequence of this line of argument is that the techniques used
in the search for xenobionts should not be narrowly conceived as if only
slightly aberrant chthonobionts were being sought. Because of its cosmic
abundance and chemical potentialities, carbon is probably an important
element in any form of life.   But, in different environments, different
groups of carbon compounds may dominate the scene.

It is possible that life is similar throughout the universe.   This
could happen because, for unknown chemical reasons, there are very
few chemical processes on which it can depend; or because early forms
were already present on the planetesimals that (according to a widely
held astronomical theory) accreted to build up planets; or because, as
Haldane suggested, cultures were sent to different planets.   In different
environments, different features of the universal biochemistry could
then become dominant.   We tend to think of biochemistry in terms of
nearly neutral aqueous solutions containing protein structures that are
unstable in other conditions.   Some organisms however maintain conditions dramatically different from these.   For example, they work at extremes of pH and manipulate phenolic compounds that would inactivate

conventional enzyme systems. It is important to note that these bizarre capacities are found in ancient rather than recently-evolved species so that it is not unreasonable to think that biochemistry was initially somewhat more catholic than it is now. We do not know what factors it is in the environment that have permitted the survival of these biochemically aberrant organisms. In another environment, mechanisms that seem aberrant here may be normal and this possibility should be envisaged in designing life-detecting equipment.

The problems raised by the study of no-longer active specimens that might give evidence for the existence of life are mainly aesthetic. There are structures in meteorites (excluding those structures that can be dismissed as terrestrial contaminants) that seem to some people to show biological organisation comparable to that found in the more ancient fossils. Their status remains uncertain. Samples brought back to Earth from elsewhere may pose similar problems. It would be well therefore if, in anticipation of this, an atlas were prepared of life-like but indubitably non-biological structures. It should include both macroscopic structures from crystallisation, diffusion and evaporation, and microscopic structures resulting in colloidal precipitates. An extraterrestrial structure that could be matched in the atlas would clearly be suspect.

There is now widespread agreement that organic compounds in extraterrestrial material need not be of biological origin: that origin would become more probable if the material showed peculiarities with which we are familiar here. Such evidence works one way. For reasons already stated, material that is unfamiliar is not necessarily non-biological. The selection of one from a pair of stereoisomers is a well known characteristic of terrestrial organisms and it is reasonable to assume that any other organisms will be selective also: otherwise there would be inefficient duplication of synthetic mechanisms within them. The recognition of optical activity in an extraterrestrial sample would therefore be of great significance, but it would not show conclusively that the sample came from an organism. Several systems are now known, or shown to be theoretically possible, that would produce one stereoisomer if there were any initial bias--and initial bias must be assumed.

It is impossible to lay-down precise rules for recognising an undefined entity. Even entities that are defined are often fuzzy at some of their limits. Thus there is now no uncertainty about whether an organism is a mammal, a bird or a fish: the evolutionary "tree" gives clear answers to questions concerned with its horizontal structure. When questioned vertically there are no sharp distinctions; species blend into one another imperceptibly. Furthermore, as soon as a definition is framed, nimble-witted people can find, or postulate, problem entities. Thus, in the early phase of organic chemistry, the aromatic category seemed clear. The growth of knowledge has produced pseudo-aromatics that would surprise the original framers of the definition. Similarly, thinking is usually accepted as an attribute of people and possibly of higher animals, but as soon as an operational definition of thinking is drawn up, a computer can be programmed to meet it. Life has not now, and may never have, even as assured an entity-status as a species, an aromatic, or a thought. The use of the word "life", like the use of the word "order" relates to the attitude of mind of the speaker rather than to the entity spoken of. The recognition of the presence or absence of these things is a matter of opinion and aesthetics.

In this article I have discussed some possible criteria for an organism in the order of their convincingness when taken in isolation. Regardless of its chemical composition, I would accept something that thinks either as being an organism or as made by an organism; similarly if its movement seems purposive. After that things become uncertain. The acceptability of apparently metabolising units would depend on their morphology, uniformity and manner of growth. When all that is recognised is a catalytic activity substantially greater than that commonly found in minerals, it will be very difficult to come to a satisfying conclusion. Another general uncertainty must be added to these: it is as true of science as it is of politics "What we anticipate seldom occurs; what we least expect generally happens." If there are xenobionts, they will be constrained by the universal principles of geometry, chemistry and physics, but within these constraints, they may have evolved mechanisms that are totally unforeseen.

Received 1 July, 1971

THERMODYNAMIC AND PHILOSOPHICAL CONSIDERATIONS

A NON-EQUILIBRIUM THERMODYNAMICAL ANALYSIS

OF THE ORIGIN OF LIFE

Ronald Forrest Fox*

Department of Physics
University of California at Berkeley
Berkeley, California 94720

INTRODUCTION

This presentation will attempt to bring together the basic concepts of non-equilibrium thermodynamics and biochemical energetics, in order to sharply define the problem of the origin of life. Having defined the problem, an assessment of its current state of solution will be presented. In addition, a number of new experiments will be suggested.

An analysis of the energetics of cellular metabolism requires two fundamental distinctions. Firstly, one must distinguish between arbitrary systems, in which the time evolution is governed by the entropy, and thermally buffered systems, in which the time evolution is governed by the free energy. Secondly, one must distinguish between systems closed with respect to energy inputs, and systems open with respect to energy inputs. Very generally, the study of the non-equilibrium thermodynamics of cellular metabolism requires the study of thermally buffered molecular systems open to energy inputs, whereas the literature of physics usually considers only the case of arbitrary systems with no energy inputs. Consequently, the natural connection between the physics and the biochemical phenomenology is not always perceived.

---

*Postdoctoral Fellow of the Miller Institute for Basic Research in Science. Present address: School of Physics, Georgia Institute of Technology, Atlanta, Georgia 30332.

Once the proper connection between the physics and the biochemical phenomenology is clarified, the details of cellular metabolism acquire a unifying overview.   The maintenance of cellular functions and the consequent long term evolution of cellular types are seen to obey traditional physical and chemical principles.   However, because the functioning of cellular metabolism involves "integrated feedbacks" to keep it going, the problem of the origin of life becomes the generation of a special, integrated, molecular feedback system when initially there is none.

The assessment of the current state of the solution to this problem is that the construction of a simple unicellular organism with evolutionary potential is possible.   In the following this claim will be made more plausible.

NON-EQUILIBRIUM THERMODYNAMICS

We shall consider a molecular system comprised of N_i atomic nuclei of species i and charge Z_i along with a complement of $\Sigma_i N_i Z_i$ electrons. For nuclei we have in mind such species as the nuclei of H, C, N, O, Na, Mg, P, S, Cl, K, Ca and Fe.   There exists a Hamiltonian for this system which determines the eigenstates of the systems and their energies (1). The jth eigenstate will be denoted by $u_j(q_1 \ldots q_N)$ which is a function of the spatial and spin coordinates of all of the constituent nuclei and electrons.   The jth energy value is denoted by E_j.   From $u_j^*(q_1 \ldots q_N)u_j$ $(q_1 \ldots q_N)$ is is possible to compute, in principle, the relative likelihoods that, for example, a carbon nucleus and a nitrogen nucleus will be specified distances apart.   In general a variety of nuclei-nuclei spatial correlations are implicit in each $u_j^*(q_1 \ldots q_N)u_j(q_1 \ldots q_N)$.   Each eigenstate is weighted by a probability coefficient $P_j(t)$ such that $\Sigma_j P_j(t)$ = 1 for all t.   It is the time evolution of the $P_j(t)$'s which determines the temporal behavior of the system (2).   Consequently, $u_j^*(q_1 \ldots q_N)u_j$ $(q_1 \ldots q_N)$ and $P_j(t)$ together provide sufficient information to determine the spatial-temporal correlations among all the constituent particles.   At any instance, t, the particular nuclei-nuclei spatial correlations provide the molecular content of the system by determining which bonds exist. The time variation of the $P_j(t)$'s determines the molecular reactions which are taking place, through temporal correlations.

Usually, the mass action laws are used to describe the time evolution of the concentrations of various molecular species in the molecular system (3).   Here we are giving more information since the $u_j^*(q_1 \ldots q_N)u_j(q_1 \ldots q_N)$'s provide information concerning the location of each

molecule of each specific type. The mass action equations provide a contracted description which is implicit in the more general $u_j*(q_1 \ldots q_N)$
$u_j(q_1 \ldots q_N)$-dependent description. The contraction results in a non-
linear description whereas the equations for the time evolution of the
$P_j(t)$'s will be seen to be linear. This situation is analogous with the contraction of the linear Liouville equation of classical mechanics into the
non-linear hydrodynamical equations (4). It should be realized that there
is no loss in generality in the linear picture which will follow since it is
given in a very many dimensional space whereas the non-linear mass action picture is given in 3-space and is in fact less general since it is a
contracted picture.

The Master equation (5) governs the time evolution of the $P_j(t)$'s by

$$\frac{d}{dt} P_i(t) = \sum_j [W_{ij} P_j(t) - W_{ji} P_i(t)] \qquad\qquad I$$

where for the arbitrary case we have

$$W_{ij} = W_{ji} \text{ and } W_{ij} \geq 0 \text{ for all i and j} \qquad\qquad II$$

while for the thermally buffered case (6) we have

$$W_{ij} = W_{ji} \exp\left[-\frac{(E_i - E_j)}{k_B T}\right] \text{and } W_{ij} \geq 0 \text{ for all i and j.} \qquad III$$

The internal energy of the system is given by

$$E(t) = \sum_i E_i P_i(t) \qquad\qquad IV$$

and the entropy is given by

$$S(t) = -k_B \sum_i P_i(t) \ln P_i(t) \qquad\qquad V$$

where k_B is Boltzmann's constant. The free energy is given by

$$F(t) = E(t) - TS(t) \text{ or}$$
$$F(t) = \sum_i E_i P_i(t) + k_B T \sum_i P_i(t) \ln P_i(t) \qquad\qquad VI$$

Using I and II, it is easily proved that the entropy given by V is a monotonically increasing function of time.

$$\frac{d}{dt} S(t) = -k_B \sum_{ij} [W_{ij} P_j(t) - W_{ji} P_i(t)] \, \ell n \, P_i(t)$$

$$= -k_B 1/2 \sum_{ij} [P_j(t) - P_i(t)] \, W_{ij} \, \ell n \left[\frac{P_i(t)}{P_j(t)} \right] \qquad\qquad \text{VII}$$

$$\geq 0$$

Using I and III, it is also easily shown that the free energy given by VI is a monotonically decreasing function of time.

$$\frac{d}{dt} F(t) = \sum_{ij} E_i \, [W_{ij} P_j(t) - W_{ji} P_i(t)] \qquad\qquad \text{VIII}$$

$$+ k_B T \, 1/2 \sum_{ij} [W_{ij} P_j(t) - W_{ji} P_i(t)] \, \ell n \left[\frac{P_i(t)}{P_j(t)} \right]$$

$$= 1/2 \sum_{ij} (E_i - E_j) + k_B T \, \ell n \left[\frac{P_i(t)}{P_j(t)} \right] \, [W_{ij} P_j(t) - W_{ji} P_i(t)]$$

$$= 1/2 \sum_{ij} k_B T \left(\ell n \left[\frac{W_{ji}}{W_{ij}} \right] + k_B T \ell n \left[\frac{P_i(t)}{P_j(t)} \right] \right) [W_{ij} P_j(t) - W_{ji} P_i(t)]$$

$$\leq 0$$

Therefore, the second law of thermodynamics for an arbitrary system, which states that the entropy of the system increases monotonically to equilibrium, becomes for a thermally buffered system the statement that the free energy of the system decreases monotonically to equilibrium. Suppose there are a finite number, N, of eigenstates. The equilibrium state for the arbitrary system is given by

$$P_j^{eq.} = \frac{1}{N} \qquad\qquad \text{IX}$$

While for the thermally buffered system the equilibrium is given by

$$P_j^{eq.} = \frac{1}{Q} \exp \frac{-E_j}{k_B T} \qquad \text{X}$$

where $Q = \exp \frac{-E_j}{k_B T}$. Equations IX and X correspond to the microcanonical and the canonical distributions of equilibrium statistical mechanics, and IX maximizes the entropy while X minimizes the free energy (7).

Putting IX into V gives for the arbitrary system

$$S^{eq.} = -k_B \sum_i \frac{1}{N} \ln \frac{1}{N}$$

$$= k_B \ln N \qquad \text{XI}$$

which is the famous Boltzmann-Planck formula (8), and is a special case, for the equilibrium state, of the fluctuation formulae of Einstein and Onsager (9)

$$W_1 \{P_i\} = W_0 \exp \frac{S\{P_i\}}{k_B} \qquad \text{XII}$$

wherein $W_1 \{P\}$ is the probability of the state $\{P\}$ occurring as a fluctuation out of equilibrium, W_0 is a normalization constant, and $S\{P_i\}$ is given by V. From XII it is seen that the equilibrium state maximizes the entropy of an arbitrary system. Consequently, low probability states, with lower than equilibrium values for the entropy, are often referred to as ordered states (10).

Thermal buffering requires a new set of relations such that the low entropy $\leftrightarrow$ order relationship, true for arbitrary systems, is no longer valid. Suppose that the energy eigenvalue equal to E_j is degenerate with degeneracy $D(E_j)$. That is, there exist $D(E_j)$ states with the same energy E_j. From VI and X it may be seen that $Q = \exp [-\frac{1}{k_B T} F (P_i^{eq.})]$. The probability of the degenerate energy level E_j, $P(E_j)$, is given by

$$P(E_j) = \frac{1}{Q} D(E_j) \exp \frac{-E_j}{k_B T} \qquad \text{XIII}$$

wherein X was used. From XI, letting $D(E_j) = N$, we get

$$P(E_j) = \frac{1}{Q} \exp \left[\frac{TS(E_j) - E_j}{k_B T} \right] \qquad \text{XIV}$$

$$= \exp \left[+ \frac{1}{k_B T} \left[F^{eq.} - F(E_j) \right] \right]$$

where $F(E_j) = E_j - TS(E_j)$. Therefore, the probability of a degenerate energy level depends upon the free energy of that level, compared with the free energy of the entire system (11). Equation XIV is a special case of the more general fluctuation formulae for thermally buffered systems (12)

$$W_1\{P_i\} = W_0 \exp \left[-\frac{1}{k_B T} F\{P_i\} \right] \qquad \text{XV}$$

Again, as in the arbitrary case, the equilibrium state is the most probable state, in this case because the equilibrium state minimizes the free energy. The improbable states, which are the ordered states, possess high free energy content relative to the equilibrium state. Therefore, in the thermally buffered case, high free energy $\leftrightarrow$ order is the correct relationship (13). Order, organization and morphological complexity are terms which may be used interchangeably to signify the same attribute.

Energy inputs may be represented by adding to I an input term

$$\frac{d}{dt} P_i(t) = \sum_j [W_{ij} P_j(t) - W_{ji} P_i(t)] + \sum_j [I_{ij}(t) P_j(t) - I_{ji}(t) P_i(t)] \qquad \text{XVI}$$

We still have $\sum_j P_j(t) = 1$ for all t. However, the equilibrium state is no longer a solution to XVI with either set of conditions, II or III. If the $I_{ij}(t)$'s are time independent, then a steady state solution exists. If the $I_{ij}(t)$'s are time dependent then no time independent asymptotic state obtains. Howovor, in oither the arbitrary system case or the thermally buffered case the energy input states are non-equilibrium states. For the arbitrary system this implies that energy inputs order the system by reducing its entropy content. For the thermally buffered case, energy inputs order the system by increasing its free energy content.

An excellent example of the preceding distinctions is given by consideration of the system comprised of the Sun and the biosphere of the Earth. The biosphere of the Earth is thermally buffered by the oceans and receives its energy inputs in the form of sunlight. The organization of the biosphere which sunlight generates is seen in the high free energy content of the biosphere. Yet, the overall system of the Sun and the Earth's biosphere shows a net increase of entropy. This results from the fact that approximately 2×10^9 times as much sunlight goes off into empty space as actually impinges upon the Earth's surface. Thus, only a tiny fraction of the overall process produces increased order, through thermally buffered free energy enhancement of the biosphere, while the greatest part of the overall process produces increased entropy, through radiation of sunlight into empty space.

It has been argued that for a thermally buffered molecular system ordered states possess high free energy content relative to equilibrium. This followed from consideration of the fluctuation formulae for fluctuations around the equilibrium state. When energy inputs exist, ordered states are generated, and in addition transitions among these states, which would be very improbable as transitions among fluctuation states, are promoted by the energy inputs. Therefore, along with recognition of ordered states, one must recognize orderly transitions among ordered states as a result of energy inputs. These two emphases will be developed in the next section of this presentation.

METABOLIC ENERGETICS AND FREE ENERGY CONVERSIONS

In reviewing the energetics of metabolism two perspectives will be emphasized. Firstly, we shall review the free energy content of particular, ordered, molecular species and the source of their free energy content. Secondly, we will discuss the various manners in which free energy is processed by free energy conversion pathways. The second perspective demonstrates the special, "integrated feedback" structure of metabolic pathways.

The necessity for using the free energy rather than the entropy as the principal thermodynamical potential arises from the experimental fact that all spontaneous processes in cellular metabolism are attended by a decrease in free energy. This results largely from the thermal buffering of metabolic processes by the heat capacity and heat conductivity of the most abundant molecular constituent of all cells, H_2O.

The ultimate source of free energy for cellular metabolism is sunlight. In photosynthetic cells, sunlight energy is converted into chemical free energy in two forms. In photoreduction NADPH is produced from $NADP^+$, excited electrons, and protons, while in photophosphorylation ATP is produced from ADP and inorganic phosphate (14). The carbon cycle utilizes the reducing potential of NADPH and the activation free energy of ATP to convert CO_2 and H_2O into carbohydrate and molecular oxygen (15). Relative to its precursor constituents, CO_2 and H_2O, carbohydrate is free energy rich and has acquired its free energy from the free energy contained in ATP and NADPH, which are free energy rich relative to their precursor constituents. Photosynthetic cells may also oxidize carbohydrates through three linked pathways of reactions called anaerobic fermentation, the tricarboxylic acid cycle, and oxidative phosphorylation by cytochrome electron transport. By this means the free energy contained in carbohydrate is converted into the free energy of phosphoric anhydride bonds in ATP (16). Non-photosynthetic cells also acquire free energy in the form of ATP by oxidizing carbohydrates, as well as fats and proteins. Again, parts of the three pathways just mentioned are used in oxidizing fats and proteins (17). The major mobile currency of free energy is contained in the phosphoric anhydride bonds of polyphosphates, principally ATP.

In addition to free energy metabolism, pathways exist for the synthesis of a variety of essential molecules with low molecular weights. These pathways yield the amino acids, the purine and pyrimidine bases, simple sugars, fatty acids, vitamins, porphyrins, and other molecules with molecular weights $\leq$ 500 Daltons (18). These pathways are connected with anaerobic fermentation, the tricarboxylic acid cycle, and the carbon cycle through a number of common intermediates such as α-keto-glutaric acid, fructose-6-P, oxaloacetic acid, pyruvic acid, and other molecules. Many of the synthetic steps require free energy activation which may, for example, be accomplished by a phosphorylation step which gets its free energy from polyphosphates like ATP (19). Other activations may be achieved by group transfers and oxidation-reduction reactions.

Overall the primary consequences of cellular metabolism are the production of a nearly universal free energy currency in ATP, and the production of a variety of low molecular weight molecules.

The processes which comprise metabolism require enzymes for catalysis of each reaction step, genes for the determination of enzyme amino acid sequences, regulatory proteins for the control of gene directed protein synthesis, and polysaccharides and lipids for structural components of membranes needed both to separate and to locate various enzymatically catalyzed reactions, and for energy storage. Proteins, in addition to their enzymatic role, are important membrane components structurally, and through the mechanism of allostery are able to contribute to the regulation of metabolism (20). Together, these various kinds of macromolecules provide the molecular basis for the vitality of the cell. Each of these species of cellular macromolecules is of high molecular weight running from several thousand Daltons up into the millions. Their synthesis follows a general mechanism of building polymers, unit by unit, out of monomers through dehydration condensations, each of which requires free energy input which ultimately derives from the free energy in phosphoric anhydride bonds in ATP or closely related polyphosphates such as GTP and UTP (21).

The monomer - -polymer transition is a free energy driven "phase" transition from low molecular weight into high molecular weight molecules. Proteins are generated from amino acids by free energy requiring dehydration condensation which forms peptide bonds. Polynucleotides are generated from nucleotides by free energy requiring dehydration condensation which forms phosphodiester bonds. Polysaccharides are generated from monosaccharides by free energy-requiring dehydration condensation,which forms glycosidic bonds. Lipids, such as the triglycerides, are free energy requiring dehydration condensates of glycerol and fatty acids wherein ester bonds are formed. The nucleotides from which polynucleotides are made are themselves free energy requiring dehydration condensates of purine or pyrimidine bases, with ribose or deoxyribose, and phosphate. Quite generally, the dehydration condensation, wherever it occurs, requires free energy which is derived from polyphosphates such as ATP (22). This results from the fact that dehydration linkages eliminate a molecule of H_2O which is thermodynamically unfavorable in aqueous environments. Therefore, relative to thermal equilibrium all of the various polymers are ordered molecules because they are free energy rich. Consequently, without a continuous source of free energy all polymers would eventually spontaneously hydrolyze into constituent monomers in an aqueous environment. It is the flow of free energy through the cell's metabolic machinery which creates and maintains its polymeric order.

The processes in the cell which lead to the synthesis of its polymeric constituents also require these polymeric constituents to enable synthesis to occur in the first place. This is an example of an integrated, positive feedback system in which some of the products of various processes are required before those processes can actually take place. The result is self-renewing cyclic processes. Several of the pathways in metabolism are of this type, such as anaerobic fermentation, the tricarboxylic acid cycle, and the carbon cycle. The problem posed by such positive feedback processes is how to initiate them when they are not yet in operation.

This is particularly difficult if the molecule which functions as the feed-back component is itself free energy rich so that its occurrence as a chance flunctuation is improbable.

In order to appreciate the nature of the preceding problem more deeply, the various kinds of free energy conversions common to metabolism will be described. As a consequence of the energy inputs, high free energy molecules are produced. Five major classes of free energy conversion pathways which result finally in the complete thermal degradation of these free energy rich molecules may be identified.

Class 1 conversions result in the direct degradation of molecular free energy content into heat. The hydrolysis of peptide bonds, for example, liberates the free energy exclusively as heat. No useful free energy is produced through coupled reactions.

Class 2 conversions result in the steady conversion of free energy in one form into free energy of another form through a series of coupled reactions. Both photoreduction and photophosphorylation are examples of this. In each the free energy content of sunlight is converted into chemical free energy in NADPH and ATP. Other examples include oxidative phosphorylation during which the free energy implicit in NADH and O_2 is converted into ATP free energy through phosphorylation of ADP, and the tricarboxylic acid cycle in which the free energy of acetyl-CoA is converted into NADH free energy content. In each of the first three examples a series of coupled reactions with no feedbacks is involved.

Class 3 conversions are positive feedback processes in which free energy of one form is converted into another form autocatalytically. The autocatalytic nature of the process requires a cycle of coupled reactions. Anaerobic fermentation is an example of this type of conversion. During fermentation glucose free energy is converted into ATP free energy through phosphorylation of ADP. However, the pathway involved requires two activation steps which utilize the free energy of 2 ATP's, before 4 ATP's are finally produced later in the pathway. See Fig. 1. Thus,   an overall net gain of 2 ATP's is achieved each time a glucose molecule is processed. Consequently, if we start with an excess of all the necessary enzymes and with a large amount of glucose, two ATP's will initiate conversion. The conversion will show a self-accelerating production of ATP as each passage down the pathway doubles the amount of ATP available for activation of the next passage. Another example which is complementary to the preceding one is given by the carbon cycle. Fig. 2. In

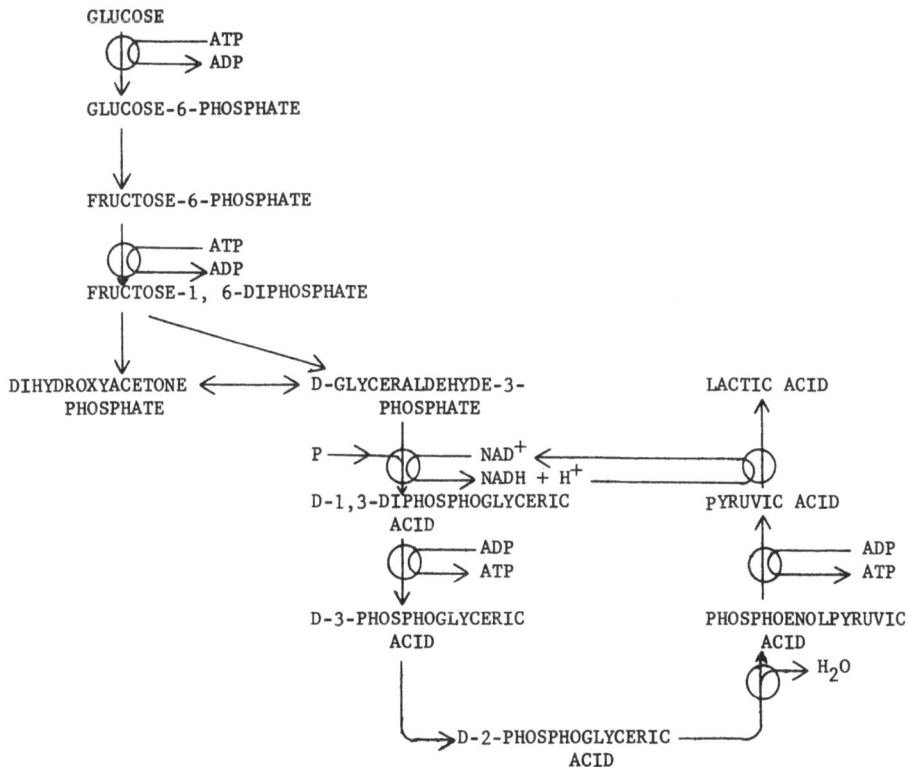

Fig. 1. Anaerobic fermentation

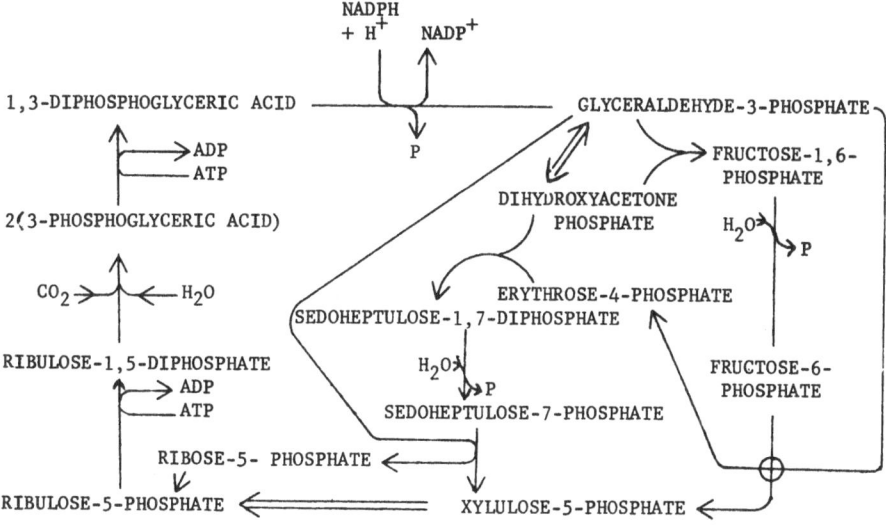

Fig. 2. Carbon cycle

this case the source of free energy is NADPH and ATP, which gets converted into carbohydrate free energy. Starting with an excess of all the required enzymes and with large amounts of NADPH and ATP, ribulose-5-phosphate will be produced in geometrically increasing amounts. As in the preceding case, this pathway requires activation by its product, ribulose-5-phsophate. In both cases the feedback component is free energy rich relative to the thermal equilibrium state and is, therefore, an improbable result of a fluctuation.

Class 4 conversions are also positive feedbacks, but the molecular component which creates the feedback is restricted to being a polymer. These conversions are called mutable positive feedback free energy conversions. As an example suppose that one has an abundance of the nucleotide triphosphates, an excess of Kornberg's polymerase enzyme, and one double helical DNA template. The free energy conversion will generate copies of the initiating template by converting phosphoric anhydride free energy into phosphodiester bond free energy. The feedback component in this case is the DNA helix. The population of DNA copies will grow exponentially. However, if a copying error is made, the resulting mutated double helix will still function in the feedback process as a satisfactory template. Thus, a mutable, positive feedback, free energy conversion is achieved. Spiegelman has extensively studied the evolution of an RNA system of this class (23).

Class 5 conversions are integrated feedback conversions which involve both positive and negative feedbacks integrated together so that the overall, complete process is a regulated positive feedback, and is mutable. A cell in a free energy rich media provides an example. The overall positive feedback process is seen in the exponential phase of cell population growth. In order to regulate and integrate the whole of the cell's metabolism, which in part involves some positive feedback processes as we have already seen, negative feedback mechanisms must be employed. These may be of various types. Two examples are, end produce inhibition of a synthetic pathway such as is seen with histidine, tryptophan, and cytidylic acid, for examples (24), and repression of operons such as in lac operon repression by the lac repressor (25). In addition, regulatory mechanisms may involve positive feedbacks as well as negative feedbacks.

Note that as in Class 3 conversions, Class 4 and Class 5 conversions involve feedback components which are free energy rich relative to full thermal equilibrium and which, therefore, are improbable as fluctuations. Consequently, the initiation of these conversions without an initiator at the beginning poses a difficult problem. In the Class 5 case one requires an entire functioning cell to initiate the conversion. The problem of the origin of life may be phrased to be the problem of initiating a mutable positive feedback free energy conversion process in a thermally buffered molecular system which has a free energy source available. Therefore, we need a Class 4 process, at least, when initially there is not one. Chance fluctuations out of equilibrium are much too improbable to have been efficient initiators. A Class 5 process could evolve from a Class 4 process or could be the first process to develop. This second possibility only reemphasizes the difficulty in initiating the process. Class 3 processes, while more simple and of the positive feedback type, do not have evolutionary potential since their feedback components are not polymers and are, therefore, not mutable. Moreover, in classes 2, 3, and 4 we have assumed the existence of the requisite enzymes, which are products of a living cell. From the point of view of the origin of life a source of catalytic potential will also be required as part of the solution to the initiation problem. In the following section of this presentation, a plausible solution to these problems of initiation will be presented.

TEMPERATURE SHIFT FREE ENERGY GENERATION AND PROTEINOID MICROSPHERES

The use of various energy inputs in simple molecular systems to generate biologically relevant, low molecular weight molecules has been very successful (26). Starting with subsets of such simple precursors as CO, CH_3, NH_3, HCN, CO_2, H_2O, H_3PO_4 and H_2CO, the application of energy inputs of various kinds has yielded a production of amino acids, simple sugars, adenine $(HCN)_5$, ATP, porphyrin, and various other biologically relevant small molecules (27). The production of polymers of amino acids or nucleotides using abiotic syntheses which possess a simplicity of mechanism which could be inputed to the primitive Earth has not been as easily achieved (28 - 31). One of the few successful methods, however, follows from a very basic thermodynamical mechanism which will be referred to as temperature shift free energy generation.

Suppose we have a thermally buffered molecular system at temperature T. In equilibrium the free energy of the system is minimized. If we raise or lower the temperature to some other temperature, let the system sit at the new temperature for awhile, and return the system to its original temperature, then the system will be out of equilibrium and will relax back into full equilibrium. The non-equilibrium states achieved in this way will of necessity contain more free energy than the full equilibrium state. In many cases the relaxation back into full equilibrium will occur rapidly, primarily through Class 1 conversions of free energy into heat, or through intermediate free energy rich states such as in Class 2 conversions followed by Class 1 conversions of the products of the Class 2 conversions. Occasionally, if the molecular composition is right and if the temperature shift is sufficient, then metastable molecular species will be trapped by the temperature shift so that the resulting non-equilibrium state is free energy rich but does not rapidly relax to full equilibrium. In this case conversions of classes 3 and 4 become more probable, especially in molecularly heterogeneous systems. An initiation factor which would be very rare as a fluctuation could in this way be produced readily.

An example of the trapping of metastable states, rich in free energy, by means of temperature shift free energy generation is given by consideration of an aqueous solution of orthophosphoric acid at a temperature around $300^{\circ}K$. If the temperature is raised to above the boiling point for H_2O, then H_2O vapor will form and the orthophosphoric acid will be left in a relatively H_2O free state. Consequently, polymerization of orthophosphoric acid into polyphosphoric acid with an attendant release of H_2O which becomes vaporized proceeds spontaneously since one of the reaction products, H_2O, is removed from the reaction by vaporization. A subsequent lowering of the temperature back to the original temperature condenses the H_2O vapor and produces a solution of phosphoric acids. The phosphoric anhydride linkages of the polyphosphate are free energy rich in aqueous media and they are also relatively metastable in so far as hydrolysis is not immediate. Thus, by means of an appropriate temperature shift a free energy rich molecular state has been achieved.

This reasoning has been applied to the temperature shift polymerization of amino acids into proteinoids; polymers of amino acids of non-biological origin (32). Thermal zones on the Earth's crust have been implicated as responsible for this process on the primitive Earth (33). As in the phosphate case, the vaporization of H_2O removes the thermodynamical barrier to dehydration, peptide bond formation, and indeed makes polymerization spontaneous. The subsequent return to the initial temperature produces a solution of metastable polymers of amino acids. In general, this process of polymerization could be performed for any of the dehydration bonded polymers. Some care must be exercised in the

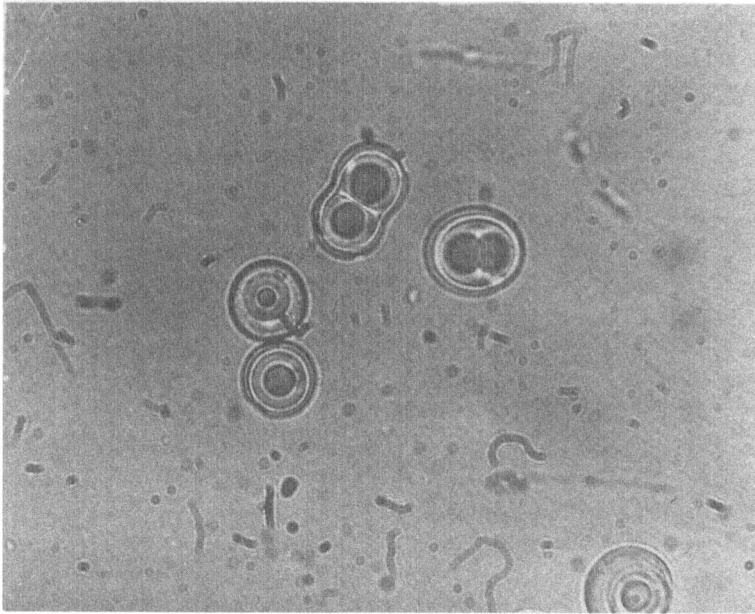

Fig. 3. Proteinoid Microspheres.
By permission from S. W. Fox.

choice of composition of the monomers so as to prevent substantial degradation of the monomers during the elevated temperature phase of the
temperature shift.   One need not envisage a primitive form of photosynthesis or oxidative phosphorylation in order to achieve the beginnings of
chemical free energy production (34).   Thermal zones work effectively
by providing temperature shift free energy generation.

The introduction of free energy into a thermally buffered system was
shown earlier to order the system.   The temperature shift method of
free energy enhancement, therefore, orders the molecular system.   In
the case of proteinoids this ordering takes on a striking morphological
manifestation. See Fig. 3. The proteinoid polymers in aqueous solution
tend to assemble into large aggregates of a few microns in diameter
and containing on the order of 10^9 polymers each. These aggregates are
called microspheres and any given preparation produces very large numbers of units of nearly identical size and shape.   Several of their physical properties have been documented (35).

The assembly of polyamino acids into aggregates is a quite general
phenomenon and has been observed particularly with the assembly of
cellular enzyme complexes from precursor subunit polyamino acids.

The assembly process which makes microspheres out of proteinoid is
qualitatively similar.   This process is often called self-assembly, but
we shall see that this is a misnomer.

In assembly processes the confusion between entropy and free energy is highlighted.   Consider two states: a) the polyamino acids are in
solution, and b) the polyamino acids are aggregated together. The aggregation process going from a) into b) is spontaneous when the ionic conditions in the solution are appropriate. At first glance it would appear that
the entropy of the polymers has decreased and that this is the sole effect
of assembly, a blatant violation of the 2nd law of thermodynamics. Assembly appears to be a spontaneous ordering process.   In fact, it is the free
energy which governs the process.   The free energy of the entire system
decreases when assembly occurs and the free energy change is dominated
by the entropy term.   Recall that $\Delta F = \Delta E - T \Delta S$. Because the
assembly of polyamino acids involves multiple weak interactions among
the residues of the polyamino acids, and the interaction of these residues with H_2O in solution are also weak at about the same strength, the
assembly process involves a small ΔE change compared with the
$T \Delta S$ term.   The ΔS change is large because assembly frees many
H_2O molecules which were bound to the polymer before the polymers
aggregate together, and because relatively few polymer molecules become aggregated. This is because each polymer molecule is large enough
to bind, in solution, a large number, N, of H_2O molecules.   Therefore,
assembly produces a negative ΔS change for the polymers alone, and a
positive ΔS for the H_2O molecules such that the positive ΔS change
for H_2O is much larger than the negative ΔS change for the polymers,
if N is sufficiently large.   The net result is an overall ΔS increase
which makes $\Delta F \simeq -T \Delta S < O$. Thus the free energy decrease during
assembly is "entropy driven" by the H_2O contribution to the overall process.   The apparent ordering of the polymer components is more than
compensated for by the disordering of H_2O.   The polymer aggregate, of
course, does not simply "self-assemble," but assembles because the
overall process involves $\Delta F < O$.

The assembly of electrolytic polymers is, therefore, a naturally
occuring consequence of the formation of the polymers. The emergence
of a morphological manifestation of the ordering process, as in proteinoid microspheres, has significance for the emergence of a potentially
Class 5 conversion process.

CONSTRUCTION OF A SIMPLE CELL

The temperature shift method of free energy enhancement provides a method for producing a chemical free energy source in polyphosphoric acid (36). It also provides for the formation of polyamino acids which have been shown to possess a variety of weakly catalytic potentials (37). Together, particularly when proteinoid microspheres are formed, these two species of molecules provide a molecular environment for free energy driven ordering because both a free energy source and a catalytic background exist. The addition of amino acids, nucleotides, and other small molecules could provide the polyphosphate-proteinoid microsphere system with the building materials for the production of more proteinoid, subsequent assembly of that proteinoid into more microspheres, and for the production of simple metabolic pathways. The presence of nucleotides permits the possibility of the formation of polynucleotides as well as proteinoids. The simultaneous polymerizations of amino acids and of nucleotides should lead to a coupling of these two processes and the emergence of a primitive regulation of one polymerization by the other. It would be of especial interest to promote simultaneously amino acid polymerization and nucleotide polymerization, to look for the emergence of polymer aggregates comprised of polymers of both species, and to look for feedback effects of these aggregates on the overall polymerization process. This may be attempted either in the context of polyphosphate - proteinoid microspheres or directly by means of the temperature shift method.

The simple case of proteinoid synthesis for amino acids which have been activated with polyphosphate through the catalytic influence of proteinoid microspheres provides a solution to the problem of generating a Class 4 free energy conversion. The question is whether or not microspheres can catalyze the phosphorylation of amino acid carboxyl groups, and thereby activate the amino acids so that polymerization will follow spontaneously. In such a system one must then think of an entire distribution of polymer compositions and sequences having a positive feedback effect on the entire distribution. A more specific feedback such as a residue by residue feedback as is seen in DNA replication and in RNA transcription requires polynucleotides (38,39). The control of sequence during polynucleotide synthesis on a template confers an added stability to the process. It enables a particular sequence to persist much longer than the half-life for hydrolysis of any particular polynucleotide. However, polynucleotides cannot promote metabolism through direct catalysis. While the proteinoids can promote metabolism weakly, they are not good at ensuring the persistence of a particular catalytic potential for longer than the half-life for hydrolysis of any particular polyamino acid. Only if an entire distribution of polyamino acids possesses a positive feedback effect on the synthesis of that same entire distribution, will persistence of catalytic potentials occur. During the initial developmental stages this is probably what happened, and only later did a coupling between the production of polyamino acids and polynucleotides occur which then greatly enhanced the possibilities for the persistence of particular catalytic potentials, or at least produced improved variations. Consequently, polyphosphate, amino acids, and proteinoid microspheres provide a basis for a Class 4 conversion, while the introduction of nucleotides provides for the subsequent development of the proteinoid microsphere system into a potentially Class 5 conversion with a more sophisticated feedback mechanism.

It has been suggested that heterogeneous organic molecular systems which possess a source of free energy be observed to see how the conversion of the available free energy orders the molecular system. In particular, polyphosphates and the monomers of amino acids provide the potential for Class 4 free energy conversions and a subsequent evolution of these conversions into Class 5 types, perhaps only after nucleotides have been introduced. Other sources of free energy have been proposed for the early stages, such as HCN and its polymers (40). Evolution has favored polyphosphates. The production of free energy rich molecular systems in general is very easily achieved by means of temperature shift free energy generation. Once a Class 4 or Class 5 process is initiated, subsequent evolution, which is implicit in the mutability of these conversions, could provide the process with the ability to utilize more sophisticated free energy metabolism such as in photosynthesis or in oxidative phosphorylation.

SUMMARY

Energy flow through a thermally buffered heterogeneous molecular solution generates ordering processes within the solution. Levels of organization may be recognized. Initially, there is the level given by all the atomic nuclei and electrons. Then there is the level of very simple small molecules such as CO_2, H_2O, HCN, H_3PO_4, etc. Then there is the level of small biologically relevant molecules of molecular weight $\leq$ 500 Daltons which includes amino acids, nucleotides, coenzymes, porphyrin, etc. Next there is the level of polymers, or macromolecules, with molecular weights in the many thousands of Daltons.   Finally there is the level of polymer aggregates such as in proteinoid microspheres. Each level emerges from the preceding one by synthesis.   The steps from very small molecules to small biologically relevant molecules, and from small biologically relevant molecules to polymers take place as energy flow generates increased order.   The initial step and the final step are spontaneous.   Thus, very small molecules will form from the elements with a release of free energy, and polymer assembly into polymer aggregates is also spontaneous.   The step from monomers to polymers requires free energy inputs to promote dehydration condensation, while polymer assembly also "eliminates" H_2O but requires no energy input.   These variations emphasize the variety of relationships, between one level and the next, which occur.

In principle, therefore, these levels of order manifest the stages of transition from the atomic elements to polyphosphate free energy driven proteinoid microspheres.   The problem at present is to see if such a paradigm will provide the means to study the emergence of the genetic code, the emergence of cellular level control mechanisms, and the emergence of cellular organelles.   If progress can be made in modeling these developments, perhaps the problem of multicellularity is also open to the synthetic, constructionistic approach.

REFERENCES

1. Schiff, L. I., "Quantum Mechanics," p. 227.   McGraw-Hill, New York, 1955.
2. Huang, K., "Statistical Mechanics," pp. 202-204.   John Wiley & Sons, New York, 1963.
3. Moore, W. J., "Physical Chemistry," p. 169.   Prentice-Hall, Englewood Cliffs, 1962.
4. Uhlenbeck, G. E., and Ford, G. W., "Statistical Mechanics," American Mathematical Society, Providence, 1963.
5. Van Hove, L., in "Fundamental Problems in Statistical Mechanics" (E. D. G. Cohen, ed.) p. 157.   North-Holland, Amsterdam, 1962.
6. Reif, F., "Fundamentals of Statistical and Thermal Physics," p. 553. McGraw-Hill, New York, 1965.
7. Huang, K., op. cit., p. 143 and p. 156.
8. Planck, M., "Theory of Heat," Macmillan, New York, 1932.
9. Onsager, L., Phys. Rev. 38, 2265 (1931).
10. Dickerson, R. E., "Molecular Thermodynamics," pp. 387-422. Benjamin, New York, 1969.

11. Tolman, R. C., "Principles of Statistical Mechanics," pp. 636-640. Oxford University Press, London, 1939.

12. Onsager, L., op. cit.

13. Loewy, A., and Siekevitz, P., "Cell Structure and Function," pp. 6-9. Holt, Rinehart, and Winston, New York, 1963.

14. Mahler, H., and Cordes, E., "Biological Chemistry," p. 491. Harper and Row, New York, 1966.

15. Mahler, H., and Cordes, E., loc. cit., p. 499.

16. Baldwin, E., "Dynamic Aspects of Biochemistry," Cambridge University Press, New York, 1965.

17. Baldwin, E., "The Nature of Biochemistry," p. 79. Cambridge University Press, New York, 1965.

18. Mahler, H., and Cordes, E., op. cit.

19. Mahler, H., and Cordes, E., op. cit., pp. 666-682.

20. Mahler, H., and Cordes, E., op. cit., pp. 316-320.

21. Krebs, H., and Kornberg, H., Ergeb. der. Physiol. Biol. Chemie und Expt. Pharm. 49, 212 (1957).

22. Ingram, V., "Biosynthesis of Macromolecules," Benjamin, New York (1966).

23. Spiegelman, S., in "The Neurosciences," (Quarton, Melnedink, and Schmitt, eds.), Vol. II, Rockefeller Press, New York, 1970.

24. Conn, E., and Stumpf, P., "Outlines of Biochemistry," pp. 331-339. John Wiley and Sons, New York, 1963.

25. Watson, J., "Molecular Biology of the Gene" (2nd ed.), pp. 435-466. Benjamin, New York, 1970.

26. Calvin, M., "Chemical Evolution," Oxford University Press, New York, 1969.

27. Hodgson, G., and Ponnamperuma, C., Proc. Natl. Acad. Sci. U.S.A. 59, 22 (1968).

28. Fox, S. W., and Harada, K., J. Am. Chem. Soc. 82, 3745 (1960).

29. Matthews, C., and Moser, R., Nature 215, 1230 (1967).

30. Schwartz, A., and Fox, S. W., Biochem. Biophys. Acta 134, 9 (1967).

31. Weinmann, B., Lohrmann, R., Orgel, L., Schneider-Bernlochr, H., and Sulston, J., Science 161, 387 (1968).

32. Fox, S. W., and Harada, K., op. cit.

33. Fox, S. W. Nature 201, 336 (1964).

34. Lipmann, F., "Wanderings of a Biochemist," pp. 212-226. Wiley Interscience, New York, 1971.

35. Fox, S. W., Nature 205, 328 (1965).

36. Lipmann, F., op. cit.

37. Fox, S. W., and Wang, C., Science 160, 547 (1968).

38. Crick, F., J. Mol. Biol. 38, 367 (1969).
39. Orgel, L., J. Mol. Biol. 38, 381 (1969).
40. Calvin, M., op. cit., pp. 121-144.

Received 12 July, 1971

THERMODYNAMICS OF SELF ASSEMBLY: AN EMPIRICAL

EXAMPLE RELATING ENTROPY AND EVOLUTION

John R. Jungck

Department of Biology, Merrimack College
North Andover, Massachusetts 01845

> A violent order is disorder; and a great disorder
> is an order. These two things are one.
>
> Wallace Stevens
> "Connoisseur of Chaos"

Although Sidney Fox is best known for his pioneering contributions in protein chemistry and molecular evolution, I think it should also be recalled that his professional origins were in the thermodynamics of biology. His Ph.D. dissertation (1) reported the measured thermodynamic properties of the compounds which participate in the urea cycle. Hugh Huffman, Fox's mentor and coauthor of a salient treatise: "The Free Energies of Some Organic Compounds" (2), commented that Fox's urea work was the first research to unequivocally demonstrate a biochemical reaction requiring linking to another under in vivo conditions (3). It was a clear illustration of "a new point of view which the application of the second law of thermodynamics to physiological questions has introduced" (3).

In the mainstream of molecular biology, many investigators were motivated by Erwin Schroedinger's little essay: "What is Life?" (4) to pursue the nature of the genetic apparatus in a search for new physical laws. Delbruck, Luria, and Stent are perhaps most representative of this group. For Stent, at least, molecular biology has been unsatisfying (5) in that no new physical laws were discovered and that the only scientific frontier where such hopes might be fulfilled is in delving into what is consciousness. Although Schroedinger's negentropy has permeated many a contemporary discussion of biology and the second law, the dilemma posed by Bridgeman (6) in 1941: "Does it mean that living organisms do or may violate the second law of thermodynamics?" was essentially solved as early as 1928 (cf. 7) with the realization that organisms are open systems. The significance of this salient fact has been ignored by or escaped the attention of contemporary antireductionists such as Koestler (8). Whereas in principle, an adequate treatment of irreversible thermodynamics in open systems should suffice to illustrate how the apparent fantastic organization of biological systems was able to evolve, Ronald Fox (9) has noted that "a non-linear theory of non-equilibrium thermodynamic processes... does not exist, so that it is not possible to quantitatively discuss the fascinating time evolution and ordering of biological systems in the manner presented here (sic) for the linear case."

In the absence of theoretical thermodynamic constructs to base an empirical view of the relationship of entropy and evolution, let us examine the available calorimetric data on biochemical systems to gain some insight into how experimental evidence may allow us to proceed to a new, workable paradigm on how apparent order is evolved. This examination of the nature of self assembly phenomena can be carried out on two fronts, both of which are resolved by the same mechanism. First, how is it that a highly ordered structure forms spontaneously; i.e., does not that violate the second law? Secondly, what is the energetic parameter which drives self-assembly reactions of proteinaceous polymers?

In 1952, Dobry and Sturtevant (10) measured the heat of reaction between the enzyme trypsin and the polypeptide, soybean trypsin inhibitor, by calorimetry. The surprising result was that the heat of reaction was very small and the equilibrium appeared to be independent of temperature. [For more up-to-date measurements, see the work of Berezin, Levashov, and Martinek (11)]. If the only process occurring was an association of the two polymers, a large drop in entropy would be expected; however, the data indicated just the opposite, i.e., this spontaneous reaction was being driven by a large increase in entropy. Two alternatives were conjured up to explain these data. Either the reaction was liberating a substantial number of ions or water bound to the polymers, or the polymers were unfolding so that they were becoming configurationally less ordered.

Sturtevant (12) also investigated another protein-peptide interaction, namely an antibody-antigen reaction, and the data collected seemed to support the idea that the small free energy change of the reaction must be in large part due to an entropy increase. Evans (13) has recently come to the same conclusion.

Although calorimetric data are not yet available, some thermodynamic insight has been gained into the labile association of mitotic spindles through a van't Hoff plot of the equilibrium constant, as determined by birefringence as a function of temperature (14). The van't Hoff plot yielded a straight line relationship which indicated a very small enthalpy but a very large entropy increase during association. Again this is suggestive of self assembly being concomitantly linked to a randomizing of water structure to compensate for the configurational ordering of the associated polymers. The low free energy of the reaction indicates that rather weak bonds such as hydrophobic interactions are responsible for self assembly.

The scope of self assembly is ubiquitous across the full panel of phylogeny. Kushner's (15) recent review describes the self assembly of simple viruses, spherical viruses, complex phages, bacterial flagella, pili, microtubules, ribosomes, collagen, actin-myosin muscle fibers, membranes, extracellular layers, and fibrin clots. The mere length of this list and the levels within organismic structure-formation which involve self assembly testify to their key role in any quantitative description of biological organization. However, it also follows from the data summarized to date that self assembly reactions contain a macroscopic paradox. Thus, the specific answer to the first question posed as to whether these reactions disobey the second law is a resounding no. We do not even have to resort to open systems to gain this insight. When viewed at the molecular level instead of the macroscopic level, the paradox is resolved by noticing that the relative entropy increase in the aqueous environment is always more than enough to account for the configurational ordering of the polymers associating. This observation, in essence, answers the second question, i.e., a release in the water structuring is responsible for the driving energy.

Two additional examples are worth describing here for reiterating the importance of entropy-driven processes in biochemical systems. First, and perhaps the most thoroughly studied self-assembly system in the thermodynamic sense, is the formation of tobacco mosaic virus (TMV) rods from TMV A-protein. This homopolymerization of A-proteins through non-covalent bonds has been extensively studied by Lauffer and his coworkers. As early as 1958, Lauffer et al. (16) were able to indicate the presence of a large entropy increase during polymerization of A-proteins using simple viscosity and optical-density measures. In 1965, Stevens and Lauffer (17) obtained osmotic data which indicated that on the order of 800 moles of water are dissociated from each mole of TMV A-protein during the process. These buoyancy changes could also be correlated with quantitative light scattering transitions (18). Recently, Stauffer et al. (19) have obtained calorimetric data on the association of TMV A-proteins which give a value of 300 cal/mole of water released. These studies thus clearly vindicate the axiom that a spontaneous process that absorbs heat demands an increase in entropy. Since the self-assembly processes fall clearly into this category, it seems that the intuitional paradox resulting from a violation of the second law by biological organization is only apparent and not real.

In the context of molecular evolution, the self assembly of proteinoids into microspheres could be very analogous to the association of TMV A-proteins. This spontaneous formation of microspheres may be an entropy-driven process due to the affinity of lyophilic side chains of proteinoids enhanced by the presence of water. Thus, the idea of "order out of chaos" would seem to be phenomenologically circumvented both at the prebiological and the biological level.

A second example which warrants treatment in the total context of entropy and evolution is a consideration of the role of ATP hydrolysis in vivo. This question has been clearly focused by Banks and Vernon (20). They showed "that (a) the hydrolysis of ATP is a forbidden reaction in intermediary metabolism, (b) that the Lipmann concept would be appropriate for a closed system containing energy-linked reactions (of which there are no known examples in biochemistry) and (c), most importantly, that since real organisms are open and not closed systems even the direction of flow of matter through a particular step cannot be predicted from the associated standard free energy change but only from the properties of the whole system making up the steady state. " These considerations as well as the above discussion of entropy and self assembly amply illustrate the lacunae in the free energy conceptualization of chemical reactions (cf. 21 - 23).

Strong and Halliwell (23) emphasize that the energy redistribution that is associated with any change, which must take a whole system into consideration, is much more important for gaining an intuitive conceptualization of what is taking place during a process than any change in the magnitude of energy. This led them to re-derive the classical Gibbs free energy equation for teaching purposes to the alternative expression:

$$\Delta S_{tu} = \Delta S_e + \Delta S_s = -\Delta H_s/T + \Delta S_s$$

where ΔS_{tu} is the totally unopposed total entropy change for a process occurring at constant temperatures and pressure (as in most biological reactions), ΔS_e is the change in the entropy of the environment, ΔS_s is the change in the entropy of the system, and ΔH_s is the change in enthalpy in the system. From this it easily follows that "a reaction proceeds when there is a process available that leads to an increase in total entropy, and the reaction stops when no way is open to total entropy increase" (23).

Using this alternative mode of thinking about energy changes during processes, it is well to reassess the reactions of ATP. The entropy data for these reactions are shown in Figure 1 as taken from Strong and Halliwell (23). These authors point "out that the evidence indicates that the reaction in which a phosphate bond is broken in ATP to form ADP is largely controlled by entropy change in the system rather than by effects on the environmental entropy. In other words, energy transfer and bonding are not the main controlling features of the reaction. Hence this is a significant argument from the evidence against the use of the term 'high energy phosphate bond' since it can easily mislead students and possibly even some biochemists. The great significance of biological systems is the tremendous organization they exhibit which in thermodynamics is described by entropy." Since the free energy of hydrolysis of ATP is used so heavily in the biochemical literature, the ramifications of these statements cannot be underscored enough.

In summary, an appreciation of how simple thermodynamic principles can be applied to biological questions has been discussed. A resolution of the paradox of development of order through spontaneous processes has been offered based on a review of the empirical data on biological organization. First, as biologists, we do not need to always resort to open systems in which energy is provided from an external source. Some processes are definitely spontaneous and proceed with a large increase in entropy. Secondly, inasmuch as many biochemical reactions are linked to ATP hydrolysis, which is entropy-driven, entropy not only does not counteract the evolution of biological organization and metabolic flux, it may in fact be the predominant force in the origin and maintenance of biological structures and functions.

J. R. JUNGCK

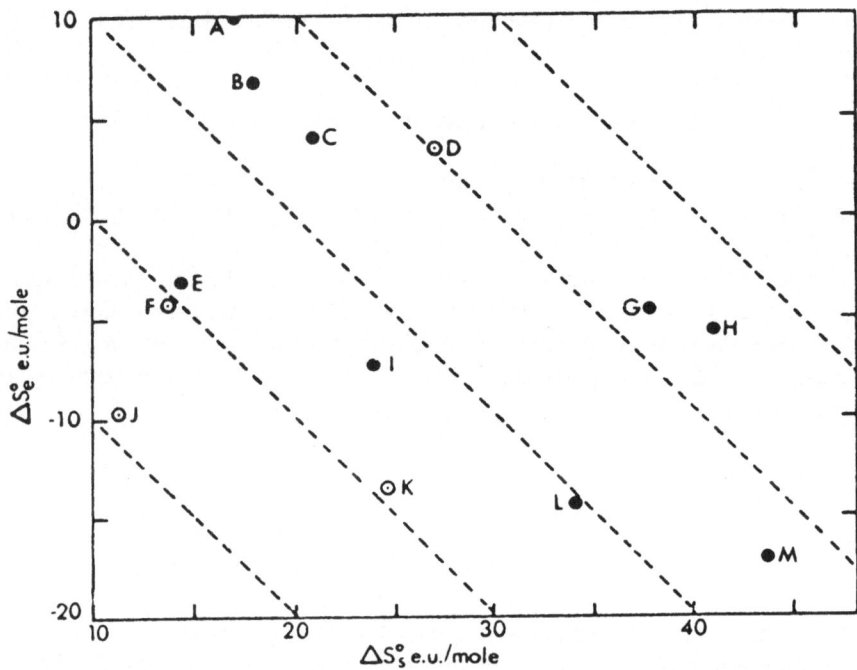

Fig. 1. Entropies of reaction for ATP, ADP, and related ions in aqueous solution at 25°C. The dotted diagonal lines represent constant ΔS_{tu}^{o} values of 0-40 eu in 10 eu increments. The points represent the values for the following reactions:

(A) $ATP + H_2O$ $\rightarrow$ $ADP + HPO_4^{-}$
(B) $H^{+} + (ADP)Mg^{-}$ $\rightarrow$ $(ADP)HMg$
(C) $H^{+} + (ATP)Mg^{2-}$ $\rightarrow$ $(ATP)HMg^{-}$
(D) $H^{+} + HPO_4^{2-}$ $\rightarrow$ $H_2PO_4^{-}$
(E) $Mg^{2+} + (ADP)H^{2-}$ $\rightarrow$ $(ADP)HMg$
(F) $H^{+} + H_2PO_4^{-}$ $\rightarrow$ H_3PO_4
(G) $H_3^{+} + ADP^{3-}$ $\rightarrow$ $(ADP)H^{2-}$
(H) $H^{+} + ATP^{4-}$ $\rightarrow$ $(ATP)H^{3-}$
(I) $Mg^{2+} + (ATP)H^{3-}$ $\rightarrow$ $(ATP)HMg^{-}$
(J) $Mg^{2+} + HPO_4^{2-}$ $\rightarrow$ $MgHPO_4$
(K) $Mg^{2+} + SO_4^{2-}$ $\rightarrow$ $MgSO_4$
(L) $Mg^{2+} + ADP^{3-}$ $\rightarrow$ $(ADP)Mg^{-}$
(M) $Mg^{2+} + ATP^{4-}$ $\rightarrow$ $(ATP)Mg^{2-}$

Taken from Strong and Halliwell (23) with permission from the Journal of Chemical Education.

For heuristic purposes, the argument presented here has been presented in utmost simplicity without qualifications. The author is aware that qualifications do exist but that these do not substantially affect the paradigm presented. Two reservations, however, should be mentioned. At least one case of protein-peptide association, namely the haemoglobin-haptoglobin reaction, that has been thermodynamically studied is driven by "enthalpy and not entropy, suggesting that hydrogen bonding (reinforced by the groups being in a hydrophobic region) is involved" (24). A second reservation is that the explanatory discussion of what is occurring during entropy-driven processes assumed axiomatically that water associated with macromolecules is structured to some degree. This axiom is still involved in some controversy (25). Even taking these qualifications into consideration, this paper has given credence to the power of entropy-driven processes in a variety of biological systems.

In conclusion, entropy will not be the nemesis of evolution; on the contrary, the selection of entropy-driven processes in biological systems has been responsible for the evolution of the sophisticated organization of contemporary biota.

ACKNOWLEDGEMENTS

The author wishes to acknowledge the influence of many critical discussions with Dr. James C. Lacey, Jr., University of Alabama, and Dr. W. Dale Snyder, University of Miami. Thanks are also due to Mrs. Ania Mejido and Dr. Marcel Gregoire for their help in the preparation of the manuscript.

REFERENCES

1. Fox, S. W., "I. The Energy Relationships in the Ornithine-Citrulline-Arginine-Urea Cycle. II. Thermal Data for Other Biochemicals. III. A Study of Sperm Agglutination." Ph.D. Dissertation, California Institute of Technology, Pasadena, California, 1940.
2. Parks, G. S., and Huffman, H. M., "The Free Energies of Some Organic Compounds," Am. Chem. Soc., Monograph No. 60. Chemical Catalogue Co., New York, 1932.
3. Borsook, H., and Huffman, H. M., in "The Chemistry of Amino Acids and Proteins" (C. L. A. Schmidt, ed.), 2nd ed., p. 822. Charles C. Thomas, Baltimore, 1944.

4. Schroedinger, E., "What Is Life?" Cambridge University Press, New York, 1945.
5. Stent, G., Science 160, 390 (1968).
6. Bridgeman, P. N., "The Nature of Thermodynamics," Harvard University Press, Cambridge, Mass., 1941.
7. Olby, R., J. Hist. Biol. 4, 119 (1971).
8. Koestler, A., and Smythies, J. R. (eds.), "Beyond Reductionism," Hutchinson, London, 1969.
9. Fox, R. F., J. Theor. Biol. 31, 43 (1971).
10. Dobry, A., and Sturtevant, J. M., Arch. Biochem. Biophys. 37, 252 (1952).
11. Berezin, I. V., Levashov, A. V., and Martinek, K., Eur. J. Biochem. 16, 472 (1970).
12. Sturtevant, J. M., in "Experimental Thermochemistry" (H. Skinner and F. D. Rossini, eds.), Vol. 1, p. 443. Interscience, New York, 1962.
13. Evans, W. V., in "Biochemical Microcalorimetry" (H. D. Brown, ed.), p. 120. Academic Press, New York, 1969.
14. Inoue, S., and Sato, H., J. Gen. Physiol. 50 (6, Part 2), 259 (1967).
15. Kushner, D. J., Bact. Rev. 33, 302 (1969).
16. Lauffer, M. A., Ansevin, A. T., Cartwright, T. E., and Brinton, C. C., Jr., Nature 181, 1338 (1958).
17. Stevens, C. L., and Lauffer, M. A., Biochemistry 4, 31 (1965).
18. Smith, C. E., and Lauffer, M. A., Biochemistry 6, 2457 (1967).
19. Stauffer, H., Srinivasan, S., and Lauffer, M. A., Biochemistry 9, 193 (1970).
20. Banks, B. E. C., and Vernon, C. A., J. Theor. Biol. 29, 301 (1970).
21. Bent, H. A., J. Chem. Ed. 47, 337 (1970).
22. Craig, N. C., J. Chem. Ed. 47, 342 (1970).
23. Strong, L. E., and Halliwell, H. F., J. Chem. Ed. 47, 347 (1970).
24. Adams, E. C., and Weiss, M. R., Biochem. J. 115, 441 (1969).
25. Frank, H. S., Science 169, 635 (1970).
26. White, D. C. S., J. Theoret. Biol. 32, 631 (1971).
27. Lumry, R., and Rajender, S., Biopolymers 9, 1125 (1970).

NOTE ADDED IN PROOF

Since the writing of this manuscript, White (26) has criticized the presentation of Banks and Vernon (20). In addition, the editors have informed me that the paper by R. F. Fox in this collection presents examples adequately adhering to the Lipmann concept (high-energy phosphate compounds). Discussions with Henry Bent, as well as the paper of Lumry and Rajender (27) and my referee's comments, have led me to the opinion that perhaps I have too enthusiastically overstated the case for entropy. However, with the historical emphasis placed on enthalpic effects, this article may contribute to an evolving understanding of the role of thermodynamics in biology which will fairly assess the contribution of all factors.

Received 12 July, 1971

THE EVOLUTIONARY PARADOX AND BIOLOGICAL STABILITY

Albert Szent-Gyorgyi

The Institute for Muscle Research at the

Marine Biological Laboratory, Woods Hole, Massachusetts

Not long ago evolution was shrouded in an unpenetrable mystery. We thought that more complex organic molecules could be built only by living systems. But the living system itself is built of such molecules, and how could a system build its own building stones? How can a hen lay the egg out of which it has to be hatched? This paradox is now partially understood. We know that Nature can build complex organic molecules without the intervention of life. Sidney Fox has pursued this building the farthest, approaching the cellular dimension (1).

Another paradox still awaits solution. Mutation is a random process. How could a random process lead to such highly ordered structures as a multicellular living organism? The usual answer to this question is that there was plenty of time to try everything. I could never accept this answer. Random shuttling of bricks will never build a castle or a Greek temple, however long the available period. A random process can build meaningful structures only if there is some kind of selection between meaningful and non-sense mutations. A certain selection was already supposed by Darwin, who made the struggle for life responsible for selection. The selection, however, in my opinion has to take place much earlier, not only with the final product.

A possible approach is opened by the fact that there is a fundamental difference in the stability of animate and inanimate structures. Animate structures, like a motor, deteriorate if used, their maximum stability being at the maximum entropy and minimum of free energy. All systems tend towards maximum stability. Contrary to this, living systems deteriorate if unused. The better and the harder they work, the more stable they are. Better work means thus a higher stability, and living systems work best at the minimum of entropy and the maximum of free energy. The "Bowditch staircase" is a demonstration of this fact in the nutshell. If the frog heart is arrested for, say a minute, the first beat after the pause will be weaker. The beats will regain their original force gradually. The short pause was already long enough to start the deterioration. Dr. Hajdu and myself (2, 3) have studied this phenomenon and we found that the weakening was due to potassium ions diffusing into the cell, approaching random distribution, increasing entropy. The entropy is decreased, the chemical structure restored by function. Thus function makes, and inactivity destroys, structure. Mutations which lead to better working of the system and have sense must thus be more stable than senseless ones.

The inner mechanism of this inactivity deterioration is a difficult problem to understand. Lewis and Randall (4) in their book on thermodynamics suggest a possible mechanism. We could suppose that complex systems have several isomorphic forms of different stability. In inactivity the system slips to a lower level of stability, but it can be brought back to its higher level of stability by activity. A mutation that has no sense does not lead to function, and the structure formed may slip gradually to lower levels of stability till it disintegrates.

REFERENCES

1. Fox, S. W., McCauley, R. J., and Wood, A., Comp. Biochem. Physiol. 20, 773 (1967).
2. Hajdu, S., and Szent-Gyorgyi, A., Am. J. Physiol. 168, 159, and 171, (1952).
3. Hajdu, S., Am. J. Physiol. 174, 371 (1953).
4. Lewis, G. N., and Randall, M., "Thermodynamics," p. 19, McGraw-Hill, New York, 1923.

Received 2 April, 1972

CHANCE IN EVOLUTION--SOME PHILOSOPHICAL REMARKS

Linus Pauling* and Emile Zuckerkandl**

*Stanford University, Stanford, California 94305

**Departement de Biochimie Macromoleculaire, B.P. 1018,
Montpellier, France

What is chance and what is its role in evolution? Jacques Monod describes the role of chance in his interesting book on biological philosophy, "Chance and Necessity." He does not say, however, what chance is.

The two notions, chance and necessity, are commonly used for a dichotomic classification of phenomena into those that are due to the one and those that are due to the other. The implication is that the qualities of chance and necessity (determinism) are inherent in phenomena and that they are antisymmetric--that is, opposite.

In our opinion, chance is never an intrinsic quality of things, whereas necessity, as soon as we acknowledge the existence of laws of nature, must be believed to be such an intrinsic quality. The two concepts are thus quite heterogeneous and unsymmetrical. Necessity is the coherence of nature. Chance is an expression of the situation of the observer with respect to the observed; it is an aspect of nature at a much different level from that of necessity.

As an example we may consider Brownian movement, the motion of a microscopic particle suspended in a fluid. The motion of the particle is described as a random motion, resulting from the transfer of momentum to the particle by the random collisions with it of the molecules of the fluid. It was shown by Einstein and by Smoluchowski half a century ago that by the methods of statistical mechanics, with use of the completely deterministic equations of the classical laws of motion, the be-

113

havior of the particle, in a statistical sense, can be completely predicted. The exact motion of the particle--the statement of its positions and velocities as a function of the time--cannot, however, be predicted, because of the lack of information by the observer of the values, at one instant of time, of the positions and momenta of all of the molecules in the fluid (as well as of the molecules in the walls of the system, with which the fluid interacts). It is said to be chance that causes the particle to move along its unpredictable path, in the way in which it is, in one series of observations, observed to move; but it is believed that this "chance" is simply the necessary result of the molecular collisions, called chance because of lack of knowledge of the observer.

It may be contended that the Brownian movement of the particle is in fact not determined, and is indeterminable, because of the fact that it is not the classical laws of motion, but rather the quantum mechanical laws of motion, that determine the behavior of the system. In fact, however, the predictions about Brownian movement of the particle made by application of the laws of quantum mechanics are identical, for a system at ordinary temperatures, with those made by application of the laws of classical mechanics. At ordinary temperatures the density of energy levels is so great that the existence of discrete quantized states, rather than a continuum of states, has no effect on the outcome of the calculations. It is not the uncertainty of the Heisenberg uncertainty rule, the principle of indeterminacy, that introduces chance into the theory of the Brownian movement; it is the impossibility for any observer to obtain information about the state of motion of the molecules of the fluid, at a level far more coarse than the level of the differences between quantized states of motion, that introduces chance into the Brownian movement.

The question of chance versus necessity is identical with that of freedom of choice versus determinism. When I (and here "I" means either Linus Pauling or Emile Zuckerkandl) use my mind (the functioning of my brain) to make a decision about which course, of two alternative courses of action, to follow, am I exercising "freedom of choice," or am I only following a course of action that is determined by the instantaneous nature of the universe and its entire past history? I feel that I have exercised freedom of choice, that I have made a decision; and in fact I have exercised freedom of choice and made a decision. There was, however, no alternative, no matter which way the decision was made, because so far as we are aware there is only one universe, and it follows only one path through time. If there were a multitude of universes, ever increasing in number and branching off from one another whenever the choice between two alternatives is made, with one universe corresponding to one choice (one branch) and another to the other choice (the other branch), then communication between these alternative universes would give us additional information; but this communication would at the same time lead to begging the question, in that it would combine the universes into a single universe, and this single (but continually multiplying) universe would as a whole be following only a single course through time. In fact, the idea of freedom of choice is by its nature such that it is never susceptible to proof. I am satisfied to feel that I have freedom of choice--sometimes the repeated necessity for choosing becomes a burden; but I am also satisfied to believe that this freedom of choice, which it pleases my consciousness, my mind, to have, is in fact one aspect of nature, which continues to change from instant to instant in a way determined by the nature of nature, by the laws of nature, laws which we comprehend only in part, and which perhaps are such that their complete comprehension by a part of nature, a part of the universe, is impossible.

There is a level, however, at which the Heisenberg uncertainty principle presents an interesting problem. The equations of quantum mechanics are such as to show that it is impossible by any known method of investigation to determine exactly, at the same instant of time, the position and the momentum of a particle, such as an electron. The result is that processes involving small particles, and at the level where the difference between adjacent quantum states is distinguishable, are in principle unpredictable by any method known because complete information about the system at one instant of time is unobtainable. Let us discuss an example. The element radium contains nuclei that are described as consisting of 226 particles, 88 protons and 138 neutrons. The nuclei of radium decompose spontaneously with emission of a helion (alpha particle), which is described as consisting of two protons and two neutrons. It is observed that in 1590 years half of the radium nuclei in a sample of radium have undergone this spontaneous decomposition. It is impossible to predict the time at which a particular radium nucleus decomposes. For example, atoms of radium may be spread thinly over a fluorescent screen, which is then watched through a microscope. When an atom decomposes it emits a helion with enough energy to cause the fluorescent substances to send out visible light, which is observed. We may accordingly watch a radium nucleus at a particular point on the fluorescent screen, and may observe that after about one year, perhaps, it undergoes decomposition and a scintillation is seen at that point on the fluorescent screen. According to the principles of quantum mechanics, it is impossible, no matter how much information is gathered by an investigator about a particular radium nucleus, to obtain enough knowledge about the nucleus, about its instantaneous structure, the instantaneous states of motion of its constituent nucleons, to permit the prediction that this radium nucleus will undergo radioactive decomposition at a particular time in the future.   The laws of quantum mechanics permit only statistical statements to be made, statements about the probability of radioactive decomposition as a function of time.   The behavior of the particular radium nucleus, its actual decomposition at a certain time, is attributed to some aspect of chance.

We ourselves find the idea that there is a fundamental element of chance in relation to the progress of the universe through time dissatisfying.   In this feeling we are in agreement with some other people--with Einstein, deBroglie, Bohm, Vigier, and a rather few others.   We feel more satisfaction in thinking that there is something about nature, something about the radium nucleus, that is correlated with its decomposition at a certain time.   Einstein said, with reference to this point, that "God does not throw dice."   This analogy, which is, we think, useful in showing what it was that troubled Einstein, does not stand up under analysis, of course, because we believe that the turn of the die is not a matter of chance, fundamentally, but rather a matter of the operation of the precise laws of motion and of the properties of material substances.   The element of chance in the throw of a die is a reflection of our lack of knowledge about the system.

The Heisenberg uncertainty principle has no relation whatever to the question of freedom of choice versus determinism. The brain of a human being functions at the temperature $310^{\circ}K$.   At this temperature quantum restrictions have little significance.   What really is restrictive to the investigator interested in deciding between freedom of choice and determinism is the impossibility of determining precisely the structure and state of the brain, even at a classical level, without destroying the brain. There is no way of testing whether the brain functions in a predictable way because there is no way of obtaining sufficient information about it to permit a prediction to be made.   If we agree that questions which by their nature can never be answered are meaningless, then we must agree that the question as to whether a human being has freedom of choice or does not have freedom of choice is a meaningless question.

These fundamental conclusions do not mean that we are restrained from discussing the question of chance and necessity in biological evolution, at a level different from the fundamental one. In thinking about biological evolution we may be concerned with understanding how things were in the past and how things are, rather than with how they will be. Let us imagine--without worrying about the fictitious aspects of the example--

that at the time the first frog was about to develop from the first tadpole a scientist, who had the tadpole in his hand, wanted to know what it was going to become. Let us also imagine that the scientist was in possession of an enormous, as yet unattainable, knowledge of biology and of analytical techniques. This man, a tremendously able biologist who had never seen a frog, is faced with another complementary relation that is somewhat analogous to the Heisenberg complementarity rule (uncertainty principle). He must make a choice: He cannot know about the frog before it comes into existence, and have him too. In order to foresee what the frog will be like, through analyzing the tadpole, he must necessarily destroy the tadpole. In fact, if no relative of the tadpole were already well known, from which to derive hints and extrapolations, the analytical destruction of a great many tadpoles would be required before the frog could be intellectually "derived," by application of a thorough understanding of the interplay of the different molecular and cellular systems. If, on the other hand, the biologist wants to see the frog, he must give up foreseeing him. There is a mutual exclusion between seeing and foreseeing, between the intellectual reconstruction of a reality and the mere existence of this reality. Again, you may know it or you may have it, not both. One might call this alternative the Lohengrin effect.

The fictitious example of the frog could be made more realistic, though less striking, by applying it to the forecast of all the effects of a single mutation of a novel type in an organism that is otherwise known.

Whatever the example, the point is that evolution is unforeseeable because, in order to be foreseen, it would have to be dispensed with. Scientists can foresee only what they can extrapolate from analogous examples. The appearance of a unique novelty, never illustrated elsewhere, cannot be predicted, and when it looks as though it can be predicted, this look arises only because of the similarity of the novelty to the known.

If we can predict the acceleration of a falling stone, it is because the acceleration of another falling stone has been measured at some earlier time. Similarly, we could predict biological evolution if we had at our disposal a number of examples of evolutionary history, as they quite probably have occurred and are occurring on other planets throughout our galaxy and in other galaxies. Since we have no such information, evolution is in our eyes a unique occurrence, and its future is an absolute novelty and thus unpredictable; it would be predictable only if we were in a position (which in fact we are not in) to make so thorough an analysis of the present that no future evolution would then be possible.

Though many of us attribute parts of the evolutionary processes to chance (in particular mutations, as they occur spontaneously and unpredictably), the evolutionary process as a whole, if not in its fine details at least in its general outlines, is just as generally thought to be determined. Determinism is indeed very firmly associated with the field of biological evolution. If it were otherwise, it would not be a field of science.

The operation of the evolutionary process was shown in a striking way by Zamenhof and Eichhorn in their study of microbial evolution through loss of biosynthetic functions. These investigators carried out experiments involving competitive growth in a chemostat of an auxotrophic mutant (a mutant requiring a nutrilite) and a prototrophic parent in a medium of constant composition containing the nutrilite. They found that the "defective" mutant, which requires the nutrilite, has a selective advantage over the prototrophic parental strain under these conditions. For example, an indole-requiring mutant of Bacillus subtilis was found to show a strong selective advantage over the prototrophic back-mutant when the two were grown together in a medium containing tryptophan. The relative number of cells of the latter decreased a million fold in 54 generations. They also found that greater advantage to the auxotroph accompanies a greater number of biosynthetic steps that have been dispensed with, that is, an earlier block in a series of reactions, when the final metabolite is available. They further found that a mutant with a gene deletion is at a distinct selective advantage over a point mutant, in that not only the synthesis of the metabolite, but also that of the structural gene, the messenger RNA, and perhaps the inactive enzyme itself would be dispensed with, and that accordingly the mutant with a deletion would replace the point mutant in competition.

It is possible to formulate the general principle that evolution will occur when there is a change in environment such as to supply to the organism essential molecules that have previously been synthesized by the organism, the advantage of the mutant being that it is rid of the machinery of synthesis and of the requirement for providing energy for the synthesis.

We may state the general principle that when an essential substance becomes available to a species, an evolutionary change will occur through the development of a mutant that has abandoned the machinery for synthesis of the essential substance. The environment has been changing rapidly in recent centuries. We refrain from making predictions about evolutionary changes that should follow from this change in the environment.

As soon as animals became large enough to live by eating and digesting plants, they began to evolve in such a way as to abandon the machinery for manufacturing many essential substances, those essential substances that were manufactured by the plants and were present in the food of the animals. The animals were made simpler in this way, and were improved, by having a smaller amount of biological machinery that could get out of order. Ten years ago, in a discussion of molecular disease, evolution, and the gene, we said that every vitamin that we need today bears testimony to a molecular disease our ancestors contracted, perhaps hundreds of millions of years ago, and that these molecular diseases are controlled by our having the proper diet; that evolution is based in part on the appearance of molecular diseases whereof the environment can cure the symptoms. Because these simplifications in the machinery of our bodies are beneficial, we conclude that "Whereas in order to achieve superiority, it is not sufficient to be degenerate, it is however necessary."

There is evidence that in one respect this degeneration has not been as advantageous as it might be. It is likely that the episode in our past history that resulted in the replacement of our ancestor who had the ability to synthesize ascorbic acid by a mutant ancestor who had lost that ability was in an environment especially rich in ascorbic acid, and that as man has spread over the earth, into areas poorer in ascorbic acid, he has suffered from a chronic deficiency in this important substance. We are in a position now to rectify this situation, by supplying, by technological methods, an adequate amount of this important substance. We know, because we have now been put in a position to know, that the replacement of auxotrophic by prototrophic organisms is not entirely due to chance, and that the development of a frog from a tadpole is not entirely due to chance. We have not yet been put in a position to know (and perhaps can never be put in that position) that the emission of the helion by a radium nucleus at a certain instant of time is not due to chance. But why should we conclude differently?

Because of the coherence between phenomena and the coherence of our relation with them, it is an economical and scientifically sound choice, although a philosophical one, to believe in the existence of a "nature of things" independent of our mind, provided that we acknowledge that the word "thing," in this statement, is used in a special sense, since in its ordinary sense a thing is already, in part, the result of the working of the mind. To this nature of things our mind closely corresponds, as witnessed by the considerable success of the efforts to develop reliable physical theories, although the complete nature of things probably cannot be formulated, since any formulation introduces a dependence on the mind.

Those people who hold that the motion of an individual electron is undetermined imply that determinism as it prevails at higher levels of integration of matter is founded on indeterminism at a lower level. They also imply that the determinism of a group of elements (the statistical behavior of electrons, or of a sample of radium) is the sum of the indeterminisms of each element belonging to the group. These statements are possible ones, and perhaps justifiable. But they seem to us to be quite gratuitously contrary to common sense. Common sense may be inadequate for comprehending the universe--it quite certainly is. But when abandoning common sense does not provide any advantage in either understanding or practice, instead of turning our backs on it we might as well keep it company, in tribute to what it has achieved for us on other occasions.

The problem of the generality of determinism is a metaphysical one, but asserting the intrinsic chance behavior or "freedom" of the electron or the radium nucleus is just as metaphysical as denying it. As scientists we may propose that, whatever the nature of the general rule that presides over the relationship between phenomena--it is called causality, determinism--this rule is to be considered to apply, intrinsically, to the relation between all phenomena, until proof to the contrary is forthcoming. There is no proof to the contrary in the proof that, in relation to some phenomena, man is not, or even cannot be, in a position to verify the workings of determinism.

Whereas there should be something intrinsic about the reality of determinism, in that it belongs to the indescribable "nature of things," there is in fact nothing intrinsic about chance. Chance is secreted by the observer, necessity by the observed.

The nature of chance as a position effect will become clearer as we enumerate the different reasons for which we may be unable to predict a given phenomenon.

1. We could not predict the velocity of fall of a body before the law of acceleration of falling bodies was formulated. Thus we may not foresee a phenomenon because we lack the necessary knowledge, a knowledge that could, in principle, be acquired. In cases of lack of knowledge it was customary in former centuries to say that the phenomenon was to be

attributed to God. In our scientific era we do not say that it is to be attributed to chance (instead, we trust that, though the phenomenon is momentarily unpredictable, it is determined). So far, so good.

2. We may not foresee a phenomenon, such as the position and the simultaneous velocity of an electron, because simultaneous observation of these quantities interferes with the production of the phenomenon. Again, we may postulate, with Einstein, that the unforeseeable, unpredictable phenomenon is in fact determined. In this case many people choose to say that the phenomenon is undetermined and is due to chance. Both of these positions are metaphysical, since they are, at the present time, at any rate, beyond the demonstrable. But the assertion of determinism is in keeping with the rest of scientific knowledge, whereas the postulate of an intrinsic indeterminacy of the electron or the radium nucleus is not, and that of its "freedom" is sheer anthropomorphic nonsense. At the same time, however, it is scientifically flawless to say that the behavior of the individual electron or radium nucleus is due to chance--this is indeed only a statement about a type of relation between the phenomenon and the observer.

3. Another type of case is closely related to the preceding one. It concerns all of those phenomena on the macroscopic scale (that is, at orders of magnitude where the Heisenberg indeterminacy is no longer to be felt) that are called "fluctuations." For example, the Brownian movement of a particle, or, as another example, how high on the beach a given wave will reach. Also included here is the question of what number will be drawn in a lottery. Generally it is not doubted that each of these individual occurrences is intrinsically determined. But none of them is foreseeable, because foreseeing would require a complete control of all the details of the system. Such a control, such knowledge, would be possible only by interfering with the event to be predicted. Here again knowledge would exclude existence, here again we have a Lohengrin effect. Whereas in the case of the simultaneous determination of the position and the velocity of an electron the notions of indeterminacy and chance are held by most people to apply simultaneously, in the case of macroscopic fluctuations the common belief is that the event attributed to chance is in fact intrinsically determined. Intrinsic determinism and chance are thus sometimes associated and sometimes dissociated in the minds of people. There is no good reason for such a variation in judgment. If the individual electron is thought not to obey causality, so should the individual wave on the beach be thought not to obey causality, since in each case determinism is not directly demonstrable.

4. We may not foresee a phenomenon because it is produced by the exercise of a <u>will</u> distinct from our own.   To be confronted with another mind--it need not be a human mind, it may be that of another animal-- poses special problems with respect to the predictable. Any author should be grateful for not being obliged to go into this subject, and so are we. We note only that the uncertainty that arises in dealing with someone else's mind is again a position effect.   We may not be able to foresee a voluntary action because we cannot be inside another person's mind.   In general it is not held that the unpredictable actions of others are due to "chance."

5. Finally, we may fail to predict a phenomenon simply because we ourselves leave it outside the field of the predictable. Each time that we define a system in order to examine the causal relationships within it, some of the factors that we have not included may act on the system. The action of external elements on a system will be due to "chance" in relation to the system, however well determined the same elements may be with- in another system. It is impossible for us to include, in an effective way, the entire universe in the system under discussion.   Once again chance turns out to be a position effect--in this case not so much the position of the observer himself as that of a factor in relation to a system defined by the observer.   The following definition thus applies here: Chance is what rules the behavior of any element that acts on a given system of de- terministic relationships without belonging to this system, as it is form- ulated by us in our analysis of the problem.

We may also be able to predict a phenomenon and yet attribute it to chance--for example, when a population of elements of a given type is considered, the transformation of this population may occur randomly, and it may be possible to apply the principles of statistical mechanics, or of statistics in general, in making predictions.   Here chance again re- sults from taking a point of view, the point of view of considering events in a set of elements, rather than the fate of the single elements that make up the set.   One of us (Zuckerkandl) is among those who recently have published data showing that amino-acid substitutions, as they become fixed during the evolution of a protein, behave collectively as though they occurred at random.   A collectivity of elements behaves randomly when the most probable event occurs most often. "Randomness" or "chance," in this case, represents the conformation of a population of elements to a probability defined by the compositional properties of that population. The behavior of the population is thus determined by its own properties, and randomness results if no external constraint interferes with the mani- festation of these properties.   Randomness is thus the determinism of systems that are isolated and left to themselves, or that behave as though they were. This determinism is, of course, always that corresponding to an increase in entropy for the universe as a whole. "Chance" here is in fact not chance at all, at the level of the population of elements considered. It is determinism in collective behavior when the second law of thermodynamics and the laws of steady states of open systems, influenced by the rates of changes which may themselves be influenced by molecules with catalytic activity, prevail. In living organisms the steady-state rates are influenced by the presence of specific catalysts, enzymes, which lead to the predominance of certain biochemical reactions and molecular changes, and to the selection of one course from among alternative courses, all of which are associated with an increase of entropy of the universe as a whole and are compatible with the laws of thermodynamics.

It can be seen that we may distinguish several types of position effects relating to the occurrence of "chance." There are essential intrinsic position effects such as that of the observer with respect to the electron or with respect to the derivation of a theoretical frog from an empirical tadpole. There are nonessential intrinsic position effects such as that of the observer who happens to be intrinsically a poor scientist when he might be a good one. There are nonessential nonintrinsic position effects when the observer deals with factors acting from the outside on the system he is analyzing, such as the "accidental" failure of a laboratory experiment because of a drop of line voltage that represents pure chance in relation to the experiment, but is perfectly determined in the mind of the electrician.

Let us emphasize that all that cannot be predicted should not be deemed to be due to chance (types of unpredictability 1 and 4), and all that is due to chance, and unpredictable, is not to be deemed to be intrinsically undetermined (types of unpredictability 3 and 5). Our own position, as already stated, is to hold, in agreement with the fundamental scientific philosophy of Einstein and some others, that all that is due to chance is at the same time intrinsically determined.

There is, at any rate, no scientific problem here, but rather a philosophical choice. Therefore the question as to whether life has arisen and developed by chance or by deterministic processes is not a scientific question, and, moreover, any scientific investigation of this problem necessarily is based on an implicit answer and is bound to fail because of the impossibility of carrying out the investigation with sufficient penetration to yield an answer.

What, then, is the good question to ask? Whether the living is predictable before it arises? Whether later phases of evolution are predictable from earlier phases? We have already stated that these questions cannot be answered, as long as previous experience with other evolving living systems is not available. The good question, then, is neither whether life is due to chance, nor whether it is due to necessity, nor whether it was predictable before it existed, but rather whether its appearance was based on a unique, or at least on a very exceptional, occurrence or not. This is, in fact, a question asked by Jacques Monod in his above-mentioned book, and to him the likely answer is in the affirmative. He thinks that the first molecular system endowed with fundamental properties of life probably arose as a single event and, furthermore, like many other biologists, he believes that major features of evolution at a much higher level, mainly the appearance of a creature endowed with the fundamental mental properties of man, are equally intrinsically unique events.

The modern concept of God's extended finger through which life passes into Adam, or rather one part of this concept, is the to some people esthetically less enchanting act of biogenesis in the "thick hot soup." In many of its local expressions, the thick hot soup can be conceived as a system, and the most primitive forms of life also represent a system. At all stages of its expression, the biotic trend must have been one relating to systems of molecular interaction. Sidney Fox is one person who not only adopted this generality, but also rendered it experimental in very suggestive ways. A system never exists in one unique state. It appears likely that biogenesis is the passage from a "non-living" system existing in a large number of states to a "living" system also existing in a large number of states. Monod's contention that the appearance of life probably was a unique, nonreproducible occurrence implies that the pathway from one system existing in multiple states to another system existing in multiple states can be unique, and its opening up can be therefore so improbable that in many million years it occurred only once. Whereas the pathway between one particular state of a system to one particular state of a qualitatively different system may indeed be unique, if two systems exist in many different states and a conversion from one into the other is possible, then it is unlikely that its conversion can occur between only one uniquely defined state of each of the two systems through but a single possible pathway. Rather, multiple pathways between numerous states of the two systems are expected to exist.

The most unlikely thing about life is its being unlikely.

As to the inherent likelihood for evolution in its later phases to take the directions that it has indeed taken on earth, this likelihood can be brought out by circumstantial evidence only, since, once again, the only direct evidence would be knowledge of many other independent evolutionary histories. The probability that man can construct artificial systems and observe them over a sufficiently great length of time as to permit the independent biogenetic process to occur seems to us to be extremely small.

Like anything else, man is a mixture of essential and accessory traits--the essential traits being defined in relation to the type of interest of the observer.   Obviously, if we consider the total collection of traits that characterize man and ask what the chances are that this total collection has arisen identically on some other planet elsewhere in the universe, we can at once state that this chance is vanishingly small.   Such a conclusion is, however, of no interest.   It is vanishingly improbable that there exists anywhere another mountain exactly like Mont Blanc. This does not mean that mountains are improbable.

It is different to ask what the chances are that in a few billion years of biological evolution a creature will appear that is able, by means of a symbolic language, to communicate novel associations of ideas to his fellow creatures. There are no good reasons for supposing these chances to be very low, since we witness, on Earth, several independent developments in the direction of mental and psychic capacities, such as those of the Cephalopod Molluscs.   These developments remain far behind those of the Primates. In Primates, however, they have gone quite a way along lineages different from man's. If we think of the distance covered since the onset of evolution, the mind even of Cephalopods is very impressive indeed.) It is undeniable that the trend toward the appearance of mental faculties on Earth is by no means limited to one lineage. On the contrary, "mind," the functioning of the brain in a strikingly effective way, is a property realized again and again, in different versions and at different levels, by evolutionary convergence.   This observation appears more striking and more significant than the traits whereby man is unique. We conclude that, in biological evolution, the appearance of some type of mind is very likely, that of highly evolved forms of mind is likely (take the dog, the porpoise, the chimpanzee), and that of a human level of mind is not unlikely.

The evolution of a psyche is not the only general trend that will be manifest as a function of reproducible circumstances. Many other trends are, and thus fail to confirm the uniqueness of basic traits of biological evolution on Earth. Consider, for instance, the convergence between the marsupials of Australia and mammals elsewhere in the world. Marsupials have evolved their "rats," their "cats," their "wolves," their "bears," and so on, morphologically speaking. These different types of morphology and the different types of behavior and of habitat associated with them will therefore be produced along independent lineages, provided that environmental conditions are similar. There are a larger number of other instances of functional convergence through evolution, and they all tend to show that under similar circumstances living systems will tend to give similar results. If we find this to be true of organisms as complex as mammals, shall we doubt that it should be true also of molecules, at the time of biogenesis?

It therefore seems safer to be among those who think that the appearance and evolution of life on earth was unavoidable than to be among those who reject this thought. This does not mean that one should deny the intervention of chance, provided that the term is not misconstrued. Through original channels of its own, the human mind retraces the steps of nature. Where the mind cannot retrace these steps in detailed correspondence, or where the mind does not choose to do so, there "chance" is the master of the field.

Received 3 January, 1972

MICROMOLECULES

THE SOURCES OF PHOSPHORUS ON THE

PRIMITIVE EARTH--AN INQUIRY

Alan W. Schwartz

Department of Exobiology, University of Nijmegen
The Netherlands

INTRODUCTION

One of the problems of chemical evolution which has remained a frustration to workers over the years is the question of the mechanism by means of which phosphorus might have entered into chemical reactions on the primitive Earth. A number of publications have discussed the incorporation or activation of orthophosphate under presumed prebiotic conditions; however, the sources of phosphorus used in most of these experiments fail to meet the test of geological plausibility. In many cases concentrations of phosphate in aqueous solution have been employed which are many orders of magnitude higher than any that can be attained under natural conditions (1-4). In other, anhydrous experiments, salts of phosphoric acid have been utilized which do not naturally occur (5). Other phosphorylation experiments have depended upon the use of highly reactive condensed phosphates, the syntheses of which have not been demonstrated under geologically plausible conditions (6-9). Only a very few studies have dealt with sources of phosphate which are geologically significant, and these publications will be reviewed briefly below. Two pathways in particular may have played a role in phosphate incorporation on the primitive Earth, although rather special circumstances are involved. Finally, an alternative approach to the question will be considered to account for the prebiological availability of phosphorus. The hypothesis to be examined does not relate to a particular set of circumstances upon the primitive Earth, but rather to the process of the formation and differentiation of the Earth itself.

THE TERRESTRIAL MINERALOGY OF PHOSPHORUS

The fact that the only significant form of phosphorus on the surface of the Earth is the highly insoluble and unreactive mineral apatite has been discussed before (10). Apatites occur both as sedimentary deposits and in igneous rocks. The igneous mineral is almost always fluorapatite (11), $Ca_{10}(PO_4)_6F_2$, while the sedimentary form is somewhat more variable, with minor substitutions of chloride, carbonate, or hydroxyl for fluoride (12). In the cycle of weathering, phosphate is transported by means of the small solubility of apatite (of the order of 10^{-6}M in P) to the oceans, where gradual reprecipitation occurs. The contemporary phosphate cycle is complicated by the presence of organisms which, because of the very low inorganic levels of phosphate present in natural bodies of water, can have a substantial influence upon the total dissolved phosphate concentration. Thus, decay of the remains of organisms in the deep ocean can contribute to the total phosphorus levels because of the presence of phosphorus-containing organic compounds, and the upper levels of the oceans are relatively depleted in phosphorus as a result of biological activity (13). This effect points to a simple but important observation, that once phosphate is incorporated into a nucleotide (or for that matter, into nearly any organic compound) its solubility under geological conditions (that is, in the presence of calcium) is increased enormously. But because calcium is always present in the environment, free phosphate is reprecipitated as soon as the concentration exceeds the equilibrium value for the apatite-water system. The first step in understanding the chemical evolution of phosphate esters, and therefore of nucleic acids and the total genetic system, is to attack the question of the prebiological mobilization of phosphorus. Two approaches have been used against this problem: attempts to account for the solubilization of apatite, so that phosphorylation may proceed in aqueous solution, and attempts to utilize apatite phosphate directly.

PHOSPHORYLATION WITH APATITE

Miller and Parris reported that a conversion of ortho- to pyrophosphate occurred on the apatite crystal surface when freshly precipitated hydroxylapatite was suspended in the presence of cyanate solutions (14). With 10^{-3}M cyanate at pH 7, approximately 0.4% of the apatite phosphate was incorporated into pyrophosphate linkages (18% yield of pyrophosphate when expressed in relation to cyanate concentration). If we can imagine this reaction occurring in the oceans, then rather large quantities of pyrophosphate could be available on the surface of suspended crystals. However, the further utility of this calcium pyrophosphate in phosphorylation reactions is open to question. Even concentrated solutions of soluble sodium pyrophosphate do not produce a significant degree of phosphorylation of nucleosides (15). In the absence of water a more favorable situation exists. Neuman et al. (16) have reported the solid state phosphorylation of adenosine by sodium pyrophosphate in the presence of apatite which apparently catalyzes the reaction. Now, the activation of suspended apatite in the presence of cyanate and the subsequent dehydration and heating in the presence of nucleoside would seem to be a somewhat unlikely sequence if apatite is precipitated in the deep oceans as is usually assumed. However, it is not certain that apatite is, in fact, precipitated directly in the oceans. There is evidence suggesting that a major pathway for nonbiological phosphate deposition is the substitution of phosphate for carbonate in deep sea aragonite deposits (17), although this question is still being debated in the literature (12). The possibility of cyanate concentrations as high as $10^{-3}M$ in the primitive oceans has also been questioned (18).

Nevertheless this combination of mechanisms could have operated in the well known "evaporating pond" and is therefore of potential promise.

There has been a recent proposal (19) that prebiotic phosphorylation of nucleosides with apatite would have been possible in the presence of urea. A suspension containing apatite, uridine, ammonium chloride, ammonium bicarbonate and urea was evaporated and heated at $100^{\circ}C$, and UMP was identified in 16% yield. The data presented suggest that the observed result was due primarily to the acid conditions generated in the procedure, since the evaporation could be expected to preferentially volatilize the basic ammonium bicarbonate. That this is indeed the case is suggested by the formation of a 1.7% yield of nucleotide in the absence of urea (19). The higher yield in the presence of urea may be due to the known effectiveness of urea-phosphoric acid mixtures in phosphorylation (20). The possible role of acid conditions in the prebiological activation of apatite has been discussed elsewhere (21), and the generation of such conditions will be considered further in the following section.

A novel experiment has recently been reported in which a nucleoside was phosphorylated by means of recoil phosphorus atoms produced by the decay of ^{31}Si. A sample of SiO_2 enriched with ^{30}Si was irradiated in a neutron reactor for ten hours at a thermal neutron flux of 8.0×10^{13} neutrons per cm^2 per sec. (22), and a solution of adenosine was added to the radioactive sample. A 2% conversion to AMP was observed after a decay period of three days. It was suggested that this process could have occurred on the primitive Earth. However, the most abundant isotope of silicon is ^{28}Si, ^{30}Si only being present to the extent of 3% (23). This effect, therefore, cannot be considered a significant natural process for incorporation of phosphate.

CONCENTRATIVE AND pH EFFECTS

Evaporation or freezing of a saturated solution of apatite could result in an increase in the phosphate concentration if supersaturation occurs, if some equilibrium can be shifted to produce protons, or if some mechanism is available to remove calcium ions from the solution. The first possibility can only operate over rather short time scales. The second, for similar reasons, probably could not contribute to phosphate solubilization on a large scale. Many of the theoretical difficulties in apatite utilization could be solved if acid environments had ever existed on the primitive Earth. Not only would acid solution provide the necessary high levels of dissolved phosphate, but evaporation and precipitation from acid solution could also produce reactive acid salts and even condensed phosphates. Although protons could be generated by a number of mechanisms, the presence of basic rocks would inevitably buffer the system, except on the shortest time scales. Calcium-complexing has been discussed and suggested by the author as a possible contributor to phosphate mobilization on the primitive Earth (24). There is admittedly some question, however, of the significance of the phosphate levels which could be reached in this manner under neutral or basic conditions. For example: in the presence of a small amount of oxalate any saturated solution of apatite can be evaporated to a phosphate concentration of the order of 10^{-3} M, assuming equilibrium. This value is still two orders of magnitude below any that have been used successfully in the synthesis of phosphate esters. If very high proportions of apatite were forced to precipitate by means of exhaustive dehydration of the solution, then a further increase in the concentration of phosphate might be achieved by supersaturation, or as a result of the expected decrease in pH (25). As already mentioned, such a pH effect could not be expected from a solution in equilibrium with basic rocks, although the required dehydration could be achieved by freezing, which would produce isolated, interstitial pockets of solution (26). Opportunity might be available by this process for reaction of phosphate with reactive compounds in eutectic phases (27). Alternatively, a proton-generating mechanism, operating in conjunction with evaporation-sequestration, might elevate phosphate concentrations where the rate of evaporation was more rapid than the approach to equilibrium. The combination of calcium-complexing and moderately reduced pH is an extremely favorable one for apatite solubilization (21).

At the moment, then, only two geologically plausible pathways have been suggested to account for prebiological activation of apatite: A surface-catalyzed reaction involving cyanate (14), followed by catalytic, thermal phosphorylation (16); and a combination of concentrative and calcium-sequestering effects in the presence of prebiological activating reagents. There is, however, another possible mechanism which has apparently not been considered before, and the remainder of this paper will be devoted to speculation concerning quite a different source for primitive phosphorus compounds.

THE MINERALOGY OF PHOSPHORUS IN METEORITES--SIGNIFICANCE
OF LOWER OXIDATION STATES

While fluorapatite is the only important phosphorus-containing mineral on the surface of the Earth, it is of only minor importance in meteorites. The meteorites which reach the surface of the Earth represent an extremely wide range of redox conditions, and although an apatite mineral occurs in the more oxidized meteorites (chlorapatite), the most common phosphorus-containing material in these objects is whitlockite (28), $Ca_3(PO_4)_2$.

As the oxidation state decreases, phosphate minerals are replaced by the phosphides schreibersite or rhabdite, $(Fe, Ni)_3P$, which occur in the metallic phase until, in the iron meteorites, phosphorus exists exclusively as the phosphide. Because the phosphide tends to be localized in nodules, rather than uniformly dissolved in the iron, total phosphorus analyses have generally been too low. A recent analysis of selected iron meteorites shows them to contain approximately 1 weight % of phosphorus (29). It is obvious that phosphorus is concentrated within the iron meteorites relative to other materials.

The difficulties which are inherent in the physical and chemical properties of the apatites led me to reconsider a suggestion originally made by Gulick (10). The fact that the calcium salts of the lower oxyacids of phosphorus--phosphorous and hypophosphorous acids--are considerably more soluble than the corresponding phosphates suggested to Gulick that these forms of phosphorus might have been present on the primitive Earth. Citing the existence of the nickle-iron phosphide schreibersite in certain meteorites, it was proposed that hypophosphite could be produced by reaction of schreibersite with water. Miller and Urey (30), in reviewing this proposal, pointed out that phosphate was thermodynamically stable in the presence of the concentration of hydrogen which they calculated for the primitive atmosphere (1.5×10^{-3} atm.), although they remarked that "....it is possible that stronger reducing agents than hydrogen reduced the phosphate...." There are two questions which are pertinent to this issue. First; what possible contribution could meteoritic schreibersite have made to chemical evolution, and second ; could the process which has produced iron phosphides in meteorites have operated also on the primitive Earth.

We can make a rough calculation of the maximum addition of schreibersite to the surface of the primitive Earth based on the infall rate of meteoric dust. Although estimates have varied in the past, some recent values are 0.5×10^{12}g per year for the present rate (31), and 3×10^{12}g per year for the primitive Earth (32). Most of this material is apparently of chondritic composition (31), but to provide an upper limit for the reduced phosphorus content let us assume that the content of iron and stony iron meteorites in the dust is the same as the observed fall frequency of about 7% (33). Since the P content of irons may average as much as 1%, this would yield an upper limit of 0.7×10^8 moles of phosphorus per year in the form of schreibersite. The present volume of the oceans is about 1.4×10^{21} liters (34). If we assume that the oceans have accumulated linearly with geologic time, then the rate of accumulation would be $1.4 \times 10^{21}/4.7 \times 10^9$, or 0.3×10^{12} liters per year. Therefore, even if a mechanism were available to oxidize this amount of schreibersite to soluble phosphorus compounds, and ignoring accumulating loss due to precipitation, the maximum level of oceanic phosphorus contributed by this process could not have exceeded $0.7 \times 10^8/0.3 \times 10^{12}$ or 2×10^{-4}M. Obviously this is an extremely rough treatment which, nevertheless, suggests that meteoritic schreibersite could not have played a significant role in the phosphorus chemistry of the primitive Earth. [The proportion of meteoritic material of chrondritic composition is probably much higher than the 93% figure derived from the fall rate (33), and even a catastrophic event such as the hypervelocity impact of an iron meteorite with a mass of 10^6 metric tons would only add an amount of P equivalent to ten year's accumulation at the average flux].

In thinking about this problem, I found that some contemporary hypotheses relating both to the origins of the meteorites and of the Earth suggest that the same mechanism which has produced schreibersite in some of the meteorites may well have produced a variety of reduced phosphorus compounds on the primitive Earth, and that the reducing agent responsible in both cases was probably carbon.

REDUCTION OF PHOSPHATE AND CHEMICAL EVOLUTION

"One must stress that at the present time all students of meteorites agree that the metallic iron in meteorites is of secondary origin--it is a product of the reduction process which occurred in the interiors of meteorite parent bodies. This conclusion if very important for planetry cosmogony, in particular for a study of the composition of the terrestrial planets" (35).

It is well established that the iron meteorites have been formed by the slow cooling of a molten, metallic phase within a much larger "parent body" (35, 36). The histories of other meteorite types are less clear, although the highly oxidized carbonaceous chondrites would seem to represent relatively primitive material (37). It appears, therefore, that differentiation within the solar system has resulted in the reduction and concentration of phosphorus within masses of metallic iron. It is logical to ask whether a similar process has occurred during the formation of the Earth to concentrate phosphorus within the core, and what the consequences of this process might have been within the context of chemical evolution. One of the more popular views of the origins of the various classes of meteorites is that the type I carbonaceous chondrites, which contain the highest proportion of carbon (3-5 wgt. %), are either examples of primitive material, or are at least similar in composition to an even more primitive parent material. Accordingly, other classes of meteorites may have been produced by heating and subsequent reduction of oxidized components by carbon, followed in some cases by further differentiation within asteroidal parent bodies (36). The iron phosphide which is present in the metallic phase of the more highly differentiated meteorites, therefore, has apparently been formed by the reduction of $Ca_3(PO_4)_2$ by carbon. An analogous model exists for the differentiation of the Earth, a key dispute being the origin of the iron which now forms the core. Thus, the Earth may have accreted from material similar to the type I carbonaceous chondrites, the iron being formed in situ by reduction with carbon (37); or accretion may have involved material of chondritic composition, in which free iron was already present, mixed with more primitive carbonaceous material (38). Of course there are other models, but for the purposes of this argument the differences are not critical, as most would result in carbonaceous material being in contact with calcium phosphate under high temperature conditions. Now, the reduction of tricalcium phosphate by carbon is the basis for the "volatilization process" for the manufacture of phosphoric acid, the first step being (39);

$$2 \ Ca_3(PO_4)_2 + 6 \ SiO_2 + 10 \ C = 6 \ CaSiO_3 + 10 \ CO_{(g)} + P_{4(g)}$$

The function of the silica is to promote formation of a slag which can be tapped from the furnace, although it may also permit the reaction to proceed at a lower temperature. Even in the absence of silica, the phosphorus can be completely volatilized in one hour at 1325^oC (39). That the conditions on the differentiating Earth would have favored this process seems to emerge as a consequence of present theories of core formation. The range of conditions would have been quite variable, and factors such as local carbon and iron concentrations, rate of heating and the maximum temperatures reached would have influenced the final distribution of phosphorus. Consider the reactions:

$$2Ca_3(PO_4)_2 + 10 \ C = 6 \ CaO + 10 \ CO_{(g)} + P_{4(g)}$$

$$2Ca_3(PO_4)_2 + 16 \ C = 2 \ Ca_3P_2 + 16 \ CO_{(g)}$$

$$2Ca_3P_2 + 12 \ C = 6 \ CaC_2 + P_{4(g)}$$

Up to 1550° CaO is the stable product. From 1600 to 1800° some proportion of Ca_3P_2 is formed, and above 1800° Ca_3P_2 is broken down into CaC_2 with further liberation of phosphorus (40). Free phosphorus will dissolve in liquid iron, and a considerable proportion of the phosphorus volatilized would have been trapped in this manner and transported to the core.

Whatever the proportion of phosphorus in the core, there does indeed seem to be an underabundance of phosphorus in the mantle as compared with that in Type I carbonaceous chondrites (41). The amount of phosphorus volatilized and made available on the surface of the Earth would have depended upon the homogeneity of the accreting material. The problem of homogeneous versus heterogeneous accretion is one still under debate (42). Too much obviously depends upon the choice of model to allow much speculation as to the quantitative significance of volatilization. However, the process which would have followed volatilization can perhaps be described somewhat more confidently. Oxidation of phosphorus by water vapor would have produced a mixture of products, the most important being phosphorous acid and phosphine:

$$P_4 + 6 \ H_2O = 2H_3PO_3 + 2 \ PH_3 \ (43).$$

Phosphine has already been shown to react with methane, ammonia and water in a spark discharge experiment to produce some of the lower oxygen acids of phosphorus, polyphosphates, and aminoalkylphosphates (44). Even after exhaustion of the initial supply of volatilized phosphorus, phosphine might have been generated at a decreasing rate by the decomposition of reactive phosphides within the crust:

$$Ca_3P_2 + 6\ H_2O = 2\ PH_3 + 3\ Ca(OH)_2.$$

The process of phosphorus volatilization could therefore have provided a supply of both volatile and water soluble phosphorus compounds for the earliest stages of chemical evolution. Both phosphine and the lower oxyacids provide the opportunity for free radical reactions to take place, through such intermediates as:

$$PH_2,\ PH,\ P_2\ \text{and}\ PO_3^{2-}\quad (45).$$

Thus the gamma-irradiation of an aqueous solution of sodium phosphite was reported to produce sodium hypophosphate, $Na_4P_2O_6$ (46). We have independently observed a similar conversion under the influence of ultraviolet light (47), and quite probably the PO_3^{2-} radical ion is involved. We are therefore studying the possibility of carrying out photophosphorylations with the lower oxyacids.

The presence of carbon (or carbonaceous material) during the accretion and differentiation of the Earth, therefore, would probably have led to some reduction of phosphate to elemental phosphorus and possibly to the accumulation of volatile and water soluble phosphorus compounds in the primitive atmosphere and hydrosphere. The quantitative importance of this process is, of course, open to question. My intention has been to suggest that a fresh approach to the problem of the sources of phosphorus on the primitive Earth may be in order, and that the unique conditions which must have prevailed during or shortly after accretion of this planet could have been involved. There is a tendency to feel uncomfortable about "once only" mechanisms in chemical evolution theory. However, it is important to remember that the process of core formation was just such an event, and may have been of critical importance for the development of life (48). The ubiquity of carbon in the Universe suggests that opportunities may exist for the reduction of phosphate under other conditions as well. There is an accumulating amount of evidence that the reactions which took place on the primitive Earth to produce organic compounds of biological importance may occur readily elsewhere. The presence of organic compounds in meteorites and associated with

interstellar dust clouds attests to this fact.   The atoms present in organic compounds on the primitive Earth may have been through many
cycles of formation, destruction, and reformation of ancestral compounds elsewhere in the Universe.   Perhaps phosphorus also takes part
in the same universal process.

SUMMARY

Three possible scenarios have been described for the prebiological
incorporation of phosphorus into primitive reaction sequences.   These
involve apatite catalysis, concentration-sequestration, and volatilization.   The first of these schemes has been shown by experiment to constitute at least a possible sequence.   The second has been tested only by
computation, but is potentially capable of experimental verification.
The third proposal is purely speculative, although the secondary consequences (i.e., involving the reactions of lower oxidation states of phosphorus) are testable.   Clearly a great deal more work is necessary before any definitive model can be proposed to account for the sources of
phosphorus on the primitive Earth.

REFERENCES

1. Steinman, G., Lemmon, R. M , and Calvin, M., Proc. Natl. Acad.
    Sci. U.S. 52, 27 (1964).
2. Halmann, M., Sanchez, R. A., and Orgel, L. E., J. Org. Chem.
    34, 3702 (1969).
3. Ferris, J. P., Science 161, 53 (1968).
4. Lohrmann, R., and Orgel, L. E., Science 161, 64 (1968).
5. Chang, S., Williams, J. A., Ponnamperuma, C., and Rabinowitz, J.,
    Space Life Sci. 2, 144 (1970).
6. Schramm, G., Grotsch, H., and Pollmann, W., Angew. Chem. Int.
    Ed. Engl. 1, 1 (1962).
7. Ponnamperuma, C., Sagan, C., and Mariner, R., Nature 199, 222
    (1963).
8. Waenheldt, T. V., and Fox, S. W., Biochim. Biophys. Acta 134,
    1 (1967).
9. Schwartz, A., and Ponnamperuma, C., Nature 218, 443 (1968).
10. Gulick, A., American Scientist 43, 479 (1955).
11. Landergren, S., in "Geochemistry" (V. M. Goldschmidt, ed.),
    p. 454. Oxford University Press, London, 1958.

12. Gulbrandsen, R. A., Economic Geol. 64, 365 (1969).
13. van Wazer, J. R., "Phosphorus and its Compounds," p. 960. Inter-
    science, New York, 1958.
14. Miller, S. L., and Parris, M., Nature 204, 1248 (1964).
15. Schwartz, A., and Ponnamperuma, C., in "Prebiotic and Biochemi-
    cal Evolution" (Kimball and Oro, eds.) p. 78. North-Holland
    Amsterdam, 1971.
16. Neuman, M. W., Neuman, W. F., and Lane, K., Currents Mod.
    Biol. 3, 253 (1970).
17. van Wazer, J. R., op. cit., p. 958.
18. Hulett, H. R., J. Theoret. Biol. 24, 56 (1969).
19. Lohrmann, R., and Orgel, L. E., Science 171, 490 (1971).
20. Daul, G. C. and Reid, J. D., Chem. Abstr. 47, 920h (1953).
21. Schwartz, A. W., in "Chemical Evolution and the Origin of Life,"
    (Buvet and Ponnamperuma, eds.), p. 207. North-Holland, Amster-
    dam, 1971.
22. Akaboschi, M., Kawai, K., and Waki, A., Biochim. Biophys. Acta
    238, 5 (1971).
23. "Handbook of Chemistry and Physics," p. B-272. Chemical Rubber
    Co., Cleveland, 1969.
24. Schwartz, A. W., and Deuss, H., in "Theory and Experiment in
    Exobiology" Vol. I (A. W. Schwartz, ed.), p. 73. Wolters-Noordhoff,
    Groningen, 1971.
25. van den Berg, L., and Soliman, F. S., Cryobiology 6, 10 (1969).
26. Wang, S Y., Nature 190, 690 (1961).
27. Sanchez, R. A., Ferris, J. P., and Orgel, L. E., J. Mol. Biol.
    30, 223 (1967).
28. Fuchs, L. H., in "Meteorite Research" (P. M. Millman ed.), p. 683.
    D. Reidel, Dordrecht, 1969.
29. Doan, A. S., and Goldstein, J. I., ibid., p. 763.
30. Miller, S. L., and Urey, H. C., Science 130, 245 (1959).
31. Singer, S. F., in Meteorite Research, p. 590.
32. Bar-Nun, A., Bar-Nun, N., Bauer, S. H., and Sagan, C., Science
    168, 470 (1970).
33. Keil, K., in "Handbook of Geochemistry" (K. H. Wedepohl, ed.),
    p. 78. Springer-Verlag, Berlin, 1969.
34. Mason, B., "Principles of Geochemistry," p. 192. John Wiley &
    Sons, New York, 1966.
35. Levin, B. J., in "Meteorite Research," p. 16.
36. Mason, B., in "Extraterrestrial Matter" (C. A. Randall, Jr., ed.),
    p. 3. Northern Illinois University Press, DeKalb, 1969.
37. Ringwood, A. E., Geochim. Cosmochim. Acta 30, 41 (1966).

38. Anders, E., Accounts Chem. Res. 1, 289 (1968).
39. Jacob, K. D., and Reynolds, D. S., Ind. Eng. Chem. 20, 1204 (1928).
40. Ershov, V. A., Chem. Abstr. 67, 83493v (1967).
41. Brett, R., Geochim. Cosmochim. Acta 35, 203 (1971).
42. Turekian, K. K., and Clark, S. P., Jr., Earth Planet Sci. Let. 6, 346 (1969).
43. Brunauer, S., and Schultz, J. F., Ind. Eng. Chem. 33, 828 (1941).
44. Rabinowtiz, J., Woeller, F., Flores, J., and Krebsbach, R., Nature 224, 796 (1969).
45. Halmann, M., in "Topics in Phosphorus Chemistry" (Grayson and Griffith, eds.), Vol. 4, p. 49. Interscience, New York, 1967.
46. Matsuura, N., Yoshimur, M., Takizawa, M., and Sakaki, Y., Bull. Chem. Soc. Japan 44, 1027 (1971).
47. Schwartz, A. W., unpublished data.
48. Anderson, D. L., Applied Optics 8, 1271 (1969).

NOTE ADDED IN PROOF

The evaporation-sequestration mechanism has now been tested experimentally in a simulated "evaporating pond." Substantial yields of nucleotides have been obtained (from initially dilute solutions) under conditions which are quite consistent with the geological constraints discussed in this article (Biochim. Biophys. Acta 281 (4), in press).

Received 30 August 1971

ABIOGENIC FORMATION OF PORPHIN, CHLORIN

AND BACTERIOCHLORIN

A. A. Krasnovsky and A. V. Umrikhina

A. N. Bakh Institute of Biochemistry
USSR Academy of Sciences, Moscow, USSR

INTRODUCTION

Oparin's hypothesis (1) maintaining that a prolonged chemical evolution preceded the emergence of life on Earth gave impetus to numerous investigations whose purpose was to reveal probable pathways of biosynthesis of complex organic compounds from simple substances which presumably existed in the primitive atmosphere and hydrosphere of the Earth.

Abiogenic formation of porphyrins in the course of chemical evolution can be studied in model systems that simulate prebiological Earth conditions. It seems likely that at a certain evolutionary stage porphyrins could arise from pyrrole precursors which, in turn, could result from ammonia and acetylene (2), acetylene and hydrogen cyanide (3), etc. Formaldehyde or other aldehydes, which may be synthesized from gases of the primary reducing atmosphere exposed to short ultra-violet irradiation or electrical discharges (4, 5), may also be involved in the synthesis of porphyrins.

In 1936 Rothemund (6) synthesized porphin from pyrrole and formaldehyde; this is a thermodynamically spontaneous reaction which proceeds during heating. Numerous experiments demonstrated a stimulating effect of water, oxygen, and different catalysts, including silica, on the synthesis of porphyrins from pyrrole and aldehyde (7). Catalytic properties of components of the Earth's crust were also described (8). Simple porphyrins can be abiogenically formed under the influence of electrical discharges and ultraviolet and gamma irradiation (9, 10). Porphyrin traces were found when methaneammonia-water mixtures were exposed to electrical discharges (11).

We detected porphyrins by measuring luminescence spectra of the reaction products of pyrrole and formaldehyde. Depending on the experimental conditions, fluorescence spectra differed essentially, thus giving evidence for a diverse composition of the products formed (7). Therefore, the products arising from the interaction of pyrrole with aldehydes required further analysis. In the present work we were mainly engaged in studying possible formation of not only porphin but also chlorin and bacteriochlorin in the pyrrole-formaldehyde system.

METHODS

The reaction was performed in sealed glass tubes heated in a boiling water bath. Solutions of formaldehyde (2%) and pyrrole (0.5%) in methanol were used in most experiments. The reaction was traced by porphin red fluorescence measurements. Fluorescence spectra were recorded on the spectrofluorometer in the range of 600 - 800 nm upon luminescence excitation by a mercury lamp DRS-250 through the filters FS-2, ZS-3, SZS-14 (mercury exciting lines 365, 404 and 436 nm). After the synthesis was completed and the fluorescence spectra were measured, reaction products were purified and partially fractionated by transferring porphyrins from methanol to benzene and subsequently to hydrochloric acid of 5, 10 and 20% concentrations. From these solutions the pigments were again transferred to benzene by neutralizing the acid with sodium acetate. To achieve final separation the pigments were chromatographed on Al_2O_3 columns (benzene solvent). At all purification stages fluorescence excitation spectra were measured. Absorption spectra could be measured only occasionally due to a low concentration of the pigment and to the presence of colored admixtures.

RESULTS AND DISCUSSION

Effect of atmospheric oxygen. The accelerating effect of atmospheric oxygen on the porphin synthesis was shown previously (7). In the air-containing tubes visible fluorescence appeared after 15-20 min heating and in the air-free tubes, several hours later.

In all the experiments the fluorescence spectrum clearly shows two peaks (Fig. 1a): in the presence of air at (I) 685 nm and at (II) 633 nm. Also found is a shoulder in the range of 620 nm (or a peak if the fluorescence at 633 nm is low) and an indistinct shoulder in the range of 760 nm. In vacuo, main maxima occur at (I) 704 nm and at (II) 648 nm. Thus, although the spectral pattern is similar in vacuo and in air, the fluorescence maxima in vacuo are shifted 15-20 nm to the long wave length region (as compared with the experiments in air). These changes may be attributed to the formation of porphyrin isomers (12). The ratio of fluorescence intensity in the maxima varies, thus indicating that different fluorescence maxima belong to different reaction products.

Effect of water.   In the presence of water in methanol the rate of porphin synthesis increased considerably.   With an increase of the water concentration to 50% the time interval between the onset of heating and the appearance of fluorescence decreased from 15-30 min to 1-2 min, but the yield of the end product was reduced.   Both in vacuo (the numbers in parentheses) and in air the fluorescence spectrum having a maximum at 633 (648) nm prevailed at the beginning of the reaction.   However, with further heating (20-30 min) the fluorescence intensity decreased significantly, especially at the maximum of 633 (648) nm, and a precipitation occurred.   Thus, the presence of water accelerates the formation of a product with a fluorescence maximum at 633 (648) nm which undergoes decomposition more rapidly than the product with a fluorescence maximum at 685 (704) nm.

Effect of foreign compounds, catalysts.   Preparations of silica, zinc oxide, iron oxides, titanium oxide, potassium chloride, and polyvinyl-pyrrolidone accelerated the synthesis of porphin.   Of the highest activity were zinc oxide (Fig. 1b) and titanium oxide. In the presence of catalysts in vacuo the time interval between the beginning of heating and appearance of red fluorescence shortened from 5-7 hours to 15-30 min. and the product yield increased as compared with the experiment without a catalyst.

The effect of electron donors and acceptors was also investigated. It was assumed that the presence of electron donors may accelerate the formation of reduced porphyrins (chlorin and bacteriochlorin), whereas the occurrence of electron acceptors may favour the formation of oxidized porphyrins (porphin).

Electron donors.   Ascorbic acid inhibited the process whereas tryptophan accelerated it actively.   In the presence of tryptophan the reaction occurred even at 3 - 5°C.   Under these conditions red fluorescence of

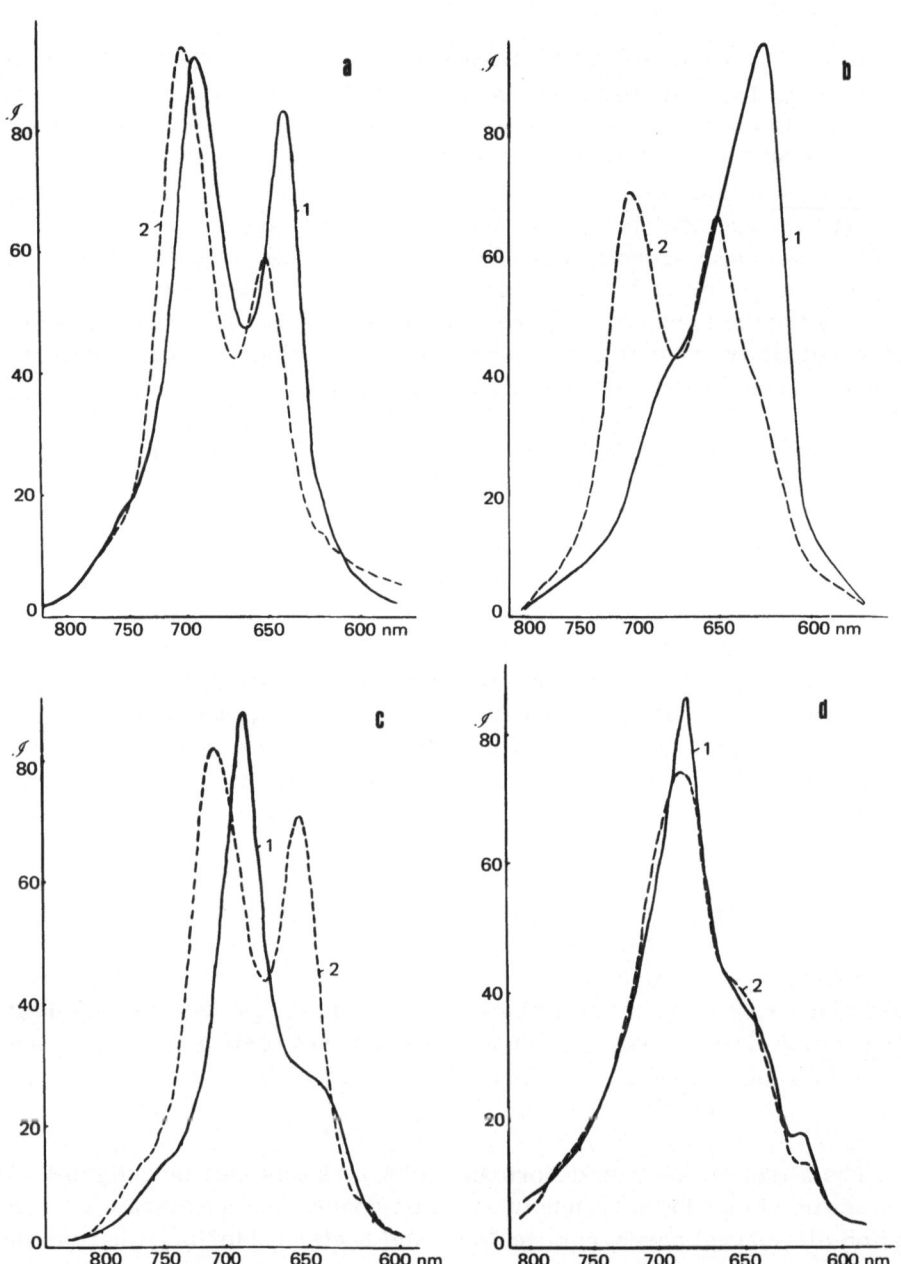

Fig. 1. Fluorescence spectra observed when heating solutions of pyrrole (0.5%) and formaldehyde (2%) in methanol (2 ml) under different conditions. a, control experiment without any additions; b, in the presence of 1 mg zinc oxide; c, in the presence of 1 mg tryptophan; d, in the presence of 1 mg p-benzoquinone (1--in the air, 2--in vacuo).

porphin appeared in a few days, while during heating it arose in 15-30 min.   In the presence of tryptophan the yield of porphins increased.   In air the main fluorescence maximum was at 685 nm (Fig. 1c) and in vacuo the maxima were 704 and 643 nm.   The heating of the solutions in the presence of excess tryptophan (10 mg per 2 ml solution) brought about, however, the formation of products with blue fluorescence, possibly dipyrrilmethenes.   The use of tryptophan instead of pyrrole did not result in the formation of porphyrins.   Other electron donors tested produced an insignificant activating effect on the porphyrin synthesis.

Electron acceptors.   Benzoquinone at a concentration of 1 mg per 2 ml of solution accelerated the synthesis of porphyrins and increased the yield of the products.   Both in air and in vacuo no compounds with a fluorescence maximum at 633 nm were formed.   Besides, no shift towards the longwavelength region was observed in vacuo (Fig. 1d) This is indicative of the synthesis of the same isomeric form of porphyrins in air and in vacuo in the presence of quinone.   The fact that a compound with a maximum at 633 nm was absent can be attributed to its oxidation by quinone.   Another oxidizing agent--ferrioxalate--inhibited the reaction.

Purification and fractionation of the products.   During the synthesis pyrrole and formaldehyde may give rise to porphin, chlorin, and bacteriochlorin.   The literature and our own data (13) suggest that the main maxima of porphin absorption in non-polar solvents occur at (I) 617, (II) 570, (III) 520, and (IV) 490 nm, the intensities being distributed among maxima 4, 2, 3, 1 (phyllotype spectrum). The luminescence spectrum exhibits two main bands at 690 and 620 nm.   In the absorption spectrum of chlorin the main maximum lies at 635 nm and the fluorescence maximum at 640 nm. The absorption spectrum of the trans-form of tetraphenylbacteriochlorin has maxima at 743 and 524 nm and that of the cis-form at 600 and 650 nm (14, 15).

Therefore, the fluorescence maxima at 685 and 618 nm in air (and at 704 and 622 nm in vacuo) which we observed in our experiments may be attributed to porphin insomers and the maximum at 633 nm (648 nm) to chlorin. The maximum at 764 nm in a concentrated pigment solution seems to occur due to the presence of bacteriochlorin.

The reaction products were pre-fractionated by extracting porphyrins by HC1 from benzene solutions.   Porphin was extracted predominantly by 5% HC1 solutions (6) and chlorin by 10 and 20% HC1 solutions.   The HC1 number of bacteriochlorin is unknown.   The pigments in 5, 10, and 20%

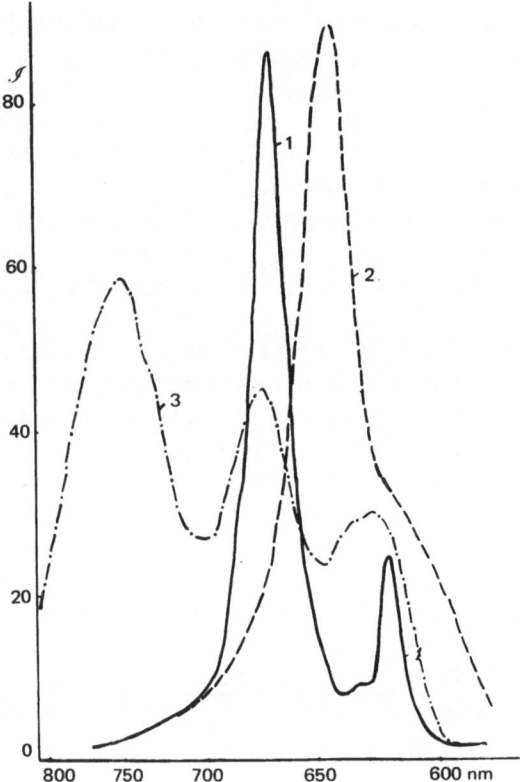

Fig. 2. Fluorescence spectra of porphyrin dissolved in benzene
following Al_2O_3 chromatography. 1, porphin; 2, chlorin; 3, mixture of pigments with bacteriochlorin in predominance.

HCl were neutralized with sodium acetate and then passed into fresh benzene; the fluorescence spectra of the resultant fractions I, II, III were
measured.   Fraction I derived from 5% HCl contained mostly porphin
(fluorescence maxima at 685 and 620 nm); however, the fluorescence spectra exhibited also a small maximum at 633 and 764 nm. Fractions II and
III derived from 10 and 20% HCl contained mainly chlorin, its fluorescence
maximum being at 633 nm.   These fractions also contained small quantities of porphin. If the starting solution had the product with a fluorescence
maximum at 764 nm, its relative content in fractions II and III was greater
than in fraction I. Thus, porphyrin fractionation by HCl solutions of different concentrations resulted only in a partial separation of the reaction
products.   To achieve final separation porphyrins were chromatographed
on Al_2O_3 columns from the benzene solutions.   When chromatographing
fraction I, the eluate in benzene contained only porphin (Fig. 2).

The first portions of the eluate obtained during the chromatographic separation of fractions II and III contained predominantly chlorin (Fig. 2), the subsequent portions contained a mixture of porphin and chlorin, and the last only porphin. Some substances showing red fluorescence (in fractions I, II, III) were firmly absorbed on Al_2O_3 columns, showed indistinct separation into zones, and were not eluted with benzene. The fluorescing zones were removed from the column in portions; the pigments were eluted with pyridine, and fluorescence spectra of the resultant fractions were measured. In the fluorescence spectrum of the upper and lower zones the main maximum was found to be at 764 nm (Fig. 2) and other maxima occurred at 700 and 640 nm. The middle zone was a mixture of porphyrins, porphin being in predominance. We so far failed to purify completely the product with the fluorescence maximum at 764 nm (presumably bacteriochlorin). Nevertheless, these experiments demonstrate that the fluorescence spectrum of both porphin and chlorin has no maximum at 764 nm (Fig. 2, curves 1 and 2) which evidently belongs to a third compound, most probably bacteriochlorin. If the eluate from Al_2O_3 in pyridine is kept in the air for several days the fluorescence intensity in the bacteriochlorin maximum at 764 nm decreases and that at 685 and 640 nm increases. This may be explained by bacteriochlorin oxidation into porphin via an intermediate formation of chlorin. There are well-known literature data indicating that bacteriochlorin pigments can be readily oxidized to form chlorin and porphin compounds (14, 16).

We showed that the luminescence excitation spectrum of a mixture of porphyrins dissolved in benzene had a maximum at 530 nm (Fig. 3) which corresponds (based on the analogy of tetraphenyl bacteriochlorin) to the second maximum of bacteriochlorin absorption. The luminescence excitation spectrum of porphin and chlorin dissolved in benzene exhibited no maximum at 530 nm.

The formation of porphyrins reduced to a different extent from pyrrole and formaldehyde can be accounted for by the fact that the pyrrole-aldehyde interaction to yield a tetrapyrrole cycle is accompanied by a release of four hydrogen atoms. The hydrogen released may either react with oxygen or other oxidants or may be utilized to reduce porphin to chlorin and bacteriochlorin. It is obvious that the reductive conditions should favour the formation of reduced porphyrins. These conditions increased the yield of reduced porphyrin forms (chlorin) and decreased the total yield of porphyrins, probably due to their secondary reduction and formation of products of non-porphyrin nature. It is plausible that the formation of reduced porphyrins may develop via a direct reaction, too.

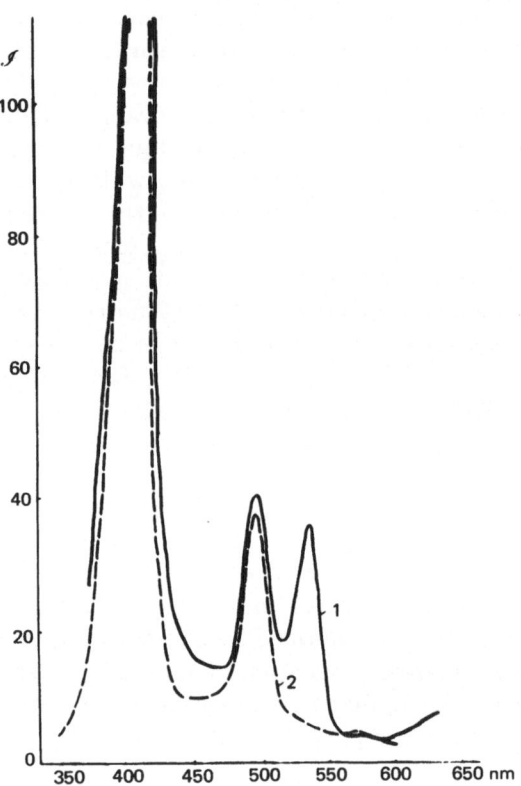

Fig. 3. Fluorescence excitation spectra of porphyrin solutions in benzene. 1, mixture of pigments; 2, porphin.

Photoreduction of porphin.   The possibility of a secondary photochemical reduction of abiogenically formed porphin aided by various electron donors should be discussed.

We examined the photochemical properties of chromatographically isolated porphin.   This pigment was readily reduced in both acidic and basic media.   For instance, porphin photoreduction by ascorbic acid in dry pyridine yielded a product with an absorption maximum at 635 nm and a fluorescence maximum at 643 nm (Fig. 4) which corresponded to chlorin.   The formed product was stable both in vacuo and in air.

Porphin photoreduction by ascorbic acid in vacuo in an acidic medium (2.5 and 5% HCl solutions in water or ethanol) included a gradual forma-

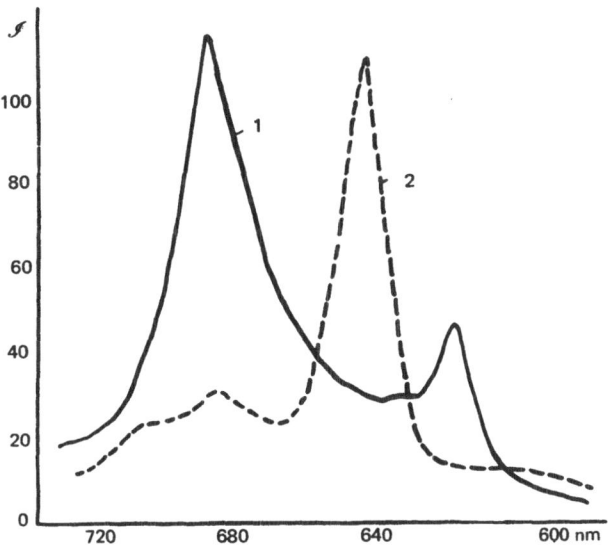

Fig. 4.   Fluorescence spectra observed during porphin photo-reduction by ascorbic acid in pyridine in vacuo. 1, before illumination; 2, after 15 min. illumination.

tion of products with absorption maxima at 440, 500, and 625 nm.   The product with the absorption maximum at 625 nm and the fluorescence maximum at 630 nm corresponded to an acid form of chlorin with respect to its spectral and photochemical properties (17).   No formation of bacteriochlorin was detected.

The results of our observations can be summarized in the following scheme:

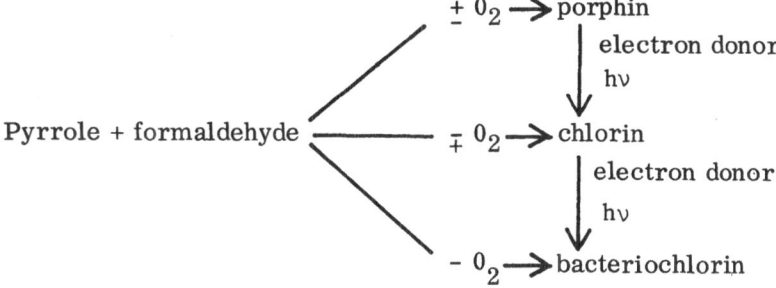

The synthesis of reduced forms of porphyrins may develop either during their synthesis from pyrrole precursors or as a result of the secondary photochemical or dark reduction by different electron donors.

The data obtained suggest that when searching for abiogenically formed porphyrins in other planets or in the depths of the Earth's crust one should take into consideration the fact that there may exist products reduced to a different extent: porphyrins, chlorins, and bacteriochlorins.

REFERENCES

1. Oparin, A. I., "Origin of Life on Earth," Izd. AN SSSR, Moscow, 1957.
2. Chichibabin, A.E., Jurn. Russk. Fiz. Chim. Obshch. 47, 703 (1915).
3. Meyer, R., Ber. 46, 3183 (1913).
4. Terenin, A. N., in "Origin of Life on Earth" (A. I. Oparin, ed.), p. 144, Izd. AN SSSR, Moscow, 1959.
5. Miller, S. L., J. Amer. Chem. Soc. 77, 2351 (1955).
6. Rothemund, P., J. Amer. Chem. Soc. 58, 625 (1936).
7. Krasnovsky, A. A., and Umrikhina, A. V., Dokl. AN SSSR 155, 691 (1964).
8. Hodgson, G. W., and Baker, B. L., Nature 216, 29 (1967).
9. Szutka, A., Hazel, J. F., and McNabb, W. M., Radiation Res. 10, 597 (1959).
10. Szutka, A., in "Origin of Prebiological Systems" (S. W. Fox, ed.), p. 245. Izd. Mir, Moscow, (1966).
11. Hodgson, G. W., and Ponnamperuma, C., Proc. Natl. Acad. Sci. US 59, 22 (1968).
12. Aronoff, S., and Calvin, M., J. Organic Chem. 8, 205 (1943).
13. Rimington, C., Mason, S. F., and Kennard, O., Spectrochimica Acta 12, 65 (1958).
14. Sidorov, A. N., Biofizika 10, 226 (1965).
15. Seely, G. R., and Talmadge, K., Photochem. Photobiol. 3, 195 (1964).
16. Krasnovsky, A. A., Drozdova, N. N., and Bokuchava, E. M., Dokl. AN SSSR 190, 464 (1970).
17. Umrikhina, A. V., Yusupova, G. A., and Krasnovsky, A. A., Dokl. AN SSSR 175, 1400 (1967).

Received 24 August, 1971

AN INVESTIGATION OF THE POSSIBLE DIFFERENTIAL RADIOLYSIS OF AMINO ACID OPTICAL ISOMERS BY ^{14}C BETAS

William J. Bernstein, Richard M. Lemmon, and Melvin Calvin

Contribution from the Laboratory of Chemical Biodynamics,
Lawrence Radiation Laboratory, and Department of Chemistry
University of California, Berkeley, California 94720

A recent report by Garay presents evidence that, in alkaline solution, ^{90}Sr β particles and/or their associated bremsstrahlung (γ-rays) cause a more rapid radiolysis of D-tyrosine than of L-tyrosine (1). The differential effect on the tyrosines was sought, but not found, in acidic solutions. The possible existence of this differential radiolysis was not tested on the crystalline tyrosine enantiomers.

A difference in the radiation susceptibility of amino acid enantiomers might provide the explanation of why modern proteins are constructed almost exclusively of L-amino acids. [The exceptions are a few percent of D-amino acids reported in the proteins of some bacteria and molds (2).] However, the Garay alkaline-solution results with tyrosine present certain theoretical difficulties. Although the circular polarization of the ^{90}Sr bremsstrahlung has been demonstrated (3), it is difficult to see how this could lead to any selective radiolysis in dilute aqueous solution. The radiolysis of organic molecules in such solutions is due to the initial production (from the water) of such active species as H atoms, OH radicals, and hydrated electrons, followed by attack of these species on the solute molecules (4, 5). It is difficult to see how any chirality of the β-rays or the bremsstrahlung could be induced in the water-derived species. However, the direct action of polarized bremsstrahlung on amino acid molecules in the solid state, in contrast to the necessarily indirect action in dilute aqueous solution, still provides an attractive hypothesis for the existence of the overwhelming predominance of L-amino acids in living systems. If it could be securely demonstrated that, under any plausible prebiotic conditions, D-amino acids were less stable than L-, we could understand the prevalence of the latter in proteins. With one exception, other theories about chirality in living systems ascribe to some chance event in the primitive Earth's history the present prevalence of one enantiomer over the other. The exception is the idea based on a minute fraction of right-circularly polarized light in sunshine (6). However, considering the very limited rotations found in the products of laboratory syntheses and/or photolyses carried out under strong sources of polarized light, the idea has gained little favor.

Earlier workers have unsuccessfully attempted to induce optical rotatory power in crystalline DL-alanine that was irradiated with the bremsstrahlung from ^{108}Ag and ^{110}Ag (7). However, in these experiments the DL-alanine was subjected to reactor neutrons and γ-radiation, as well as to the γ-radiation from the silver isotopes themselves. These additional forms of ionizing radiation may have masked any possible differential radiolysis induced by the bremsstrahlung.

This laboratory has a considerable stock of ^{14}C-labeled, crystalline DL-amino acids that were synthesized here 12 - 24 years ago. During the ensuing years these amino acids have been subjected to the radiolytic effects of their own longitudinally polarized β particles. Carbon-14 is a pure β emitter; thus, the amino acids were subjected to no nuclear γ-rays. The β particles have a low kinetic energy (maximum = 156 keV) and most of them are stopped in the sample itself. Consequently, the samples are well irradiated by both the β particles and their associated bremsstrahlung, each of which are polarized and present the possibility of a differential interaction with the two optical isomers.

The amount of beta-particle radiation that each sample received was calculated on the basis that all beta particles were absorbed by the sample--a reasonable assumption since the average range of ^{14}C betas (av. energy = 5.0×10^4 eV) in organic material is only about 0.1 mm (8). For example, the total radiation absorbed by our DL-norvaline sample, synthesized 15.0 years ago, and having a specific activity of 5.0 μ Ci/mg (=5.0 mCi/gram) is:

$(5.0$ mCi/gram$)$ $(2.2 \times 10^9$ dis/min/mCi$)$ $(5.0 \times 10^4$ eV/dis$)$

$\qquad$ x $(5.25 \times 10^5$ min/year$)$ $(15.0$ years$)$

$\qquad$ = 4.34×10^{21} eV/gram

The rad (100 ergs/gram) is equivalent to the absorption of 6.2 x 10^{13} eV/gram. Therefore, the DL-valine's radiation dose was 7.0 x 10^7 rads.

In Table I are summarized the data on our ^{14}C-labeled amino acids.

TABLE I. Amino Acid ^{14}C-Exposure Data.

DL-amino acid	Sp. Act., mCi/gram	Date Syn.	Radiation dose, Rads (x 10^{-7})	% decomp.
Norvaline	5.0	5/56	7.0	3
Alanine	3.2	9/59	3.5	26
Dopa	1.1	6/47	2.5	3
Aspartic acid	2.4	7/52	4.2	75–80
Methionine	7.3	2/56	10.4	59

The percentages of decomposition were measured on a Beckman Model 120C amino acid analyzer, and by the counting of radioactive spots on paper chromatograms of the radiation-decomposed material. The two methods were in essential agreement in all cases. The apparent small amounts of decomposition of the norvaline and dopa (dihydroxyphenyl-alanine) are surprising. However, there may have been considerably more radiolysis. If the radiolysis products were volatile, both the amino acid analyzer and the radioactive-spot counting would show less decomposition than that which had actually taken place.

The analyses for possible differential radiolysis of the optical isomers was carried out by a search for measurable optical activity in each sample. The following amounts of sample were used: norvaline (100 mg), alanine (268 mg), dopa (5.1 mg), aspartic acid (1.093 g), and methionine (185 mg). Each sample was dissolved in 10 ml H_2O and placed in a 7-ml (2-cm) "lollipop" cell and analyzed in the ORD mode of a Cary 60 spectropolarimeter. The sample of aspartic acid was crystallized from ethanol-water yielding 20% of the pure amino acid. This allowed the ORD observation to be extended further into the UV region. (In all cases the impurities resulting from radiolysis gave the samples significant absorption well into the visible.) In addition, it provided the possibility of concentrating, in the crystals, the enantiomer that might be in excess in the impure material. It is known that the aspartic acid enantiomers are less soluble in water than is the racemate (9), and there is no reason to believe that the L- (or D-) crystals are thermodynamically less stable than the DL-crystals (10).

No rotations were found within the sensitivity of the spectropolarimeter (0.002°). Using the equation of Kuhn and Knopf (11), then, an upper limit to the degree of differential decomposition could be found for each sample. [Differential decomposition = $(k_D - k_L)(/k_D + k_L)/2$, where k_D and k_L are the rates of decomposition of the D and L enantiomers, respectively]. These were: norvaline, 4.0; alanine, 0.08; dopa, 17.0; aspartic acid, 0.007; methionine, 0.5. The upper limit values for norvaline and dopa were very high due to the low degree of decomposition of these samples, and the low specific rotation for methionine also resulted in a high upper limit. However, the results for alanine and aspartic acid clearly rule out any significant differential radiolyses in these cases.

The failure to detect any differential radiolysis may be rationalized in one of two ways. First, of course, the enantiomers may react equally to polarized β and γ-rays. Second, ^{14}C betas have a low kinetic energy (156 KeV maximum), and the degree of polarization is known to be directly proportional to that energy (12). Thus, β particles from ^{14}C may not have sufficient polarization to cause experimentally observable amounts of differential radiolysis of amino acid enantiomers.

REFERENCES

1. Garay, A. S., Nature 219, 338 (1968).
2. Stevens, C. M., Halpern, P. E., and Gigger, R. P., J. Biol. Chem. 190, 705 (1951).
3. Goldhaber, M., Grodzins, L., and Sunyar, A. W., Phys. Rev. 106, 826 (1957).
4. Allen, A. O., "The Radiation Chemistry of Water and Aqueous Solutions," pp. 117-146, Van Nostrand Co., New York, 1961.
5. Hart, E. J., and Anbar, M., "The Hydrated Electron," pp. 124-169. Wiley-Interscience, New York, 1970.
6. Byk, A., Naturwissenschaften 13, 17 (1925).
7. Ulbricht, T. L. V., and Vester, F., Tetrahedron 18, 629 (1962).
8. Kamen, M. D., "Isotopic Tracers in Biology," 3rd Ed., p. 306. Academic Press, New York, 1957.

9. Greenstein, J. P. , and Winitz, M. , "Chemistry of the Amino Acids, " p. 564. J. Wiley & Sons, New York, 1961.
10. Jaeger, F. M. , "Optical Activity and High Temperature Measurement," pp. 203-214. McGraw-Hill, New York, 1930.
11. Kuhn, W. , and Knopf, E. , Z. Physic. Chem. 7B, 292 (1930).
12. Ulbricht, T. L. V. , Quart. Rev. 13, 48 (1959).

Received 1 July, 1971

SYNTHESIS OF GLUTAMIC ACID VIA CYANOETHYLATION OF

N-ACYLAMINOACETONITRILES IN LIQUID AMMONIA*

Kaoru Harada and Tadashi Okawara

Institute for Molecular and Cellular Evolution
and Department of Chemistry
University of Miami, Coral Gables, Florida

Several reports on the prebiotic formation of amino acids have been previously published. Glutamic acid was synthesized from a mixture of simple gases by electric discharge (1, 2) and by heat (3). Glutamic acid was also formed from hydrogen cyanide and ammonia (4), from formaldehyde, ammonium chloride, and ammonium nitrate under irradiation of ultraviolet rays (5), by pyrolysis of formamide (6), and by pyrolysis of a reaction product of formaldehyde and ammonia (7).

Also the organic synthesis of glutamic acid using acrylonitrile or ethyl acrylate with diethyl acetoaminomalonate (8) or ethyl acetamido cyanoacetate (9) has been reported. The latter two compounds could be considered as activated derivatives of glycine.

In the present study, the formation of glutamic acid from acrylonitrile and derivatives of aminoacetonitrile was investigated. Both acrylonitrile (1) and aminoacetonitrile (10) could be regarded as prebiological organic compounds on the primitive Earth. According to Akabori's "fore-protein" hypothesis (10, 11), the formation of proto-protein on the primitive Earth could occur by the polymerization of aminoacetonitrile followed by mild hydrolysis to form polyglycine.

---

* Contribution no. 200 of the Institute for Molecular and Cellular Evolution, University of Miami.

$$n(H_2N-CH_2-CN) \rightarrow H_2N-CH_2-C\left[\begin{array}{c} NH-CH_2-C \\ \| \\ N \\ H \end{array}\right.\left.\begin{array}{c} NH-CH_2-CN \\ \| \\ N \\ H \end{array}\right]_{n-2}$$

$$\downarrow (n-1) \quad H_2O$$

$$H_2N-CH_2-CO-(NH-CH_2-CO)_{n-2}-NH-CH_2-CN$$

The C-terminal residue of the presumed polyglycine "fore-protein" could be N-substituted aminoacetonitrile. It would be expected that the methylene group of the aminoacetonitrile residue would be more chemically reactive than that of the other methylene groups in glycyl residues of this polymer.

In this investigation, N-benzoylaminoacetonitrile was used as a model compound for cyanoethylation to form glutamic acid. Liquid ammonia was used as a solvent. Although the solvent might not be present on the primitive Earth, it may be available on the outer giant planets such as Jupiter. In addition, liquid ammonia has been used as a solvent by Furuyama et al. (12) to add acetaldehyde to glycyl residues in polyglycine to form threonyl residues.

Benzoylaminoacetonitrile was treated with an excess of acrylonitrile in liquid ammonia in a sealed tube for varying times at room temperature in the presence of sodium ethoxide as a catalyst. After evaporation of ammonia, the reaction products were hydrolyzed. The yields of glutamic acid range from 10 to 20% based on the starting N-benzoylaminoacetonitrile. Polymerization of acrylonitrile was observed during the reaction. After acid hydrolysis, the formation of β-alanine was observed. β-Alanine could be formed by addition of ammonia to acrylonitrile. Similar results were obtained when acrylonitrile, liquid ammonia, and sodium ethoxide were replaced by methyl acrylate, dioxane, and Triton B. However, the yields of glutamic acid were lower (less than 10%). Sodium hydroxide, sodium amide, Triton B and triethylamine were also used as catalysts in the liquid ammonia reaction. However, the yield of glutamic acid was very low or only a trace. The results are summarized in Table I.

A glutamic acid analogue, β-(carboxymethylamino)-propionic acid (HOOC-CH_2-CH_2-NH-CH_2-COOH, CAPA), was synthesized as a standard

TABLE I

SYNTHESIS OF GLUTAMIC ACID FROM BENZOYLAMINOACETONITRILE

$$C_6H_5CONHCH_2CN \quad \xrightarrow[\text{(Triton B, Dioxane)}]{\substack{CH_2=CHCN \\ (CH_2=CHCOOMe) \\ \text{liq. } NH_3, NaOEt}} \quad C_6H_5CONHCHCHCN \ \big(\substack{CH_2 \\ CH_2CN}\big) \quad \xrightarrow[H^+]{H_2O} \quad \text{glutamic acid}$$

$C_6H_5CONHCH_2CN$	$CH_2=CHCN$	Catalyst	Solvent	Reaction time (hr., room temp)	Glu (%)[a]	Gly (%)[a]	CAPA (%)[a]	β-Alanine (mmole)
1. 0.80 (0.005)	4.0 (0.075)	NaOEt	liq. NH$_3$	24	13	53	–	5.2
2. 0.80 (0.005)	13.0 (0.25)	NaOEt	liq. NH$_3$	24	10	22	–	b
3. 0.80 (0.005)	4.0 (0.075)	NaOEt	liq. NH$_3$	72	17	49	trace	6.7
$C_6H_5CONHCH_2CN$	$CH_2=CH-COOMe$							
4. 0.80 (0.005)	0.43 (0.005)	Triton B	Dioxane	120	9	59	11.5	–
5. 0.80 (0.005)	0.86 (0.01)	Triton B	Dioxane	120	8	17	12.2	–
6. 0.80 (0.005)	4.3 (0.05)	Triton B	Dioxane	120	0	22	c	–

[a] The yields are calculated based on the starting benzoylaminoacetonitrile.

[b] The sample was not analyzed.

[c] The analytical data was lost.

in order to identify by chromatographic comparison the unknown substance in the reaction product.   The CAPA could be formed by the cyanoethylation of the amino group of the aminoacetonitrile during the reaction. The physical properties of CAPA are very similar to those of glutamic acid.   The DNP-CAPA could not be separated from DNP-glutamic
acid by celite column chromatography and also by thin layer chromatography (alumina) using four different solvents.   However, free CAPA
and glutamic acid were separated by high voltage electrophoresis (2,000
volts) using 2 N acetic acid as a buffer.   The CAPA and glutamic acid
could also be separated with an automatic amino acid analyzer (retention
time: CAPA, 71-80; glutamic acid, 92-98).   Therefore, the yield of
glutamic acid and CAPA could be determined by the use of automatic
amino acid analyzer. In the actual cyanoethylation reactions, the yields
of CAPA were not as high as expected; they were in the same range as
glutamic acid or less.   In some reactions, we could not see the peak of
CAPA on the chart of the amino acid analyzer.

Another problem studied was the cyanoethylation of optically active
(N-benzoylaminoacyl) aminoacetonitriles.   It might be expected that the
cyanoethylation of optically active (N-benzoylaminoacyl) aminoacetonitriles would form optically active glutamic acid after hydrolysis.   N-
Benzoyl-L-alanylaminoacetonitrile, N-acetyl-L-phenylalanyl-aminoacetonitrile, N-benzoyl-L-prolyl-aminoacetonitrile, N-benzoyl-L-valyl-
aminoacetonitrile, N-benzoyl-L-leucyl-aminoacetonitrile, and N-acetyl-
D-phenylglycyl-aminoacetonitrile were prepared by coupling the optically
active acylamino acids and aminoacetonitrile with dicyclohexylcarbodiimide. The yields, physical properties, and elemental analyses of these
N-benzoyl- and N-acetyl-aminoacylaminonitriles are summarized in
Table II.

The cyanoethylation of these (N-benzoylaminoacyl) aminonitriles
was carried out in liquid ammonia at room temperature as described in
the cyanoethylation of benzoylaminoacetonitrile. After cyanoethylation,
ammonia was evaporated and the residue was hydrolyzed with 6 N hydrochloric acid. A typical example of a chromatogram of the hydrolyzate of
the reaction products using N-benzoyl-L-alanyl-aminoacetonitrile is
shown in Fig. 1. The yields of glutamic acid range from 12 to 35%. The
optical purities of resultant glutamic acid are rather low and range from
0 to 3.2%.   When L-phenylalanine and L-proline were used as asymmetric moieties, L-glutamic acid was obtained.   On the other hand, when
L-valine and D-phenylglycine were used, D-glutamic acid and L-glutamic
acid were obtained, respectively. Amino acid compositions of the hydrolyzate of the reaction products and the optical purities of the newly
formed glutamic acid are summarized in Table III.   As described ear

TABLE II

SYNTHESES OF ACYLAMINOACETONITRILES

Acylaminoacetonitriles	$[\alpha]_D^{25}$ [a]	m.p. (°C)	Yield (%)		Elemental Analysis		
					C	H	N
					Calcd. Found		
$C_6H_5CONHCH_2CN$		142-143	88	$C_9H_8N_2O$	67.49 67.52	5.03 5.08	17.49 17.37
L-$CH_3CHCONHCH_2CN$ $NHCOC_6H_5$	-10.5 (EtOH c=3.3)	161-162	69	$C_{12}H_{13}N_3O_2$	62.33 62.19	5.67 5.86	18.17 18.29
L-$C_6H_5CH_2CHCONHCH_2CN$ $NHCOCH_3$	+ 3.57 (EtOH c=3.1)	180-181	46	$C_{13}H_{15}N_3O_2$	63.66 63.61	6.16 6.15	17.13 17.13
L-[ring]$N{-}CONHCH_2CN$ $CO\phi$	-23.8 (EtOH c=1.6)	190-191	46	$C_{14}H_{15}N_3O_2$	65.36 64.84	5.88 5.88	16.33 16.35
L-$\dfrac{CH_3}{CH_3}{>}CHCHCONHCH_2CN$ $NHCOC_6H_5$	-26.9 (EtOH c=1.9)	186-187	59	$C_{14}H_{17}N_3O_2$	64.85 64.36	6.61 6.57	16.20 16.15
L-$\dfrac{CH_3}{CH_3}{>}CHCH_2CHCONHCH_2CN$ $NHCOC_6H_5$	-0.22 (EtOH c=1.8)	159-161	61	$C_{15}H_{19}N_3O_2$	65.91 65.98	7.01 7.10	15.37 15.21
D-$C_6H_5CHCONHCH_2CN$ $NHCOCH_3$	-18.9 (methyl-cellosolve c=1.6)	188-190	64	$C_{12}H_{13}N_3O_2$	62.33 61.91	5.67 5.69	18.19 17.74

[a]The specific rotations were measured by JASCO ORD/UV-5 Optical Rotatory Dispersion Recorder using a 10 mm cell.

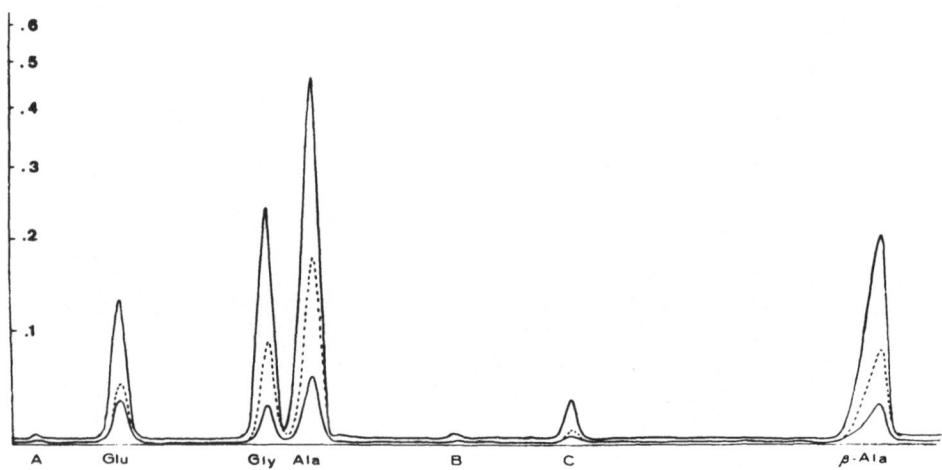

Fig. 1. Amino acid chromatogram of the hydrolyzate of the product from the cyanoethylation of N-benzoyl-L-alanylaminoacetonitrile. A: The peak corresponds to β-(carboxymethylamino)-propionic acid (CAPA). B, C: Unknown.

lier, DNP-glutamic acid and DNP-CAPA could not be separated by the use of the chromatographic technique. Therefore, the DNP-glutamic acid fraction isolated from the reaction mixture could contain a (very) small amount of DNP-CAPA, and the specific rotations shown in Table III represent the minimum values of the optical purity of synthesized glutamic acid.

It is rather difficult to discuss the possible steric course of these sterically controlled reactions and the inversion of configuration of the synthesized glutamic acid because of the limited number of experiments and the low degree of optical purity of the resulting glutamic acid. However, it might be relevant to mention that the inversion of configuration of glutamic acid by the use of valine and phenylglycine could be related to the isopropyl or phenyl groups of the asymmetric moiety. These groups are rigid, branched, and cannot be bent. Therefore, the effective bulkiness of these groups is larger than that of the benzyl group that is flexible and is not branched at the α -carbon to which this group is attached.

TABLE III

SYNTHESES OF GLUTAMIC ACID FROM OPTICALLY ACTIVE
(N-ACYLAMINOACYL)AMINOACETONITRILES

$$\text{Acyl NH} \overset{*}{\text{CH}} \text{CONHCH}_2\text{CN} \quad \overset{\text{CH}_2=\text{CH-CN}}{\underset{\text{liq. NH}_3,\text{NaOEt}}{\longrightarrow}} \quad \text{Acyl NH} \overset{*}{\text{CH}} \text{CONHCHCN} \quad \overset{\text{H}_2\text{O}}{\underset{\text{H}^+}{\longrightarrow}}$$

(with R substituent and CH₂CH₂CN side chain)

glutamic acid

$$\text{NH}_2\overset{*}{\text{CH}}\text{COOH}$$
R

Acylaminoacetonitrile[a]	Glutamic Acid			Yield of Amino Acids after Hydrolysis[d]				
	Config.	$[\alpha]_D^{25}$ of DNP-glu (AcOH)[b]	Optical purity (%)[c]	Glu (%)	Gly (%)	Optically active moiety(%)	CAPA (%)	β-Ala (mmole)
L-CH₃CHCONHCH₂CN / NHCOC₆H₅	L	slightly c=0.41	-	24	43	Ala 95	9.2	2.4
L-(CH₃)₂CHCH₂CHCONHCH₂CN / NHCOC₆H₅	DL	0 c=0.30	-	23	11	Leu 66	21.4	3.6
L-C₆H₅CH₂CHCONHCH₂CN / NHCOCH₃	L	-3.2 c=0.91	3.9	30	44	Phe 61	trace	2.1
L-[ring]CONHCH₂CN / N COC₆H₅	L	-2.7 c=0.93	3.3	12	50	Pro 91	-	5.1
L-(CH₃)₂CHCHNHCH₂CN / NHCOC₆H₅	D	+0.8 c=1.28	1.0	22	23	Val 73	-	1.0
D-C₆H₅CHCONHCH₂CN / NHCOCH₃	L	-0.12 c=0.25	0.2	35	45	PheGly 91	-	4.8

[a]N-Acylaminoacetonitrile (0.005 mole each), acrylonitrile (0.05 mole), sodium ethoxide (0.005 mole), and 25 ml of liquid ammonia were used in these reactions. The mixtures were allowed to stand at room temperature for 72 hrs.

[b]The Specific rotations were measured by JASCO ORD/UV-5 Optical Rotatory Dispersion Recorder using a 10 mm cell.

[c]Optical purity is defined as $([\alpha]_D$ observed$/[\alpha]_D$ literature$) \times 100$. DNP-L-Glutamic acid: $[\alpha]_D = -80.8°$ (AcOH), literature 14.

[d]The yields of amino acids are calculated based on the starting amount of N-acylaminoacetonitrile.

EXPERIMENTAL

The specific rotations were measured by the use of a JASCO ORD/ UV-5 Optical Rotatory Dispersion Recorder using a 10 mm cell. The melting points were measured with a Mel-Temp apparatus. The melting points are uncorrected.

N-Benzoylaminoacetonitrile

N-Benzoylaminoacetonitrile was prepared from aminoacetonitrile (18.5 g, 0.20 mole) benzoylchloride (31.0 g, 0.22 mole), and 2 N sodium hydroxide (210 ml, 0.42 mole) by the usual Schotten-Bauman procedure. Yield, 28.0 g (88%), m.p. 142-143° C (after recrystallization from ethanol). Elemental analysis is shown in Table II.

Glutamic Acid from N-Benzoylaminoacetonitrile

N-Benzoylaminoacetonitrile (0.80 g, 0.005 mole) and sodium ethoxide (0.34 g, 0.005 mole) were dissolved in 25 ml of liquid ammonia. To this solution, 3.98 g (0.075 mole) of acrylonitrile was added. The solution was sealed and was kept at room temperature for 72 hrs. Ammonia was evaporated at room temperature and the residue was evaporated under reduced pressure to minimize residual ammonia. To the residue, 50 ml of 6 N HCl was added and the mixture was refluxed for 6 hrs. The solution was evaporated under reduced pressure. A small amount of water was added and the evaporation procedure was repeated. In order to isolate glutamic acid hydrochloride, the residue was extracted with absolute alcohol. After evaporation of the alcohol under reduced pressure, the amino acid mixture was dissolved in a small amount of water and this solution was placed on a Dowex 50 column (H^+ form, 4 cm x 25 cm). The column was eluted with 1.5 N aqueous ammonia. The effluents that contained amino acids were combined and evaporated to dryness in vacuo. Since the crude amino acid mixture contained glycine, β -alanine, and other ninhydrin positive materials, the yield of glutamic acid was determined by using an automatic amino acid analyzer after appropriate dilution. The yield of glutamic acid was 17%. Also, paper chromatography and analysis on an automatic amino acid analyzer revealed that the crude amino acid mixture contained several unknown ninhydrin positive materials.

The crude glutamic acid fraction was placed on a Dowex-3 column (HCOOH form, 2.1 cm x 14 cm); the column was washed with water and then eluted with 1 N formic acid. The effluents which contain glutamic acid were combined and evaporated to dryness under reduced pressure. Part of the glutamic acid was converted to DNP-glutamic acid by reacting with DNFB (13, 14). The DNP-glutamic acid was purified on a celite column that was treated with a pH 4 phosphate-citrate buffer (15) without fractionation of optical isomers (16), m.p. 155-159°C. Anal. Calcd. for $C_{11}H_{11}N_3O_8$: C, 42.18; H, 3.54; N, 13.42. Found: C, 42.11; H, 3.77; N, 13.18.

In the reactions using methylacrylate in dioxane, it was found that as the amount of methylacrylate increased, the yield of glutamic acid and glycine decreased. This suggests that under the conditions used methylacrylate may attack the acylated amino nitrogen. The results are summarized in Table I.

β-(Carboxymethylamino) propionic acid (CAPA)

The β -(carboxymethylamino) propionic acid (CAPA) was prepared from aminoacetonitrile and acrylonitrile in dioxane in the presence of a small amount of sodium ethoxide at room temperature. m.p. 192-193°C.

Anal. Calcd. for $C_5H_9O_4N$: C, 40.81; H, 6.17; N, 9.52. Found: C, 40.39; H, 6.22; N, 9.30. DNP-CAPA, m.p. 159-164°C (DNP-glu, 155-160° C).

The CAPA and glutamic acid were separated by high voltage electrophoresis using 2 N acetic acid as a solvent (pH 2.2), 2,000 volts. After 3 hrs., CAPA migrated 10.7 cm to the anode side, glutamic acid migrated 16.8 cm in the same direction.   The retention times of CAPA and glutamic acid in the automatic amino acid analyzer (Phoenix K-5000) were found to be 71-80 and 92-98, respectively. The C-value of CAPA is very small, 0.524 (12.55 for glutamic acid).

The DNP-CAPA and DNP-glu could not be separated chromato - graphically using a celite column which was treated with pH 4.0 citrate-phosphate buffer. Similarly, these two DNP-amino acids could not be separated with thin layer chromatography (alumina) using four solvents, (a) EtOH:conc.NH_3aq:H_2O (18:1:1), (b) t-BuOH:sec. BuOH:conc.NH_3aq: H_2O (4:3:2:1), (c) benzylalcohol: EtOH (9:1), (d) nBuOH:AcOH:H_2O (3:1:1). The glutamic acid and CAPA were not separated by paper chromatography using BuOH:AcOH:H_2O (3:1:1) and EtOH:conc. NH_3aq:H_2O (18:1:1).

(N-Benzoyl-L-alanyl)aminoacetonitrile

N-Benzoyl-L-alanine (19.3 g, 0.1 mole) and free aminoacetonitrile (5.6 g, 0.1 mole) were dissolved in 200 ml of freshly purified dioxane. To this solution in an ice bath, 22.7 g (0.11 mole) of dicyclohexylcarbodiimide in 80 ml of dioxane was added slowly. After the addition was complete, the reaction mixture was kept at room temperature overnight. The precipitated dicyclohexylurea was removed by filtration and the filtrate was evaporated to dryness under reduced pressure. The residue was dissolved in ethylacetate, and this solution was washed with 0.3 N hydrochloric acid, 5% sodium hydrogen carbonate, and water. The solution was dried with anhydrous sodium sulfate and was evaporated to dryness in vacuo. The residue was recrystallized from ethylacetate and ether. Yield, 15.9 g (69%), m.p. 161-162°C. Elemental analyses are shown in Table II.

(N-Acetyl-L-phenylalanyl)aminoacetonitrile, (N-benzoyl-L-prolyl)aminoacetonitrile, (N-benzoyl-L-valyl)aminoacetonitrile, (N-benzoyl-L-leucyl)aminoacetonitrile, and (N-acetyl-D-phenylglycyl)aminoacetonitrile were prepared in a similar way. The yields, physical properties, and elemental analyses are shown in Table II.

Glutamic Acid from (N-Acetyl-L-phenylalanyl)aminoacetonitrile

(N-Acetyl-L-phenylalanyl)aminoacetonitrile (1.23 g, 0.005 mole) and sodium ethoxide, 0.34 g (0.005 mole) were dissolved in 25 ml of liquid ammonia. To this solution, 2.65 g (0.05 mole) of acrylonitrile was added. The container (a screw capped thick wall vial) was sealed and shaken a few times to make the solution homogeneous. The reaction mixture was kept at room temperature for 72 hrs. The liquid ammonia was evaporated at room temperature and then the mixture was treated under reduced pressure in a hot water bath to minimize the residual ammonia. The residue was hydrolyzed with 50 ml of 6 N hydrochloric acid for 6 hrs., and this solution was then evaporated in vacuo. A small amount of water was added and the evaporation procedure was repeated. The residue was extracted with absolute alcohol to eliminate sodium chloride and the solution was evaporated under reduced pressure. The residue was dissolved in a small amount of water and was placed on a Dowex 50 column (H^+ form, 4 cm x 25 cm). After washing with water, the column was eluted with 1.5 N aqueous ammonia. The ammoniacal solution which contained amino acids was evaporated to dryness in vacuo.

The amino acid mixture was diluted and analyzed for amino acid composition by using an automatic amino acid analyzer. The yield of glutamic acid was 29.8%. The amino acid mixture contained several unknown ninhydrin positive materials that were detected by paper chromatography and analysis on an automatic amino acid analyzer.

The amino acid mixture was then treated on a Dowex 3 column (HCOOH form, 2.1 cm x 14 cm) and glutamic acid was isolated by eluting with 1 N formic acid. The solution was evaporated to dryness under reduced pressure. The residual glutamic acid was converted to DNP-glutamic acid in the usual way by the use of DNFB (13, 14). The resulting DNP-glutamic acid was separated and purified by the use of a celite column that was treated with pH 4 phosphate-citrate buffer (15). $[\alpha]_D^{25} = -3.2^0$ (c = 0.91, AcOH0, optical purity: 3.9%.

SUMMARY

According to Akabori's hypothesis about polyglycine being the first protein on the primitive Earth, the C-terminal residue of the "fore-protein" could be aminoacetonitrile. The methylene group of the aminonitrile residue would be expected to be more reactive compared with other methylene groups of glycyl residues in the "fore-protein." In this investigation, cyanoethylation of acylaminoacetonitrile in liquid ammonia was studied. a) Benzoylaminoacetonitrile and b) several optically active (N-acylaminoacyl)aminoacetonitriles were used as model compounds for cyanoethylation. The yield of glutamic acid was in a range 10-35%. In some of the latter reactions (b) using asymmetric moieties, weakly optically active (~4%) glutamic acid was obtained.

ACKNOWLEDGMENTS

This work was supported by Grant no. NGR-10-007-052 of the National Aeronautics and Space Administration. The authors wish to express their thanks to Mr. Charles R. Windsor for amino acid analyses and to Dr. Kazuo Matsumoto for valuable discussion.

REFERENCES

1. Miller, S. L., J. Am. Chem. Soc. 77, 2351 (1955).
2. Grossenbacher, K. H., and Knight, C. A., in "The Origins of Pre-

biological Systems," (S. W. Fox, ed.), p. 173.   Academic Press, New York, 1965.

3. Harada, K., and Fox, S. W., Nature 201, 335 (1964).

4. Oro, J., and Kamat, S. S., Nature 190, 442 (1961); Lowe, C. U., Rees, M. W., and Markham, R., Nature 199, 219 (1963).

5. Pavlovskaya, T. E., and Pasynskii, A. G., in "The Origin of Life on Earth," I.U.B. Symposium Series I, p. 151.   Pergamon Press, New York, 1959.

6. Harada, K., Nature 214, 479 (1967).

7. Fox, S. W., and Windsor, C. R., Science 27, 984 (1971).

8. Morrison, D. C., J. Am. Chem. Soc. 77, 6072 (1955).

9. Moe, O. A., et al., U.S.P. 2,551,566 (1951).

10. Akabori, S., Kagaku 25, 54 (1955).

11. Akabori, S., "The Origin of Life on the Earth," I.U.B. Symposium Series I, p. 189.   Pergamon Press, New York, 1959.

12. Furuyama, T., Sakiyama, F., and Narita, K., Bull Chem. Soc. Japan 36, 903 (1963).

13. Sanger, F., Biochem. J. 39, 507 (1945).

14. Rao, K., and Sober, H. A., J. Amer. Chem. Soc. 76, 1328 (1954).

15. Perrone, J. C., Nature 167, 513 (1951); Court,A., Biochem. J. 58, 70 (1954).

16. Harada, K., and Matsumoto, K., J. Org. Chem. 32, 1794 (1967).

Received 4 February, 1972

MACROMOLECULES

PROTEINS AND NUCLEIC ACIDS IN PREBIOTIC EVOLUTION

J. C. Lacey, Jr. and D. W. Mullins, Jr.

Laboratory of Molecular Biology
University of Alabama in Birmingham
Birmingham, Alabama

INTRODUCTION

The "central dogma" of molecular biology, as set forth by Watson and Crick in 1953 (1), and by Crick in 1958 (2), allows for the bidirectional transfer of genetic information between DNA and RNA, as well as from RNA to protein. Nevertheless, it came as no little surprise to many when Temin (3) and Baltimore (4) reported RNA-dependent DNA polymerase activity from viruses. That genetic information can, in fact, flow from RNA to DNA has now been solidly incorporated within the framework of the "central dogma" (5), and has resulted in a great deal of excitement in several areas, notably cancer research (6). The discovery that RNA can serve as a template for the synthesis of DNA, however, has not altered the major prohibitory tenets of the "central dogma," namely, that genetic information cannot flow from proteins to nucleic acids, nor from proteins to proteins. Other work presented in this volume by Lipmann, however, demonstrates that information does flow from proteins to polypeptides in contemporary systems.

We do not propose to take exception with the tenets of the "central dogma" merely because they exist, however, nor does our proposal necessarily affect its validity with respect to contemporary systems; our approach might be viewed as somewhat indirect in that it relates to early evolutionary systems. Basically, we believe that there exists a body of experimental evidence, together with a logical rationale, to suggest the following hypothesis: Proteins were essential both in the establishment

171

and early evolution of biological systems, and evolution could have proceeded, for a time, without nucleic acids. Consequently, early informational exchange between these two major macromolecular systems may have been from proteins to nucleic acids. That is, proteins may have served as templates for the production of nucleic acids. This hypothesis, while controversial, is certainly not original, and has been discussed by a number of investigators in the fields of molecular evolution and the origin of life (7 - 13).

PARAMETERS OF MOLECULAR EVOLUTION

Before presenting our arguments for the hypothesis stated above, we will discuss what we believe to be the essential characteristics of prebiotic molecular evolution. We will restrict our discussion, however, to only those aspects of evolution that would most likely be equated with a unidirectional transition from primordial to contemporary life systems. This unidirectional evolutionary transition is one of increasing functionality. Increasing functionality is here interpreted to mean not just an increase in the number of functions, but also an increase in the relative efficiency of each function. This direction in evolution can thus also be characterized by an increase in complexity and independence of the environment (Figure 1).

Fig. I Evolutionary direction

Evolution is characterized by the persistence of individual entities and the non-persistence of others. Since persistence can be equated with natural selection, contemporary organisms can be regarded as the end result of a sequential line of persistent entities. Persistence of individual molecules, however, would prove to be too harsh a judge of fitness for natural selection; neither proteins nor nucleic acids, for example, would be expected to persist in an aqueous environment for long periods of time. Persistence in an evolutionary sense, therefore, must also require a net synthesis of new material relative to old, before the latter ceases to persist. A line of persistent entities, therefore, can be viewed as depicted in Figure 2.

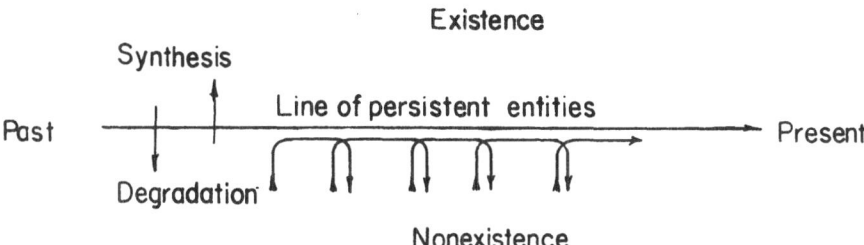

Fig.2 Informational overlap of entities leading to the long-term persistence of selected functions.

The above scheme requires that there be a coupling agent, or process of duplication, between the information content of the old and that of the new, lest the information in the former be lost. Consequently, an essential feature of evolution is that at some point in time such a process of duplication must have appeared.

Error-free duplication, however, would not have appeared, nor would it have been desirable, since it would have tended to retard the rate of evolutionary change by decreasing the amount of variability. Duplication errors, therefore, would have constituted one means of introducing new functions into evolving systems by producing variations in prior functions.

A subset of all living systems, the living cell, is the ultimate focal point of all our considerations. This morphological entity, the cell, is a common microsystem in the set of living things. It did not arise, however, through the fragmentation of a larger, pre-existent entity, but rather through the coalescence of less complex materials. An essential and critical step in prebiological evolution, therefore, must have been the appearance of isolated chemical microsystems. Professor Fox has emphasized this point repeatedly and has demonstrated experimentally how such microsystems might have arisen (14).

Aside from the fact that the very existence of cellular structures in contemporary organisms merits considerations as to their origin, there are other arguments which suggest that the appearance of biological microsystems was an inevitable step in evolution. These arguments, in turn, relate to still another aspect of evolutionary development from the primordial to the contemporary, namely, that successive generations of persistent individual entities must have exhibited increasing independence of the terrestrial environment. This means that any one entity displayed, with time, an increased probability of persistence in the face of increased numbers and kinds of environmental assaults. It is not difficult to see that a single molecule would be incapable of exhibiting such persistence, and that what is required is a collection of molecules with different specific functions (i.e., a collection of functions). It is largely irrelevant, at present, what these precise functions might have been; what is important is that such a collection of functions resulted in the persistence of an isolated microsystem to contain them.

The formation of such isolated microsystems can be viewed as a partitioning of the terrestrial environment as depicted schematically in Figure 3. Subsequent to this partitioning, the microsystem is then seen capable of interacting with, and responding to, both the terrestrial and universal environments, and is thus made subject to the processes of natural selection.

As previously discussed, those isolated microsystems which were able to persist over relatively long periods of time would be those 1) capable of self-duplication at a rate in excess of terrestrial environmental pressures for elimination and, 2) which contained functions capable of dealing with such pressures. In essence, the terrestrial environment was selecting for functions, and the criterion for the selection of micro-systems must have been functionality.

In summary, we believe the following statements are true of early molecular evolution:

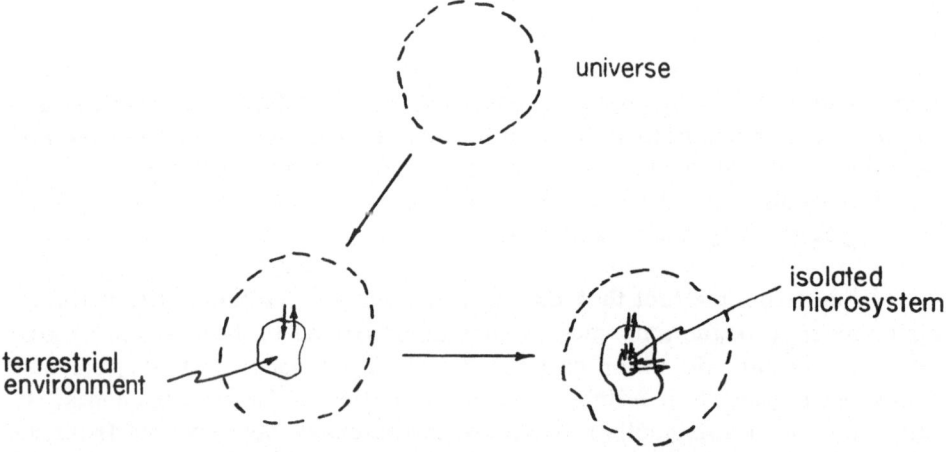

Fig. 3 Successive partitioning in the evolutionary development of life.

1) The unidirectional transition from primordial to contemporary was characterized by increases in functionality, complexity and independence.

2) Persistence can be equated with natural selection.

3) A process of duplication of functions must have appeared.

4) An essential and critical step in prebiological evolution must have been the appearance of isolated chemical microsystems.

5) Natural selection by the terrestrial environment was for specific functions.

If it is assumed that the above factors are the essential ones in early evolution, we can proceed with the major objective of asking whether proteins or nucleic acids alone could have satisfied them.

PROTEINS, NUCLEIC ACIDS AND MOLECULAR EVOLUTION

In Figure 2 we suggested that informational overlap is a requirement for the maintenance of a line of individual persistent entities. Once a rudimentary self-copying process had developed, information was certainly passed from persistent entity to persistent entity, while indirect environmental control was maintained through natural selection. However, prior to the development of such a self-directed duplicatory system, information for the repeated synthesis of monomeric and polymeric molecules could only have been derived from the environment and the physiochemical properties of the participating reactants themselves. Consequently, although it is conceptually not of major importance which came first in the evolutionary development of life, proteins or nucleic acids, it is of interest to ask the question in this way: If the initial flow of information was directly from the environment to the isolated microsystem, which macromolecular group, proteins or nucleic acids, would have best served to express this environmental information?

Natural Occurrence

We will not relate here the experimental evidence concerning the question of the origin of proteins and nucleic acids, as it is adequately summarized elsewhere (15, 16). Suffice to say, however, that all available experimental evidence at this point strongly favors the prebiotic formation of polypeptides over nucleic acids. Amino acids have been synthesized from simpler materials under a wide variety of geologically relevant conditions, and these have been shown to be capable of thermal condensation to protein-like materials. Moreover, the amino acids which have been produced by such procedures, and their relative proportions, are similar to those found in contemporary systems.

While similar experiments with the nucleic acids have shown that all of the relevant purine and pyrimidine bases, as well as ribose, can be produced by such procedures, different conditions are generally required for the synthesis of each; furthermore, although the conversion from nucleosides to nucleotides has been observed to occur, the production of nucleosides from ribose and purines or pyrimidines has, thus far, not been modelled. Finally, attempts to polymerize nucleotides under a variety of conditions have yielded only oligomeric fragments composed of five or less monomeric units. Consequently, the most logical evolutionary scheme available to us is one which exhibited in its early stages the terrestrially-directed synthesis of proteinoids.

The Development of Isolated Microsystems

The discussion above has emphasized the fact that molecular evolution could be expected to result in the formation of isolated biochemical microsystems. It might be asked, at this point, which of the two major macromolecular species, nucleic acids or proteins, would be best suited for this role of environmental partitioning or micro-system formation? This question can be approached in two ways:

(1) Through an examination of contemporary biological systems to determine what materials have been selected in evolution to form membraneous structures (reductionistic approach).

(2) By considering data relative to the abiological formation of membrane-like structures from proteins and nucleic acids (constructionistic approach).

Contemporary systems. The plasma membranes of higher order contemporary biological systems are composed primarily of proteins and lipids, although minor components are also present and the precise arrangement of participating units is still a subject of contention. The present picture seems to indicate that the general biological membrane is a mosaic of proteins with enzymatic functions being represented in the matrix. The cell membranes of bacteria also appear to be composed primarily of lipoprotein.

Some cell membrane types are modified by RNAase (17) and may contain RNA. However, RNA is not an integral and indispensable part of all bio-

logical membranes, whereas proteins are. Perhaps the closest function-
ally related structure to contain a nucleic acid-like polymer in large
amounts is the bacterial cell wall.   Here the techoic acids (18) appear to
be essential to the maintenance and function of these extracellular structures. There are two basic types of techoic acids, polyglycerol phosphate
and polyribitol phosphate. Both are apparently linked via short oligopep-
tides to other polysaccharides in the cell wall. The composition of techoic
acids suggests that they might be viewed as either "relatives" or perhaps
evolutionary precursors of the ribose phosphate backbone of polynucleotides.   Nevertheless, the fact remains that nucleic acids have not been
shown to be integral parts of biological membranes.

There are structures, including the chromatin, ribosomes, and viruses, in contemporary biological systems which do contain nucleic acids.
All of these are also known to contain protein complexed with the nucleic
acid but none exhibits a membraneous organization.

Thus, the bulk of evidence from contemporary systems suggests that
proteins are universally employed in the formation of biological membranes, while nucleic acids are not.   One cannot, however, legitimately
employ the observations made by a reductionistic approach alone to establish firm conclusions with regard to structures in prebiological systems. Nevertheless, we feel that these data, coupled with those presented
in the following section, strongly implicate the exclusion of nucleic acids
from any membraneous components of isolated biochemical microsystems
in early evolution.

Constructionistic experiments.   Professor Fox and his coworkers
have, through the years, demonstrated unequivocally that some types of
thermal proteinoids will form protein microspheres when dissolved in
warm water, and then cooled (14).   Furthermore, by altering the pH of
solutions containing these microspheres, Fox has demonstrated that these
particles can be caused to empty themselves and exist as membrane-bound
structures; these membranes appear to have a double-layered morphology
when observed with electron microscopy.   The microspheres themselves
have been observed to grow and form buds which detach and increase in
size, apparently by accretion.   Again, this form of growth requires that
the terrestrial environment furnish the raw materials and polymers needed
for growth and duplication; the microsphere itself still lacks internally
directed synthesis and a self-duplicatory mechanism.   Thus, while the
persistence of these microspheres does depend upon externally directed
synthesis, they do represent a microenvironmental isolate which serves
to collect and enclose primitive functions. We know of no similar experi-

ments employing nucleic acids which show these molecules to be capable of forming membraneous structures.

Another set of experimental data relating to the formation of isolated microsystems is available, and concerns studies made on the relative affinities of nucleic acids and proteins for air-water interfaces. It has been well established that proteins are, in general, highly surface active (19), and this fact is quite significant with regard to the problem at hand. Since high surface activity results from a greater affinity of a solute for the surface phase as opposed to the solution phase, a protein solution caused to form microdrops the size of cells (e.g., by spraying) would yield droplets which contained protein at their surfaces. One can easily imagine that through partial oxidation and evaporation, a proteinaceous surface film might be developed. Indeed, experiments similar to this have been performed in which protein-lipid mixtures have been demonstrated to give rise to isolated microsystems (20).

Although there is some evidence to indicate that DNA is present at air-water interfaces (21), we have found (22) that neither mononucleotides nor polynucleotides (DNA or poly A) exhibit significant surface activity. If it is true that nucleic acids are not normally found at interfaces to any measureable degree, then they would not be expected to form membraneous structures, but rather exist in solution in the interior of protein-bound microdroplets. This, of course, is precisely what is found in contemporary cells.

The above data are consistent with the physicochemical properties of the macromolecules themselves (see below), and indicate that proteins are readily capable of forming membraneous structures with or without lipids, but that nucleic acids are not.

Functionality

We believe, as pointed out above, that functionality is the basis for natural selection in biological evolution, and that only those functions which contribute to the persistence of a microsystem will be so selected. It is, therefore, pertinent to ask, "How does a microsystem acquire functions?" The simplest answer, of course, is that new functions arise through the acquisition of new proteins. Why proteins? Again, by taking the reductionistic approach, we can turn to contemporary systems and observe what various functions have been assigned to proteins and nucleic acids.

An enumeration of functions for nucleic acids proves to be quite short:

(1) Information storage (DNA, or RNA, in the case of some viruses.)
(2) Information transfer (via mRNA) to proteins.
(3) Formation of ribosome structure.
(4) Acceptance of activated amino acids (by tRNA) prior to their incorporation into proteins.

While there is no question but that the above functions are absolutely essential in contemporary systems, it is also obvious that all these functional roles are involved with but a single parameter of molecular evolution, namely, the duplication of functions.

It would be unfeasible, on the other hand, to attempt an enumeration of the various functions associated with proteins in contemporary organisms--to do so would yield a list of unwieldy proportions. Primarily these functions are enzymatic in nature, but also included are structural roles in virtually all intra- and extracellular organelles (ribosomes, mitochondria, membranes, flagella, etc.), gene regulation, biological transport, hormonal activity, and others. In short, proteins appear to be responsible for nearly all cellular functions except those reserved solely for the nucleic acids (above). Furthermore, those functions given to nucleic acids are also rendered an essential assist from proteins in contemporary systems.

Whether functions similar to those enumerated above can arise spontaneously and concomitantly with the abiological synthesis of proteins is an essential question in any discussion of biochemical evolution, and one which has been repeatedly answered in the affirmative by Professor Fox and his associates. A wide variety of enzyme-like activities, as well as some hormonal activity (23), has been found associated with thermal proteinoids.

Evidence at hand thus indicates that proteins are required over nucleic acids for major functionality in isolated microsystems, and that both proteinoids and associated functionality can arise spontaneously.

It is perhaps of interest to digress momentarily and consider the physiochemical basis for functionality in the two macromolecular species being considered. Why do proteins have many and varied functions associated with them, and the nucleic acids but a few? Part of the answer, of course, lies in the fact that there are twenty amino acids, but only four nucleotides, resulting in $(20)^2=400$ possible dipeptide sequences, but only $(4)^2=16$ possible dinucleotide sequences. These values, and the difference between them, are magnified, of course, for longer sequences of the two systems. For example, a tripeptide has 8000 possible sequences but a trinucleotide only 64.

Although the possible number of proteins of a specified length is astronomical when compared to the possible number of nucleic acid molecules of similar length, the immense diversity of the former is made still greater by the fact that the amino acid side chains are, themselves, so diverse. Thus, there exist acidic, basic and hydrophobic residues, as well as those with intermediate polarity, such as serine. Too, the polypeptide backbone may be either helical or in a random coil configuration, and thus contribute even more to the tremendous functional variability found associated with these particular macromolecules.

The nucleic acids on the other hand, have a polyanionic ribose phosphate backbone which, when associated with cations, establishes some helical rigidity to the entire molecule. In addition, however, the organic base portions, four in number, differ only slightly from one another in physiochemical properties, each being either a purine or pyrimidine. One particularly important aspect with regard to the purine and pyrimidine components of nucleic acid molecules is that they tend to base stack with one another. That is, the plane of the ring structure of a purine or pyrimidine base in a polynucleotide tends to lie in a slightly offset, but parallel, configuration with respect to adjacent bases (24). As a result of this interaction, nucleic acids tend to approach a rigid rod-like form in solution. In essence, the nucleic acids have both limited sequence variability and limited conformation and display a monotonous polyanionic character. Such molecules would not, by their very nature, be expected to possess many and varied functions but rather exhibit limited and quite specific functional roles in biological systems.

A Duplicating Function

Since the major function of nucleic acids in contemporary biological systems is that of information duplication and transfer, and because this function is essential to a unidirectional pattern of evolution, a key question with regard to the present hypothesis is whether proteins alone could carry out this duplicatory role in the absence of nucleic acids (i.e., could proteins duplicate proteins?).

Calvin (11) has introduced the idea of 'growing end control' of polypeptides in which a metal-complexed catalytic end of a nascent polypeptide chain generates a new oligopeptide fragment from the former; this oligopeptide then acts as an initiator for the residue-directed synthesis of a second, identical, or nearly-identical, polypeptide. While this idea is conceptually sound, such a mechanism would appear to exhibit rather low specificity and control, and tend toward the production of repeating sequences and, hence, low functionality.

Orgel (8) has discussed the possibility of the existence in primitive cells of a protein copying enzyme, although such a mechanism of protein duplication would have necessitated the simultaneous appearance of two such proto-enzymes, one to copy the other. Orgel considers this to be a low probability event.

Perhaps as a consequence of Orgel's thinking, we propose to ask, at this point, a more generalized question: How probable is it that two proteinoid molecules having identical functions would arise simultaneously in the same isolated microsystem? In an attempt to answer this question, we can point to the work of Rohlfing and Fox (25) with thermal proteinoids having the ability to hydrolyze p-nitrophenyl acetate. This function was shown to be dependent upon the presence of histidine residues in the proteinoid, and it was observed that different preparations of proteinoid material employing varying amounts of histidine exhibited varying activity. Nevertheless, repeated preparations did exhibit p-nitrophenyl acetate hydrolysis activity, indicating that this catalytic function, at least, was able to arise repeatedly and spontaneously--apparently, at least one "active site" was synthesized per proteinoid preparation. Moreover, a preparation of proteinoid material weighing several milligrams was shown to retain the above mentioned catalytic activity when multiple samples were removed and assayed. This demonstrates that several "active sites" were produced in each preparation, suggesting statistical reproducibility of the function. A proteinoid microsphere weighing 1 μ gm and composed of individual proteinoids with an average molecular weight of 10,000 would contain about 6×10^{13} molecules. The probability of having two "active sites" per microgram would seem to be quite high if the amino acid input to the formation of proteinoid had a composition consistent with the "active site" of whatever enzymatic function was being considered.

Even today, there are hundreds of sites on the Earth's surface having a temperature sufficiently high to allow the synthesis of thermal proteinoids; therefore, it appears obvious that such an event was, prebiotically, quite frequent. Furthermore, different geographical sites might be expected to produce variations in the amino acid content of synthesized proteinoids and result in the spontaneous origin of a variety of functions. When viewed in this perspective, it seems quite probable that two copies of a protein-copying enzyme could have arisen simultaneously in the same microenvironmental system somewhere on the Earth's surface.

If it is to be proposed, however, that a protein-copying enzyme did arise spontaneously, some consideration must be given as to the mechanism by which such a catalyst might function. The principle requirements for this proto-enzyme are that it recognize and bind to a pre-existent polypeptide molecule, move along that molecule's backbone and simultaneously synthesize a new polypeptide that is related, in some way, to the template protein. Would such a newly synthesized polypeptide be identical with, or similar to, the old protein, or complementary to it-- that is, would a positive or negative copy be produced?

In nucleic acid synthesis, partial recognition is apparently achieved through base pairing between the incoming monomers and their template complements, and a newly synthesized strand is complementary to an old one. Similarly, the recognition process for a protein copying enzyme might operate on the same basis (i. e., by side chain interactions). However, while a positively charged side chain in the template protein could be expected to result in the incorporation of a negatively charged (complementary) residue in the daughter molecule, hydrophobic groups, on the other hand, would result in the incorporation of other hydrophobes. The newly synthesized polypeptide would thus be complementary (i. e., a negative copy) with respect to charged residues, but non-complementary (i. e., a positive copy) with respect to hydrophobic amino acids. Two rounds of copying would be required to approximately duplicate a function.

The internally directed synthesis of proteins as described above would require the presence of ATP for the activation of participating amino acids. This latter stipulation is necessary because the formation of the peptide bond is an energy-requiring event. While there are some data (26) to indicate that ATP, divalent cations, and amino acids in solution will spontaneously give rise to activated amino acids (aminoacyl adenylates), the abiological formation of ATP has, unfortunately, been incompletely modelled in pre-biological systems, particularly with regard to the adenine-adenosine transition. We would propose, however, that the protein copying enzyme had the ability to aid in the formation of the aminoacyl adenylate of each incoming amino acid. If so, it might represent an evolutionary precursor of the contemporary aminoacyl-tRNA synthetases, enzymes which also employ ATP in the formation of activated adenylates.

It has been well established that aminoacyl adenylates of mixed amino acids polymerize spontaneously in solution to yield peptides (27, 28). The present proposal merely suggests that, in prebiotic systems, such an event might have been caused to occur selectively on a protein template under the direction of an enzyme. In this regard, it has been shown (29) that certain clays can serve to greatly enhance peptide bond formation from alanyl adenylate.

While the above discussion appears reasonable in its proposals and conclusions, there are, at present, no data to indicate that a protein copying enzyme is possible. The only contemporary system that might involve an analogous process is the causative agent of Scrapie, a disease which infects sheep (30, 31). This "organism" appears to be completely devoid of nucleic acid, yet is still capable of intracellular reproduction. It is possible, however, that this agent's replication is carried out through means employing the sheep's genetic information. Lipmann's work with gramicidin S does answer the question conceptually. It is possible for information to flow from proteins to polypeptides.

Although a protein-copying enzyme would allow the duplication of functions, and hence, of the microsystem itself through an increase in mass, there is a major disadvantage inherent in a "polypeptide dependent polypeptide polymerase" system--that is, that natural selection processes could not be permitted to select for one or the other of the two polypeptides (template or product polypeptide). Selection must be for both, since both are required for polypeptide duplication.

Since it is entirely possible that two complementary proteins might have conflicting functions, any microsystem which contained a set of such catalysts would exhibit decreased efficiency. Natural selection pressures, consequently, might demand a system in which the complementary sequence of a functional polypeptide was nonfunctional (32). As we have already indicated, nucleic acids were prime candidates for satisfying such a requirement. Nevertheless, as a result of this fact, evolution was faced with the immense problem of transferring information present in existing polypeptides to nucleic acid molecules. A solution to this problem, however, has been proposed by Lacey and Pruitt (12), who have suggested that such a process of informational transfer might have taken place through the formation of a mononucleotide helix around the alpha-helix of a polypeptide template, with the subsequent polymerization of the monomers to form a polynucleotide. Corey-Pauling models suggested a 3:1 nucleotide to amino acid ratio, and that ribonucleotides might have been preferred over deoxyribonucleotides. Although there is ample evidence to suggest that negatively charged mononucleotides interact with positively-charged polyamino acids in a selective fashion (12, 33, 34), the precise nature of such interactions appears to be based primarily on a cooperativity of electrostatic interaction and base stacking.   Wagner and Wulff (35) have also demonstrated that GMP induces helicity in polylysine, although genetic code relationships have not been observed in any of these experiments.   If such a model were feasible, the nucleotides might be coupled by some form of carbodiimide derived from cyanamide (36).

Although, as mentioned, Corey-Pauling models implicate a 3:1 nucleotide to amino acid ratio in helical complexes between these two systems, it is possible that such reverse translation might have passed through a stage in which the coding ratio was 1:1, and subsequently evolved to the contemporary value.

Needed Experiments

We feel that the collection of facts and rationale underlying the here-proposed protein → protein and protein → RNA informational flow are self-consistent.   Nevertheless, both tenets lie well within the bounds of hypothesis and considerably more experimental evidence is needed before acceptance, abandonment or modification of the theory is possible.

One of the most obvious pieces of data required is that concerning the interaction of amino acid side chains.   It is known, for example, that in proteins there is a strong tendency for hydrophobic amino acids to form specific clusters with each other and for oppositely charged side chains to interact electrostatically--the tertiary structure of most globular proteins is, in fact, determined in large part by these very forces. It is not known, however, whether monomeric amino acids (or their adenylates) will so interact with specific side chains in existent polypeptides.   Experiments involved in establishing the relative solubility of amino acids in the presence of various specific peptides might help to answer such a question. Other approaches might involve a study of the induction of helicity in polypeptides of known composition by various amino acids, as well as an analysis of the composition of peptides synthesized from mixed aminoacyl adenylates in the presence of pre-existent peptides. A further elucidation of the mechanism involved in Scrapie infections might also prove helpful.

Similarly, the suggestion that informational flow once occurred in a protein → RNA direction also needs considerable study.   Specific complexes between proteins and nucleic acids are known to exist in nature, as between rRNA and ribosomal protein or DNA and chromatin. Allfrey's group (37) is presently involved in demonstrating tissue and species specific binding of acidic nuclear phosphoproteins to DNA, while Kurland and his associates (38) have recently demonstrated that 16s rRNA contains specific binding sites for four ribosomal proteins. What remains to be determined in all of these studies, however, is the precise nature of these binding specificities; mainly we need to know if they are in any way code related. While it is certainly true that chromosomal and ribosomal proteins do not serve as templates for their associated nucleic acids, studies of how specificities are manifested in chromatin and ribosomes may suggest how information could flow from proteins to nucleic acids. Perhaps a more direct approach to this problem would be to study the influence of pre-existent peptides on the composition of oligonucleotides synthesized abiologically in their presence. The principal drawback to such experiments is that a really satisfactory and geologically relevant method of synthesizing oligonucleotides larger than about five subunits is not yet available, although considerable progress employing imidazole catalysis is being made (39, 40, 41).

SUMMARY

We have herein presented those available facts, together with a rationale that we believe to be consistent with these facts, in support of the hypothesis that it was possible for life to have originated as an isolated microsystem containing proteins, and to evolve, for a time, without nucleic acids. This hypothesis necessitates proposing the spontaneous appearance, in time, of a protein-copying enzyme to facilitate the transmission of biological information from proteins to proteins, and the subsequent appearance, due to increased selection pressure, of nucleic acid molecules complementary to protein. Both protein → protein and protein → RNA informational transitions can be envisioned as requiring an alpha helical polypeptide template.

The basic arguments favoring the preminence of protein in early evolution and the subsequent appearance of nucleic acid are as follows:

(1) An early requirement for the appearance of life was the development of isolated microsystems. Thermal proteinoids are easily produced under a variety of geologically relevant conditions, and do spontaneously form such microsystems. Nucleic acids are, on the other hand, not easily synthesized under simulated prebiological conditions and do not form membraneous structures.

(2) The natural selection of microsystems must have been on the basis of functionality. Proteins exhibit high functionality because of the properties inherent in their component side chains and secondary and tertiary structure, whereas nucleic acids display limited functionality as a consequence of the monotonous nature of their polyanionic backbone and the stacking interactions of their organic bases. Furthermore, nucleic acids, even if they had arisen early in the development of living systems, could not be selected on the basis of informational storage and transmission until transcriptional and translational mechanisms of high fidelity had appeared.

(3) The evolutionary development of contemporary systems from prebiotic structures required functional duplication so that new microsystems could arise which were similar to previous ones. Nucleic acids could not have participated in this duplication of functions until the related processes of transcription and translation had themselves appeared and been refined. Furthermore, these processes are also dependent upon functional proteins (polymerases, synthestases, etc.). A much simpler evolutionary scenario proposes the initial development of a protein-copying enzyme, followed by the appearance, in time, of an enzyme system for transmitting information from proteins to nucleic acids. We feel, as does Hanson (7), that the functions commonly associated with nucleic acids in contemporary organisms appeared much later than proteins, so that these latter functional molecules would have nonfunctional, but complementary, counterparts.

Much as music must exist before it can be recorded on a tape and played back, so functional proteins must have existed before their nucleic acid "tapes" could evolve and be "played back" through processes of transcription and translation.

ACKNOWLEDGMENT

This work was supported by National Institutes of Health General Research Support Grant No. 5-S01-RR-05300-10.

REFERENCES

1. Watson, J. D., and Crick, F. H. C., Nature 171, 737 (1953).
2. Crick, F. H. C., Symp. Soc. Exp. Biol. 12, 138 (1958).
3. Temin, H., and Mizutani, S., Nature 226, 1211 (1970).

4. Baltimore, D., Nature 226, 1209 (1970).
5. Crick, F. H. C., Nature 227, 561 (1970).
6. Spiegelman, S. et al, Nature 227, 563 (1970).
7. Hanson, E. D., Quart. Rev. Biol. 41, 1 (1966).
8. Orgel, L. E., J. Mol. Biol. 38, 381 (1968).
9. Mueller, H. J., Science 121, 1 (1955).
10. Sagan, C., Evolution 11, 40 (1957).
11. Needham, A. E., Quart. Rev. Biol. 34, 189 (1959).
12. Lacey, J. C., and Pruitt, K. M., Nature 223, 799 (1969).
13. Calvin, M., "Chemical Evolution," Oxford University Press, New York, 1969.
14. Fox, S. W. in "The Origin of Prebiological Systems (S. W. Fox, ed.), p. 361. Academic Press, New York, 1965.
15. Fox, S. W., et al., Chem. and Eng. News 48, 80 (1970).
16. Lacey, J. C. and Fox, S. W. in "Theory and Exp. in Exobiology," Vol. 2, (A. Schwartz, ed.), Walters-Noordhoff, Groningen, 1972, p. 35.
17. Weiss, L., in "Permeability and Function of Biological Membranes" (L. Bolis, A. Katchalsky, R. D. Keynes, W. R. Loewenstein, and B. A. Pethica, eds.), p. 95. North-Holland/American Elsevier, New York, 1970.
18. Archibald, A. R., et al., "Advances in Enzymology" (F. F. Nord, ed.), p. 223. Inter-science Publishers, New York, 1968.
19. Adamson, A. W., "Physical Chemistry of Surfaces," p. 152. Inter-science Publishers, New York, 1960.
20. Goldacre, R. J., in "Surface Phenomena in Chemistry and Biology" (Danielli, J. T. et al., eds), p. 276. Pergammon Press, N. Y., 1958.
21. Frommer, M. A., and Miller, J. R., J. Phys. Chem. 72, 2862 (1968).
22. Lacey, J. C., PhD. Thesis., University of Alabama (1969).
23. Rohlfing, D. L., and Fox, S. W., Adv. in Catalysis 20, 373 (1969).
24. Bugg, C. E. et al., Biopolymers 10, 175 (1971).
25. Rohlfing, D. L., and Fox, S. W., Arch. Bioch. Biophys. 118, 127 (1967).
26. Ryan, Jack, PhD. Thesis, University of Miami (1970).
27. Krampitz, J., and Fox, S. W., Proc. Natl. Acad. Sci. 62, 399 (1969).
28. Nakashima, T., et al., Naturwissenschaften 57, 67 (1970).
29. Paecht-Horowitz, M. et al., Nature 228, 636 (1970).
30. Gibbons, R. A., and Hunter, G. D., Nature 215, 1041 (1967).
31. Griffith, J. S., Nature 215, 1043 (1967).
32. Eigen, M., UMSCHAU 24, 777 (1971).
33. Woese, C. R., Proc. Nat. Acad. Sci. 59, 110 (1968).

34. Wagner, K. G., and Arav, R., Biochemistry 7, 1771 (1968).
35. Wagner, K. G., and Wulff, K., Biochem. Biophys. Res. Comm. 41, 813 (1970).
36. Sulston, J. et al., Proc. Nat. Acad. Sci. 60, 409 (1968).
37. Teng, C. S., Teng, C. T., and Allfrey, V., J. Biol. Chem. 246, 3597 (1971).
38. Schaup, H. W., Green, M., and Kurland, C. G., Molec. Gen. Genetics 109, 193 (1970).
39. Pongs, O., and T s'o P. O. P., Biochem. Biophys. Res. Comm. 36, 475 (1969).
40. Ibanez, J., Oro', J., and Kimball, A. P., Southwestern Regional Meeting, Am. Chem. Soc., Tulsa, Okla. (1969).
41. Jungck, J. R., and Fox, S. W., Southeastern Regional Meeting, Am. Chem. Soc., New Orleans (1970).
42. Fuller, W. D., Sanchez, R. A. and Orgel, L. E., J. Mol. Biol. 67, 25 (1972).
43. Sanchez, R. A., and Orgel, L. E., J. Mol. Biol. 47, 531 (1970).

NOTE ADDED IN PROOF

Fuller et al. (42) have recently reported the successful synthesis of adenine nucleoside isomers from ribose and adenine under prebiotic conditions (dry heat). Similar results have been obtained with guanine. Sanchez and Orgel (43) have also previously reported the successful synthesis of the naturally occurring pyrimidine nucleosides utilizing D-ribose, cyanoacetylene and cyanamide.

Received 14 July, 1971

MODEL EXPERIMENTS ON THE PREBIOLOGICAL

FORMATION OF PROTEIN

by Shiro Akabori and Mihoko Yamamoto

Protein Research Foundation, Ina, Minoo, Osaka, Japan

Since it was emphasized by Oparin (1) in 1957 that the formation of coacervate of Bungenberg de Yong (2) in the primitive ocean might have been the most important step of chemical evolution of life, the theory has been widely accepted by the workers in the field (3). The prerequisite for the formation of coacervate particles is the presence of high-molecular organic sustances together with other biochemical substances. Among those high-molecular substances a long chain polypeptide must have been one of the most important ones.

In 1957 Akabori (4) proposed a hypothesis that the "Fore-protein" could have been formed by polymerisation of amino-acetronitrile followed by hydrolysis to polyglycine and then by the introduction of various side chains onto methylene group of polyglycine.

$$n \; NH_2-CH_2-CN \longrightarrow (-NH-CH_2-\overset{\|}{\underset{NH}{C}}-)_n$$

$$\xrightarrow{\overset{+H_2O}{\quad}} (-NH-CH_2-CO-)_n + n \; NH_3.$$

$$\downarrow$$

$$-(NH-\underset{R}{\overset{|}{C}}H-CO-)_n$$

This hypothesis may well explain why natural proteins are all composed of α-peptide bonds. Experimental evidences supporting this hypothesis are reported by Akabori and Hanafusa (5), Okawa, Sato and

Akabori (6), Sakakibara (7), and Losse and Bohm (8).

On the other hand Fox (9) has taken a position that the prebiological protein could have been formed by thermal polycondensation of preformed amino acid mixtures. Fox and his associates (10) carried out a series of experiments on the thermal polycondensation of amino acids and obtained protein-like substances which they called "proteinoids. They vary in their properties according to the ratio of amino acid components and conditions of polymerization. It is highly interesting to observe their proteinoid microspheres suspended in dilute salt solutions resembling unicellular microorganisms.

However, by the theory of thermal polycondensation, it is difficult to understand why β-carboxyl group of aspartic acid, γ-carboxyl group of glutamic acid and ε-amino group of lysine did not participate in the prebiological polypeptide formation, if we assume abiogenically accumulated amino acids were composed of all kinds of amino acids which are contained in the present day protein.

Concerning the prebiological formation of glutamyl residues in polypeptide chains the following mechanism was suggested by Akabori (4):

$$-NH-CH_2-CO- \longrightarrow \begin{array}{c} -NH-CH-CO- \\ | \\ CH_2-CH_2-CN \end{array} \longrightarrow \begin{array}{c} -NH-CH-CO \\ | \\ CH_2-CH_2-COOH. \end{array}$$

$$+H_2C=CH-CN$$

An interesting experiment supporting this possibility was reported by Dzugaj, Siemion and Ojrzynski in 1966 (11). They obtained glutamic acid by hydrolysing the reaction product of phenylazlactone with acrilonitrile:

$$NC-CH=CH_2 + H_2C \underset{N \;\; O}{\overset{CO}{\rule{0pt}{1em}}} \longrightarrow NC-CH_2-CH_2-CH \underset{N \;\; O}{\overset{CO}{\rule{0pt}{1em}}}$$

$$C_6H_5 \qquad\qquad C_6H_5$$

$$\overset{+H_2O}{\longrightarrow} \quad HOOC-CH_2-CH_2-\underset{NH_2}{CH}-COOH \quad +C_6H_5-COOH.$$

The most difficult problem is to find a reasonable chemical explanation for the prebiological formation of lysyl residues, because all lysyl residues in proteins are contained only in α-peptide form, leaving the ε-amino group free. In 1957 Akabori (4) proposed an apparently factitious speculation on the prebiological formation of lysyl residue as follows:

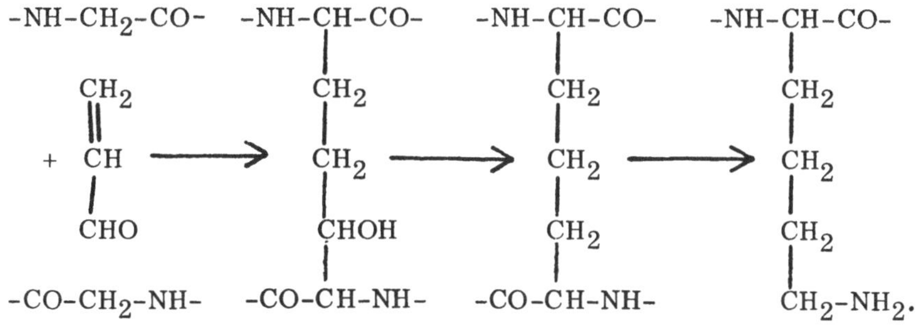

$$
\begin{array}{c}
-NH-CH_2-CO- \\[4pt]
CH_2 \\
\| \\
+ \quad CH \\
| \\
CHO \\[4pt]
-CO-CH_2-NH-
\end{array}
\longrightarrow
\begin{array}{c}
-NH-CH-CO- \\
| \\
CH_2 \\
| \\
CH_2 \\
| \\
CHOH \\
| \\
-CO-CH-NH-
\end{array}
\longrightarrow
\begin{array}{c}
-NH-CH-CO- \\
| \\
CH_2 \\
| \\
CH_2 \\
| \\
CH_2 \\
| \\
-CO-CH-NH-
\end{array}
\longrightarrow
\begin{array}{c}
-NH-CH-CO- \\
| \\
CH_2 \\
| \\
CH_2 \\
| \\
CH_2 \\
| \\
CH_2-NH_2.
\end{array}
$$

If such a cross-linked bilateral condensation of acrolein should actually occur, we could be able to isolate diaminopimelic acid by the reductive hydrolysis of the condensation product of acrolein with polyglycine. By the hydrolysis or reductive hydrolysis of the reaction product of polyglycine and acrilonitrile or acrolein the formation of glutamic or diaminopimelic acid was very small and further confirmation was impossible. Therefore, we used N·N'-dibenzoyldiketopiperazine (DBDP) as a model substance in place of polyglycine. The reaction scheme is shown as follows:

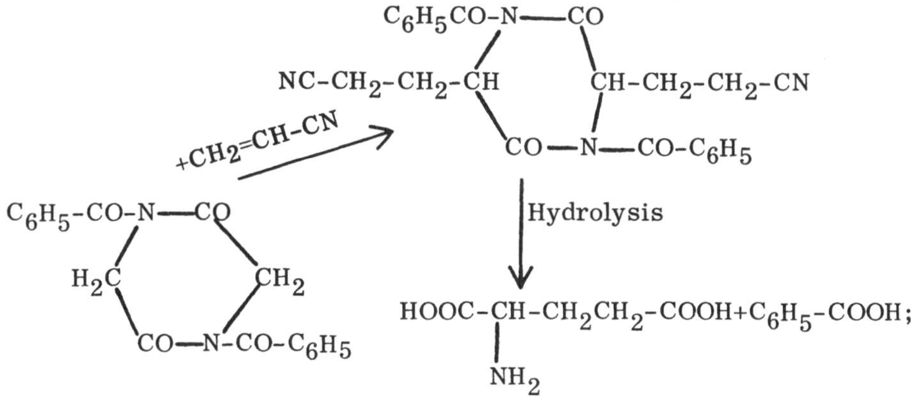

$$
\begin{array}{c}
C_6H_5CO-N\!\!-\!\!-CO \\
NC-CH_2-CH_2-CH \qquad CH-CH_2-CH_2-CN \\
CO\!\!-\!\!N\!\!-\!\!-CO-C_6H_5
\end{array}
$$

$$
C_6H_5-CO-N\!\!-\!\!-CO
$$
$$
H_2C \qquad CH_2
$$
$$
CO\!\!-\!\!N\text{-}CO-C_6H_5
$$

$$\xrightarrow{+CH_2=CH-CN}$$

Hydrolysis

$$HOOC-CH-CH_2CH_2-COOH+C_6H_5-COOH;$$
$$\qquad\quad NH_2$$

$$\overset{\displaystyle N\text{----}CO}{\underset{\displaystyle CO\text{----}N-}{-CH}} \quad \underset{}{CH\text{-}CH_2\text{-}CH_2\text{-}CH\text{-}CH} \quad \overset{\displaystyle N\text{----}CO}{\underset{\displaystyle CO\text{----}N-}{CH-}}$$

$$\downarrow \text{Reductive hydrolysis}$$

$$HOOC\text{-}\underset{NH_2}{CH}\text{-}CH_2\text{-}CH_2\text{-}CH_2\text{-}\underset{NH_2}{CH}\text{-}COOOH \quad + \ C_6H_5\text{-}COOH$$

It is very interesting that a small amount of lysine was formed besides diaminopimelic acid.

EXPERIMENTAL

1) Dibenzyol-diketopiperazine (DBDP). This compound was synthesized according to Sasaki and Hashimoto (12); m.p. 225-227° C.

2) Reaction of DBDP with acrylonitrile. Ten g of DBDP were dissolved in 120 ml dimethylformamide, and 10 ml each of acrylonitrile and triethylamine were added. The mixture was kept at 105° in an oil bath for 46 hrs in a sealed bottle. After the solvent and volatile matter were distilled off <u>in vacuo</u> the remaining residue was boiled with 300 ml of 6 N hydrochloric acid for 42 hrs. Hydrochloric acid was driven off in vacuo, and the residue was dissolved in water. The solution was run through Dowex 50 x 8 (H^{+} form) column (4 x 45 cm), washed with water and then eluted with 2 N ammonia. The eluate was concentrated and subjected to amino acid analysis. The results of the analysis are shown in Fig. 1.

Peak 1 in Fig. 1 corresponds to that of aspartic acid, but it was found in our laboratory that aspartic acid and isoglutamic acid cannot be distinguished by ordinary chromatography; therefore, the acidic amino acid fractions were converted to the trifluoracetyl amino acid ethylester and analyzed by a Shimadzu gas chromatograph GC-4A-PF using a column (3 m x 4 mm) charged with NGS (neopentylglycol-succinate). By this gas chromatography we identified the amino acid of peak 1 as iso-glutamic acid HOOC-CH$_2$-NH-CH$_2$-CH$_2$-COOH.

The glutamic acid fraction was treated with DNFB to obtain DNP-derivatives, and the product was crystallized; m.p. 167-169°C. No depression of m.p. was observed by mixing with an authentic sample of DNP-DL-glutamic acid, m.p. 167.5-170°. Infrared spectra of both samples of DNP-glutamic acid are shown in Fig. 2.

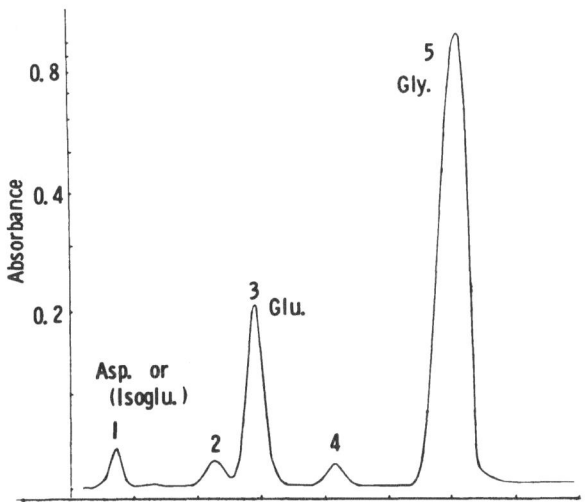

Fig. 1. Amino acid analysis in the effluents from Dowex 50 x 8 column.

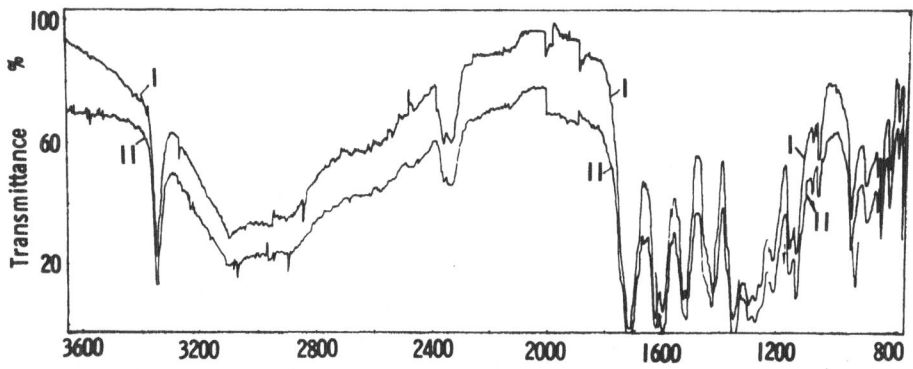

Fig. 2. Infrared spectra of authentic and synthesized DNP-Glu. I; Authentic DNP-Glu. II; Synthesized DNP-Glu.

The amounts of amino acids found in the reaction mixture were: glutamic acid, 0.109g; isoglutamic acid, 0.168g; and glycine, 1.145g.

3) Reaction of acrolein with DBDP. DBDP, acrolein and triethylamine were dissolved in acetyldimethyl amide and kept at various temperatures for various lengths of time. The reaction product was subjected to reductive hydrolysis. An example of the experiments is described below.

Two hundred mg of DBDP, 0.05 ml of acrolein and 0.1 ml of triethylamine were dissolved in 3 ml of acetyldimethylamide and kept in a sealed tube at 27° for 44 hrs. After the solvent and volatile matter were driven off from the reaction product, to the residue was added 7 ml of 52% hydroiodic acid and 30 mg red phosphor, and the mixture was heated at 100° for 24 hrs. The hydrolysate was diluted with water. Red phosphor was filtered off, and liberated iodine was removed by extraction with toluene. The solution was run through Dowex 50 x 8 (H^+ form) column 1.5 x 10 cm. The amino acid mixture was eluted by 2 N aqueous ammonia and applied to an automatic amino acid analyzer. The yield of diaminopimelic acid was 16.7 mg. A small amount of lysine (0.23 mg) was also detected.

Further identification of the product was carried out by converting it to ethyl ester of the trifluoroacetyl derivative and analyzing by gas chromatograph with an authentic sample prepared from diaminopimelic acid as a standard (Fig. 3).

DISCUSSION

Experimental evidences on the prebiological formation of peptides have been reported by man workers, as mentioned in the introduction. The present work may be of some value as an additional evidence in favour of the theory of Akabori.

Results of our previous work on the formation of leucine or isoleucine was very unsatisfactory. Highly interesting work was recently reported by Elad and Sperling (13). These workers succeeded in modifying glycyl residues in the peptide chain to leucyl, isoleucyl and phenylalanyl residues, respectively, by irradiation with ultraviolet light in the presence of 1-butene, isobutene and toluene, respectively, using acetone as photosensitizer. Their experiments clearly demonstrate the possibility of the transformation of polyglycine to a polypeptide containing various

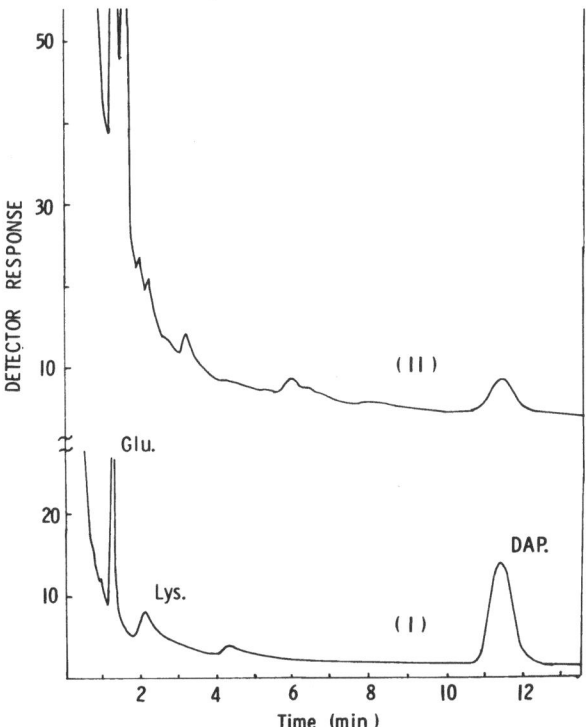

Fig. 3. Gas chromatography of authentic
(I) and synthesized α, ε-diamino pime-
lic acid (II).

amino acid residues, under prebiological conditions on the Earth. The present author assumed the presence of acrylonitrile and acrolein on the surface of the prebiological Earth to explain the non-biological formation of glutamyl and lysyl residues. This assumption may be considered to be too factitious to most biochemists. However, we should remind ourselves that, in the present-day peterochemistry, acrylonitrile is produced in a large scale by the catalytic dehydrogenation of the mixture of propylene and ammonia.

$$H_2C=CH-CH_3+NH_3 \longrightarrow H_2C=CH-CN+3H_2$$

Acrolein was also produced in an industrial scale by an incomplete combustion of propylene.

$$H_2C=CH-CH_3+O_2 \longrightarrow H_2C=CH-CHO+H_2O$$

Another apparently factitious hypothesis is the mechanism of the prebiological formation of diaminopimelic acid residues by bilateral condensation of acrolein with glycyl residue, followed by the dehydroxylation.

Diaminopimelic acid (DAP) was first discovered by E. Work in 1949 (14), and soon it was found that this diamino-dicarboxylic acid is widely distributed among various microorganisms with the exception of Gram-positive cocci by E. Work and D. L. Dewy (15). Holdsworth (16) discovered that diaminopimelic acid in Corynebacterium diphtherie was mostly located in the cell wall. In more recent studies of the bacterial cell wall, DAP has been shown to be a main amino acid components of cell wall mucopeptide of various bacteria. This fact is now applied to chemical taxonomy of bacteria (17). Kingan and Ensign concluded that DAP in cell wall mucopeptide of Bacillus thuringiensis (18) was mainly involved in cross-linkage (up to 66% of total DAP), based on their chemical analysis of the mucopeptide.

Considering those studies described above it could be accepted that DAP had a role in maintaining the rigidity of cell walls in the primitive microorganism or protocell.

ACKNOWLEDGEMENT

The authors are greatly indebted to Dr. S. Sakakibara of the Protein Research Foundation for his kind advice throughout our study.

REFERENCES

1. Oparin, A. I., "Die Entstehung des Lebens auf der Erde," Deutsche Verlag der Wissenschaften, Berlin, 1957; "The Origin of Life on the Earth," Pergamon Press, New York, 1959.
2. Bungenberg de Yong, H., Protoplasma 15, 110 (1932); " La Coaservation," Hermann, Paris, 1936.
3. Smith, A. E., Bellware, B. T., and Silver, J. J., Nature 214, 1038 (1967). Liebl, V., and Lieblova, J., J. Br. Interplanet. Soc. 21, 312 (1968).

4. Akabori, S., I. U. B. Symposium Series, 1, 189 (1959).
5. Akabori, S., Hanafusa, H., Bull. Chem. Soc. Japan 32, 626 (1959).
6. Akabori, S., Okawa, K., and Sato, M., ibid. 29, 608 (1956).
7. Sakakibara, S., ibid. 34, 205 (1961).
8. Losse, G., and Böhm, R., J. Prakt. Chem. 38, 69 (1968).
9. Fox, S. W., Science 132, 200 (1960); Fox, S. W., Harada, K., and Kendrick, J., Science 129, 1221 (1959).
10. Fox, S. W., and Harada, K., "A Laboratory Manual of Analytical Methods of Protein Chemistry," Vol. 4, 129. Pergamon Press, New York, 1066; Fox, S. W. and Yuyama, S., Ann. N. Y. Acad. Sci., 108, 487 (1963).
11. Dzugaj, A., Siemion, I. Z., and Ojrzynski, Z., Roczniki. Chem. 40, 1329 (1966).
12. Sasaki, T., and Hashimoto, T., Ber. Beutchen, Chem. Gesells. 54, 2688 (1921).
13. Elad, D., and Sperling, J., J. Chem. Soc. (C), 1969, 1579 (1969); Sperling, J., and Elad, D., J. Am. Chem. Soc. 93, 967 (1971).
14. Work, E., Biochem. Biophys. Acta 3, 400 (1949).
15. Work, E., and Dewey, D. L., J. Gen. Microbiol. 9, 394 (1953).
16. Holdsworth, E. S., Biochem. Biophys. Acta. 9, 19 (1952).
17. Hamaguchi, T., J. Bacteriol. 89, 444 (1965).
18. Kingan, S. L., and Ensign, J. C., J. Bacteriol. 95, 724 (1968).

Received 26 October, 1971

ON THE ELECTROPHORETIC BEHAVIOR OF THERMAL

POLYMERS OF AMINO ACIDS

Klaus Dose and Horst Rauchfuss

Max-Planck-Institut für Biophysik

Frankfurt (Main), Germany

INTRODUCTION

The limited heterogeneity of a number of thermal polymers of amino acids, also called proteinoids (1), has been established in several reports (2-8). Particularly, if the polymers were subjected to electrophoretic analysis they could be separated into only a small number of definable fractions, mostly three or less. Often electrophoresis even signified true-near-homogeneity. In some instances the single fraction appeared to have a higher degree of homogeneity than even purified organismic proteins. Examples are the acrylamide gel electrophoresis of an unfractionated, but amidated 1:1:1* proteinoid (8) and the gel electrophoresis of a hemoproteinoid (molecular weight about 18.000) which possesses peroxidase-like activity (6-7). However, substantial evidence indicates that electrophoreses is inferior in sensitivity to fractionation on DEAE-Sephadex, DEAE-cellulose, and other cellulose ion exchangers. Fox and Nakashima (8) separated on DEAE-cellulose the amidated 1:1:1 proteinoid into at least three major fractions which all

---

*1:1:1 proteinoids are polymers prepared from equal proportions (by weight) of aspartic acid, glutamic acid, and an equimolar mixture of all 20 amino acids common to protein.

showed identical electrophoretical mobilities.   A number of related results which have been obtained in our laboratory are published here.

MATERIALS AND METHODS

Synthesis of Thermal Polymers of Amino Acids

Polymer A 61:   The amino acid mixture (see Table I for composition) was heated at 178° for 7 hours under N_2.   The raw product was
dispersed in water and dialyzed for 3 days against distilled water and
lyophilized. The dialyzed product was free of amino acids and low molecular weight peptides. The product could be precipitated by 10% trichlo-
racetic acid.   The precipitate gave a typical reaction with biuret reagent
(9).   The optical density, however, yielded only 52% the value produced
by the same amount (by weight) of human serum albumin.   Without previous precipitation the corresponding biuret value was 66%.   The ash
contents of the product was 2.5%; Kjeldahl nitrogen was 14.8%.

Polymer A β :   Two parts of an amino acid mixture (see Table II for
composition) and one part of sodium polyphosphate were heated for 48
hours at 125° under N_2. The raw product was dispersed in water at 80°.
The dispersion was cooled to 0° under stirring, filtered, dialyzed against
distilled water for 4 days and lyophilized.   Biuret values and contents in
Kjeldahl nitrogen were similar to A 61.   The ash content was slightly
higher.

Polymers A 96 a-d: Two parts of four different amino acid mixtures
(containing increasing amounts of aspartic acid according to Table III)
were each heated with one part of sodium polyphosphate at 120° for 48
hours under N_2.   The raw product was dispersed in distilled water at
40°, stirred and cooled to 2^O.   The sediment was homogenized in a Star
Mix and filtered.   The filtrate was dialyzed for 5 days against distilled
water until the contents in free amino acids (determined by paper electrophoresis and ion exchange chromatography) was less than 3%.   Biuret
values and nitrogen contents are shown in Table III .

Polymers A 98 a-d: Two parts of an amino acid mixture as shown in
Table III were mixed with one part of sodium polyphosphate.   To this
mixture variable amounts of 1 N sodium hydroxide or HCl were added so
that the pH of 10 mg. reactant mixture/ml H_2O after 30 min at 120° was
obtained as shown in Table III .   To polymerize the amino acids the re-

actant mixtures were heated 72 hours at 120° under nitrogen. Finally the raw product was dispersed in water, cooled to 2°, homogenized in a Star Mix and filtered. The filtrate was dialyzed 4-5 days until all free amino acids were removed. The relative biuret values and contents in Kjeldahl nitrogen are shown in Table III.

Analytical Methods

Amino acid analyses were carried out according to Gundlach et al. (10). The standard procedure for acid hydrolysis was: 1 mg polymer per 2 ml 6 N HCl at 108° C for 18 hrs under vacuum.

The number of free carboxylic and amino groups of the various proteinoids were determined with a Radiometer Titrator essentially according to Cannon (11) and Alberty (12). For each assay 40 mg of dialyzed proteinoid was dissolved in 2 ml water (deionized and bidistilled). The pH value of the resulting solution was taken as the isoionic point. Titrations were carried out with 1 N HCl.

The number of carboxylic groups per mg proteinoid was calculated from the number of meq. of HCl added to change the pH from 5.5 to 2.0; the number of basic groups (assumed to be amino groups) was accordingly calculated from the number of meq. of HCl added during the titration from the isoionic point to pH 2.0. The values were corrected for the amount of meq. of HCl which reacted with the water.

Molecular weights were estimated by gel filtration on various Bio Gel columns (13).

Acrylamide gel electrophoresis was carried out at pH 8.9 (Tris-glycine buffer; gel concentration was 7%) according to Maurer(14).

Electrophoresis on cellulose acetate films (Selekta Elektrophorese-Folien from Schleicher and Schull, Einbeck) was performed by use of a Pherograph from Hormuth, Heidelberg. The solutions, being about 1-5% in protein(oid), were applied with a micropipette. The potential gradient was about 20 volts x cm^{-1}; the time for one run was about 40 min; the temperature of the Pherograph was set at 5-7°. Mostly a barbital-barbital sodium buffer of pH 8.6 was used. After the electrophoretic separation the fractions were localized by staining with amido black 10 B in methanol: glacial acetic acid = 9:1. The excess of the dye was removed by repeated treatment with methanol: glacial acetic acid = 9:1. In order to produce transparent films the wet strips were carefully transferred to microscopic slides, sprayed with acetic acid and dried at 100°.

The IR-spectra were taken with a Perkin-Elmer 021; 1-2 mg of the proteinoid and 250 mg KBr were well ground, mixed and pressed at a pressure of 10 atmospheres.

For the chromatographic fractionation DEAE-SS-cellulose from Serva, Heidelberg, was used. The material was equilibrated with 0.005 M phosphate buffer at pH 7.5. Columns, 30 x 200, were charged with 150-300 mg of the proteinoid. The standard procedure was to eluate with 0.02, 0.05, 0.1, 0.2, and 0.5 M phosphate buffer at pH 7.5, 0.1 N HCl and 0.1 N NaOH, in this sequence. An LKB-Uvicord Ultraviolet Absorptiometer (at 280 nm) was used. The flow rate was 25-30 ml per hour. The separated fractions were dialyzed for 1 to 10 hours until they were practically free of electrolyte. Then they were lyophilized. This procedure was followed by amino acid analysis and/or characterization by electrophoresis.

Some proteinoids were also fractionated by precipitation at various concentrations of ammonium sulfate. For this purpose 100 mg of the proteinoid (e.g., A β) were dissolved in 10 ml 0.02 M tris/HC at pH 7.2. Solid ammonium sulfate (p.a., Merck, Darmstadt) was added until the required degree of saturation (25, 50, 75, and 100%) was reached. The precipitations were carried out at 0°. The precipitates were collected by sedimentation (1500 x g for 20 min) about 1 hour after the desired amount of ammonium sulfate was added. After sedimentation the precipitates were redissolved, dialyzed for 10-24 hours against distilled water and finally lyophilized.

RESULTS

Polymer A 61 represents the dialyzed fraction of a product obtained by heating a given amino acid mixture (Table 1, column b) for 7 hours at 178°. The amino acid composition of the unfractionated polymer is shown in Table 1, column c. The biuret value relative to bovine serum albumin is 66%. The hydrolysis of 1 mg. polymer yields about 0.5 mg amino acids. The polymer is largely homogenous according to the anaiysis by gel chromatography with Bio Gel. The Rf-values indicate an approximate molecular weight of 5000. The results of electrophoretical analysis suggest that the material has about the same net charge per molecule

TABLE I.   Polymer A 61.   Amino Acid Composition of Reactants and Dialyzed Product and Fractions Obtained by DEAE-Chromatography.   All values in gm %; tr corresponds to values below 0.5%.

a amino acid	b composition of reactants	c composition of polymer	d F_1	e F_2	f F_3	g F_4	h F_6	i F_7
Asp	8.4	5.0	3.7	6.0	5.2	6.1	6.4	6.5
Thr	2.5	tr	0.8	0.7	tr	0.8	tr	0.7
Ser	7.2	0.9	1.0	1.1	1.7	1.5	1.9	0.8
Glu	15.0	29.4	25.6	30.5	28.0	29.4	25.3	27.0
Pro	2.5	tr	2.8	tr	tr	tr	tr	3.2
Gly	1.7	3.9	3.1	3.8	3.7	4.1	4.3	3.6
Ala	2.0	6.2	5.6	5.7	6.6	6.3	6.5	6.3
Val	2.5	1.2	1.2	1.2	tr	1.4	tr	1.3
1/2 Cys	5.3	tr	tr	tr	tr	tr	tr	tr
Met	3.1	1.4	1.6	1.5	tr	0.7	tr	0.8
Ile	2.8	tr	tr	tr	tr	tr	tr	tr
Leu	2.8	1.9	1.8	1.8	tr	2.0	2.0	2.2
Tyr	7.6	1.2	1.1	tr	tr	tr	tr	1.2
Phe	3.7	7.4	5.1	5.8	8.7	6.1	8.1	7.0
NH_3	–	–	2.0	2.0	4.4	2.3	4.6	2.7
Lys	20.5	36.0	37.1	35.7	34.5	34.5	36.5	32.5
His	6.3	2.2	3.2	1.3	2.4	2.3	2.3	2.5
Trp*	4.3	tr	tr	tr	tr	tr	tr	tr
Arg	1.6	2.3	3.3	2.1	2.0	2.0	2.2	1.9
α -ABA	–	1.0	0.9	0.7	tr	1.0	tr	tr
β -Ala	–	–	tr	tr	tr	tr	tr	tr

* Tryptophan is largely destroyed by acid hydrolysis.

inasmuch as the material migrates in a single major fraction at pH 8.6 on cellulose acetate (Fig. 1a) and at pH 8.9 in acrylamide (Fig. 1b). The tailing of the electrophoretic fraction (Fig. 1a) is caused by an interaction between proteinoid and cellulose acetate. The isoionic point of polymer A 61 is 6.2. The titration experiments showed 1.53 meq. carboxylic groups and 1.60 meq. amino groups per gm. polymer. At pH 8.6 (on cellulose acetate) the electrophoretic mobility (anodic) is 1.35 times that of bovine serum albumin.

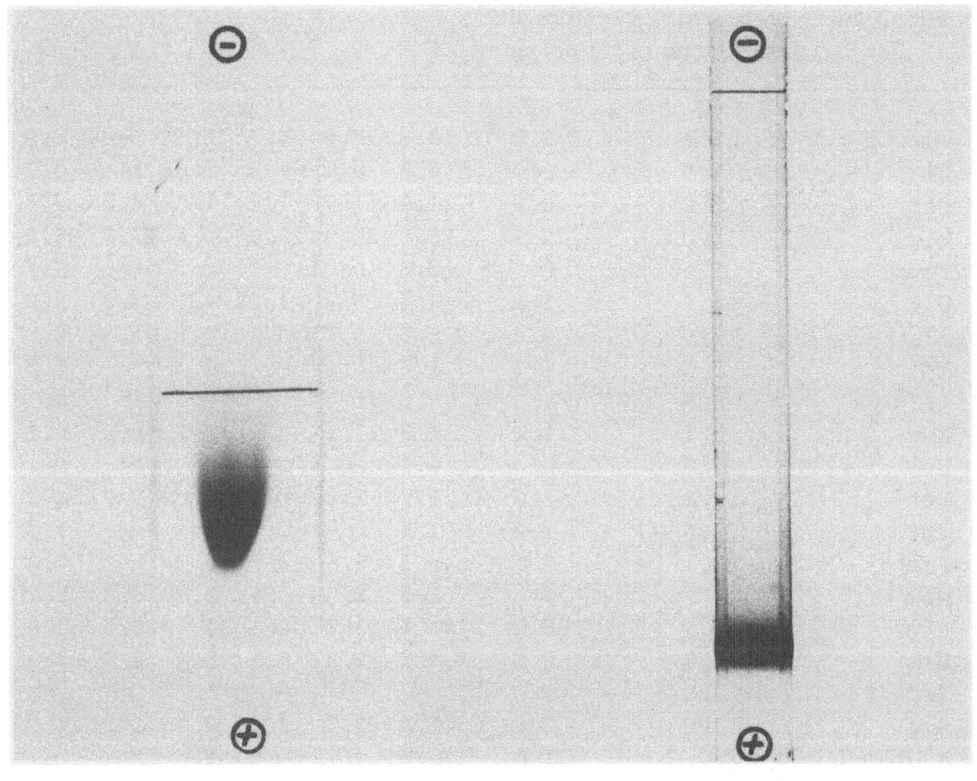

1a   1b

Fig. 1a. Electrophoresis of polymer A 61 on a cellulose acetate film at pH 8.6. The anodic movement of the main fraction is 1.35 times that of bovine serum albumin.

Fig. 1b. Electrophoresis of polymer A 61 in acrylamide gel at pH 8.9. The amount of tailing and other disturbances can be greatly reduced by appropriate choice of the supporting material.

In figure 2 the IR-absorption spectrum of A 61 is compared with the corresponding spectrum of bovine serum albumin. The similarity of the spectra appears remarkable. Related similarities between IR-spectra of proteins and proteinoids have been found in all instances so far analyzed. However, in spite of many similarities, differences are also evident, e.g., in the amide I and II range. Presently no firm basis has been found to discuss in more detail these differences.

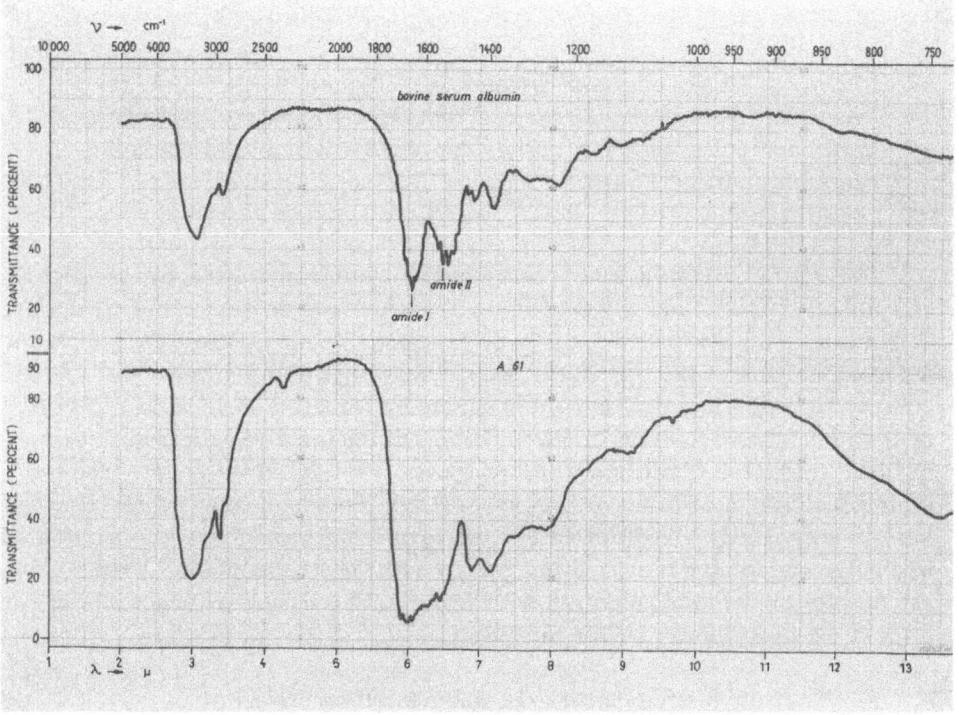

Fig. 2. The IR absorption spectrum of polymer A 61 is compared with the spectrum of bovine serum albumin. In spite of a general agreement of both spectra a close comparison shows several differences, e.g., in the range of the amide I and II absorption. These differences have not yet been explained.

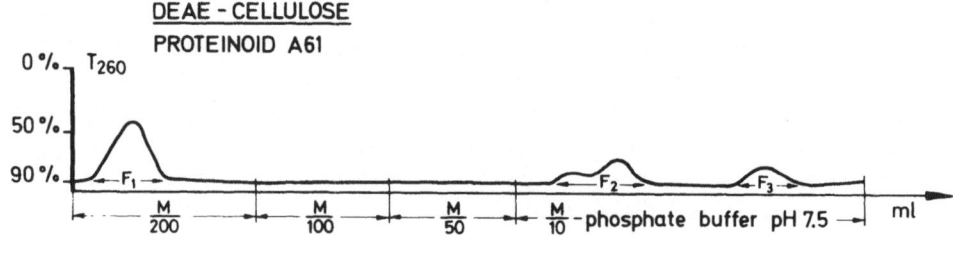

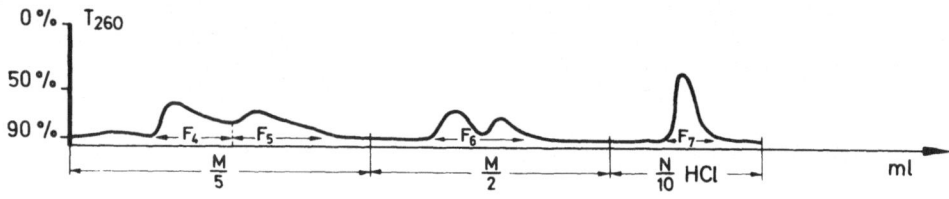

Fig. 3.   DEAE-chromatography of polymer A 61.   See text
(materials and methods) for more details. The fractions which
have been collected and further analyzed (amino acid composition) are marked F_1-F_7.

Polymer A 61 was also fractionated by chromatography on DEAE-
cellulose.   More than seven major fractions were separated (Fig. 3).
The amino acid composition of these fractions is only little different from
the composition of the unfractionated material (Table I , column d-i). On
contact with DEAE-cellulose in phosphate buffer at pH 7.5 or in 0. 1 $\underline{N}$ HCl
at 20°, proteinoid A 61 undergoes a conversion into two electrophoretically distinguishable fractions which migrate cathodically (Fig. 4). The possible mechanism of this conversion will be discussed later. This conversion is only partly responsible for the larger number of fractions obtained
by DEAE-chromatography, if compared with electrophoresis. Experimental results of the kind obtained with proteinoid A 61 are characteristic for
almost all proteinoids so far analyzed:

1. Near-by homogeneity according to electrophoresis,
2. partial or complete convertibility into one or two electrophoreti-
    cally distinguishable fractions by interaction with polar agents
    (ion exchange resins and/or electrolytes),
3. separability into a larger number of fractions by DEAE-chroma-
    tography.

⊖

⊕

a   *b*   *c*   *d*   *e*

Fig. 4. Electrophoretic control of the conversion of polymer A 61;
  4a: untreated material;
  4b: 24 h, 20°, 0.1 N HCl
  4c: 24 h, 20°, 0.1 N HCl then 24 h, 20°, 0.1 N NH$_4$OH;
  4d: 24 h, 20°, 0.1 N NH$_4$OH, then 24 h, 20°, 0.1 N HCl ;
  4e: typical electrophoretic pattern of A 61 after elution from DEAE-
      cellulose (mixture of fraction F$_1$-F$_4$); no cathodic material could
      be eluted.
Cellulose acetate electrophoresis pH 8.6.

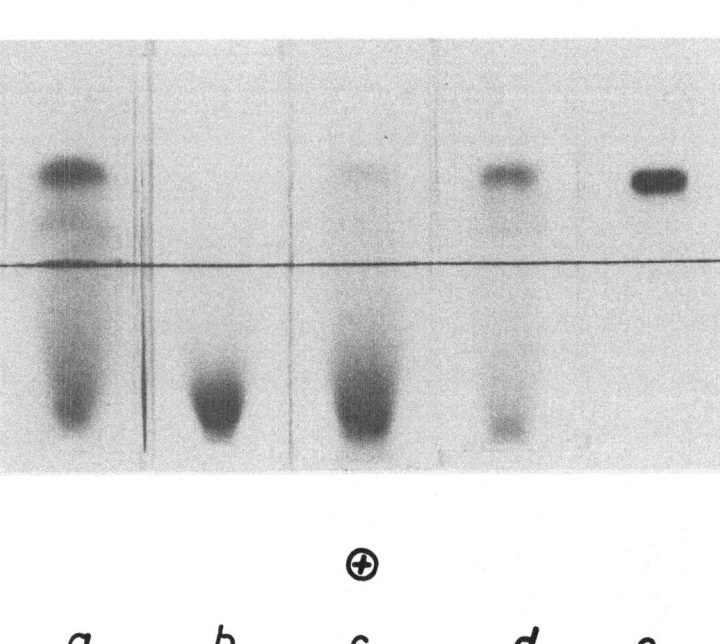

Fig. 5.   Electrophoretic control of the fractionation of
polymer A β with ammonium sulfate. The electropherograms, b, c, d and e represent the analysis of the materials precipitated at 25, 50, 75, and 100% saturation with
ammonium sulfate.   The mobility of the acidic fraction is
about 1.2 times that of bovine serum albumin. The cathodic mobilities of the upper two zones are 0.45 times and
0.2 times that of lysozyme.   (pH 8.6; cellulose acetate).

The results obtained with proteinoid A β are typical for the fractionation of proteinoids by ammonium sulfate precipitation. Figure 5 a shows
the electropherogram of polymer A β in the unfractionated state.   The
electropherograms of the precipitates isolated after 25, 50, 75, and 100%
saturation with ammonium sulfate are shown in Fig. 5b-5e, respectively.
The biuret value of the unfractionated material is 68% relative to bovine
serum albumin.   The nitrogen content (Kjeldahl) is 13%.   In Table II are
summarized the data on amino acid composition of the reactants (column b),

TABLE II. Polymer A 61. Amino Acid Composition of Reactant Mixture, of A β , and Fractions Obtained by Precipitation with $(NH_4)_2SO_4$. All values in gm.%; tr corresponds to values below 0.5%.

a amino acid	b reactant mixture	c A β unfract.	d F_1	e F_2	f F_3	g F_4
Asp	1.4	4.1	4.7	4.3	4.7	3.8
Thr	1.2	tr	tr	tr	tr	tr
Ser	27.0	4.7	3.8	3.7	6.0	6.3
Glu	9.8	19.9	23.9	22.2	24.7	15.3
Pro	1.2	tr	tr	tr	tr	tr
Gly	0.8	3.0	3.9	4.1	2.6	2.2
Ala	0.9	14.2	20.4	8.1	18.2	10.5
Val	1.2	1.4	tr	1.9	1.6	0.7
1/2 Cys	18.5	3.8	tr	4.7	6.4	4.1
Met	1.5	0.9	tr	1.4	tr	tr
Ile	1.3	tr	tr	tr	tr	tr
Leu	1.3	1.0	tr	2.4	tr	tr
Tyr	1.9	tr	tr	tr	tr	tr
Phe	1.7	1.5	tr	2.8	tr	tr
NH$_3$	–	3.7	tr	tr	tr	tr
Lys	17.5	30.4	31.5	31.9	20.7	42.1
His	9.8	5.9	6.0	6.0	6.5	6.4
Trp*	2.1	tr	tr	tr	tr	tr
Arg	1.8	1.9	1.5	1.5	1.8	2.7

* Tryptophan is largely destroyed by acid hydrolysis.

the unfractionated polymer (column c), fraction F_1 (25% saturated $(NH_4)_2SO_4$, column d), and fraction F_2 (50% saturated $(NH_4)_2SO_4$, column e). Fraction F_1 contains little less than 30% acidic and 40% basic amino acids, but its anodic mobility at pH 8.3 is 1.2 times that of bovine serum albumin. Fraction F_4, on the other hand, migrates towards the cathode (about half as fast as lysozyme) in qualitative agreement with the predominance of basic amino acids in its hydrolyzate. The relatively fast anodic migration of F_1, however, clearly disagrees with the amino acid composition of its hydrolyzate. This result is characteristic for proteinoids with relatively high contents in lysine.

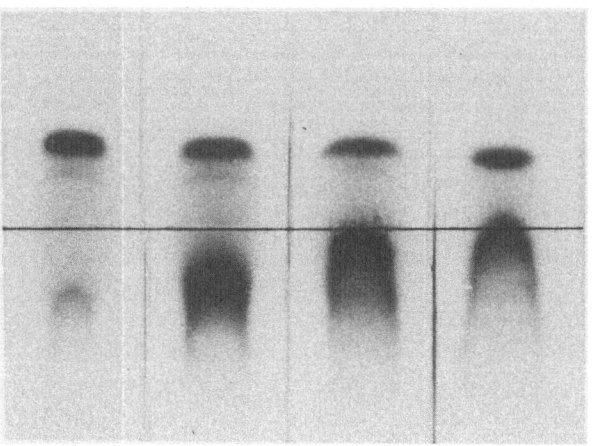

A96a A96b A96c A96d

Fig. 6. Electrophoretic properties of polymers
A 96 a-d containing increasing amounts of aspartic acid (Table III).   The major cathodic
fraction migrates about 20% as fast as lysozyme.
The anodic fraction moves about as fast as bovine serum albumin.   (pH 8.6; cellulose acetate).

In Fig. 6 are compared the electropherograms of the thermal polymers A 96 a-c.   These polymers were obtained by heating two parts of a
mixture of amino acids containing varied amounts of aspartic acid and 1
part of sodium polyphosphate.   The data on the amino acid composition of
the reactant mixture and of the 4 unfractionated polymeric products are
summarized in Table III.   The cathodic fraction migrates mostly due to
electroosmosis. Its mobility relative to lysozyme is 0.22 at pH 8.3. The
anodic fraction of A 96 a migrates about as fast as bovine serum albumin.
Aqueous solutions of polymers A 96 c and A 96 d show (average)isoionic

TABLE III. Amino Acid Composition of Polymers (A 96 a–d) and Corresponding Reactant Mixtures with Increasing Amounts of Aspartic Acid (0.5 gm Na polyphosphate/gm). All values in gm % (aspartic acid excluded; the percentages for aspartic acid are given relative to the total of all other amino acids); tr corresponds to values below 0.5%.

a amino acid	b reactant mixture (asp omitted)	c A 96 a 0.013 gm asp/gm	d A 96 b 0.213 gm asp/gm	e A 96 c 0.413 gm asp/gm	f A 96 d 0.813 gm asp/gm
Asp	1.3	11.0	32.4	46.2	53.5
Thr	1.1	tr	tr	tr	tr
Ser	26.5	8.1	5.6	3.6	2.6
Glu	10.0	22.0	17.4	17.0	16.4
Pro	1.1	tr	tr	tr	tr
Gly	0.7	2.2	2.3	3.0	1.4
Ala	0.7	10.6	12.0	17.8	13.2
Val	1.0	tr	tr	tr	tr
1/2 Cys	18.1	6.1	4.5	6.9	3.3
Met	1.6	tr	tr	tr	tr
Ile	1.2	tr	tr	tr	tr
Leu	1.2	tr	tr	tr	tr
Tyr	1.8	tr	tr	tr	tr
Phe	1.7	tr	tr	tr	tr
NH_3	–	2.8	3.3	2.8	2.8
Lys	18.1	33.0	34.1	32.2	39.3
His	10.1	7.3	12.3	8.5	11.5
Trp*	2.0	tr	tr	tr	tr
Arg	1.7	1.9	2.2	1.7	2.4
other aa	–	<1.0	<1.3	<1.6	<2.0
aa recovered, μ gm/mg (after hydrolysis)		366	396	461	507
pH of 10 mg reactant/ml after 30' at 120°		6.0	5.5	5.0	4.5
relative biuret value, % (bovine serum albumin =100%)		55	55	53	40
N-Kjeldahl, %		13.2	12.3	12.8	13.3

* Tryptophan is largely destroyed by acid hydrolysis.

TABLE IV. Amino Acid Composition of Polymers (A 98 a–d) and the Corresponding Reactant Mixture (gm %). All values in gm %; tr corresponds to values below 0.5%.

a amino acid	b reactant mixture	c A 98 a	d A 98 b	e A 98 c	f A 98 d
Asp	1.3	3.1	3.5	1.9	1.8
Thr	1.1	tr	tr	tr	tr
Ser	26.7	tr	4.2	8.0	8.3
Glu	10.1	19.5	19.1	8.0	5.3
Pro	1.1	tr	tr	tr	tr
Gly	0.7	1.4	1.5	2.2	2.1
Ala	0.7	11.4	10.6	8.7	6.9
Val	1.0	tr	tr	tr	tr
1/2 Cys	18.2	2.8	6.8	5.7	0.9
Met	1.5	tr	tr	tr	tr
Ile	1.2	0.8	tr	tr	0.9
Leu	1.2	tr	tr	tr	tr
Tyr	1.7	tr	1.2	0.9	tr
Phe	1.6	tr	1.1	1.0	tr
NH$_3$	–	1.6	3.1	2.4	1.6
Lys	18.2	42.0	36.4	50.1	62.1
His	10.1	9.5	5.7	3.7	2.2
Trp*	2.0	tr	tr	tr	tr
Arg	1.6	2.8	2.6	3.1	3.6
aa recovered, μ gm/mg (after hydrolysis)		320	301	361	200
pH of 10 mg reactant/ml after 30' at 120°		3.0	7.0	9.0	11.0
relative biuret value, % (bovine serum albumin=100%)		48	57	57	40
N–Kjeldahl, %		15.7	14.5	13.9	12.8

* Tryptophan is largely destroyed by acid hydrolysis.

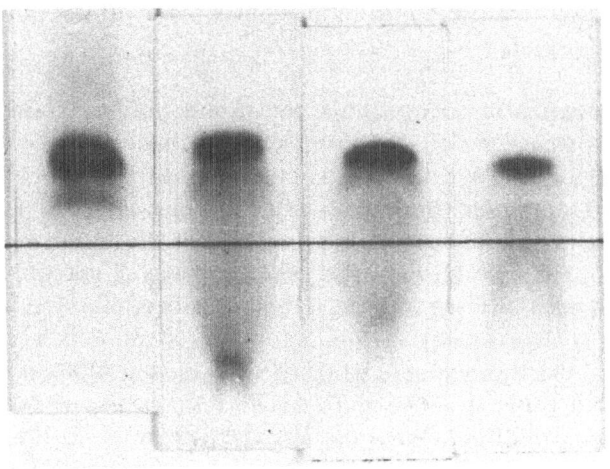

A98a A98b A98c A98d

Fig. 7. Electrophoresis of polymers A 98 a-d
(Table IV). The mobilities of the two basic
fractions relative to lysozyme are 0.46 and 0.2.
The mobility of the acidic fraction is 1.5 relative to bovine serum albumin. Applied amounts:
A 98a: 200 μgm.; A 98b: 300 μgm.: A 98c:
400 μgm.; A 98d: 600 μgm. (pH 8.6; cellulose
acetate).

points of 4.22 and 4.6, respectively. The number of meq. (per gram of
material) of carboxylic groups are 3.0 and 3.6, the meq. of amino groups
are 1.63 and 3.10, respectively. These and related results obtained with
other acidic proteinoids are in qualitative agreement with their electrophoretic behavior. The increasing amounts of aspartic acid which are
incorporated into the polymers, however, do not correlate with the quantity and the mobility of the acidic fraction. Each of the polymers A 96
a-d contains significant amounts of a slightly basic fraction. Its proportion only slightly decreases from A 96 b to A 96 c (Figure 6), in spite of
a significant increase in aspartic acid (Table III). As can be seen (Table

III), the increase in aspartic acid is in part compensated by additional incorporation of basic amino acids. In addition, incorporation of serin, cysteine, and glutamic acid is significantly lowered with increase in aspartic acid.

The incorporation of amino acids into a polymer can also be controlled by the presence of inorganic acids or bases among the reactants as shown for polymers A 98 a-d. These polymers were obtained by polymerization of an amino acid mixture containing relatively high proportions of cystine. Prior to the polymerization the mixtures were acidified or alkalified by addition of small amounts of sodium hydroxide or hydrochloric acid (for the pH of 10 mg reactant mixture in 1 ml water see Table IV). The amino acid composition of the 4 unfractionated products (A 98 a-d), their respective biuret values and N-contents are also summarized in Table IV. The incorporation of lysine and serine increases under alkaline conditions, whereas the incorporation of histidine, aspartic and glutamic acid decreases. Optimal incorporation of cystine appears to proceed when the reactant mixture is close to neutral pH. The polymers appear to be less polypeptide-like when they are synthesized under alkaline conditions. Stainability with amido black 10 B, biuret value, and N-Kjeldahl value are significantly decreased under such conditions. In Figure 7 are compared the electropherograms of polymers (A 98 a-d). An acidic fraction is present in all polymers of this series even if the overall contents in lysine are increased to 64% (polymer A 98 d, Fig. 5). Related results as described for this series have been obtained also in other instances. The general inference is that the incorporation of lysine is strongly increased when the reactant mixture is alkalified; in spite of increasing incorporation of lysine, the isoelectric points of the basic fractions thus obtained are close to pH 8.5 or lower.

DISCUSSION

The results presented here show that thermal polymers of amino acids can be separated in electrophoretically definable fractions. These fractions are biuret-positive and stainable with amido black 10 B, though to a lesser degree than proteins. The amino acid composition of the polymers can be controlled by the composition of the reactant mixture (proportions of amino acids and other additives). If the reactant mixture is alkaline, higher amounts of lysine but not histidine are incorporated. However, most of the amino groups of lysine are chemically bound. The nature of a significant proportion of these bounds is not peptidic for stoichiometric reasons. Even in proteinoids with predominance of lysine the titratable amino groups only slightly exceed the carboxylic groups.   The number of amino groups released after hydrolysis, however, strongly exceeds the number of carboxylic groups. Lysine-rich polymers, therefore, are less basic than comparable proteins or synthetic polymers. In thermal polymers which contain more than 20% lysine the total number of titratable amino groups (groups with pK above 8.0) is smaller than expected from the amino acid composition of the hydrolyzate. With increasing amounts of lysine being incorporated into proteinoids the number of titratable groups with pK of about 7 increases, whereas the number of groups with pK-values in the neutral range could indicate an increased incorporation of histidyl residues.   But the results of amino acid analysis do not bear on this assumption. We therefore suggest that imidazole-type residues are formed by condensation of amino groups with two adjacent peptide bonds during the thermal condensation according to the equation below. More detailed research is required, however, to resolve the reaction mechanism.

$$
\begin{array}{ccc}
\underset{\underset{O}{\|}}{-C}-NH-\underset{\underset{\underset{O}{\|}}{C}}{\overset{\overset{R_1}{|}}{CH}}-NH- & -H_2O & R_1-C=C-NH- \\
+ & & \\
\underset{\underset{R_2}{|}}{NH_2} & +H_2O &
\end{array}
$$

Polymers which contain high proportions of aspartic acid or glutamic acid exhibit to a lesser degree an acidic character than comparable proteins.   This phenomenon is likely related to the reaction of the β-carboxylic group of aspartate with a neighboring peptide link, to give an imide compound (16).

Electrophoretically homogenous polymers can be further fractionated, e. g., by chromatography on DEAE-Sephadex or precipitation with $(NH_4)_2SO_4$. The evaluation of such fractionations, however, is sometimes complicated by secondary reactions.   One example is the conversion of proteinoid A 61 into two electrophoretically more basic fractions on contact with DEAE-cellulose or by 0.1N HCl. We assume that the hydrolysis of reactive bonds, perhaps according to the back reaction of the above equation, is responsible for the increase in electrochemically active amino groups. Correspondingly, the titrigraphic results (15) show that lysine-rich proteinoids are easily converted into more basic polymers under mild hydrolytic conditions (e.g., 2 hours at pH 4.2 and 100°). An overall reaction as indicated by the above equation could account for this effect.

SUMMARY

1. Thermal polymers of α-amino acids have been synthesized in the absence and in the presence of sodium polyphosphate. In some instances the amino acid composition of the polymeric end products has been varied by condensation under acidic, neutral or alkaline conditions or by adding increasing amounts of aspartic acid to the reactant mixture.

2. The thermal polymers were subjected to electrophoresis in cellulose acetate or acrylamide gel. In all instances the polymers migrated in three or less fractions. The electrophoretic behavior was found to be in agreement with the number of titratable acid and basic groups. However, the isoionic points are mostly lower than anticipated on the basis of the amino acid composition of the polymers.

3. DEAE-chromatography reveals a larger number of proteinoid fractions than electrophoresis. This increase in the number of separable fractions is partly caused by a hydrolytic liberation of polar groups during chromatography. The various fractions obtained by DEAE-chromatography of a given proteinoid sample usually represent a family of polymers with respect to their amino acid composition. Stronger differences have only been found for fractions which are clearly distinguishable by electrophoresis.

4. The fractionation of proteinoids by precipitation with $(NH_4)_2SO_4$ is applicable for the production of electrophoretically homogenous fractions.

5. The electrochemical properties, particularly of basic proteinoids, do not well agree with their amino acid composition. A mechanism which explains the reversible disappearance of free amino groups is proposed.

REFERENCES

1. Hayakawa, T., Windsor, C. R., and Fox, S. W., Arch. Biochem. Biophys. 118, 265 (1967).
2. Fox, S. W., and Nakashima, T., Biochem. Biophys. Acta 140, 155 (1967).
3. Vestling, C., in S. W. Fox and K. Harada, J. Am. Chem. Soc. 82, 3745 (1960).
4. Usdin, V. R., Mitz, M. A., and Killos, P. J., Arch. Biochem. Biophys. 122, 258 (1967).
5. Fox, S. W., Harada, K., Woods, K. R., and Windsor, C. R., Arch. Biochem. Biophys. 102, 439 (1963).
6. Dose, K., and Zaki, L., Z. Naturforsch. 26b, 144 (1971).
7. Dose, K., and Zaki, L., in "Chemical Evolution and the Origin of Life" (R. Buvet and C. Ponnamperuma, eds.), Molecular Evolution, Vol. 1, North-Holland Publishing Company, Amsterdam, 1971.
8. Fox, S. W., and Nakashima, T., unpublished results.
9. Weichselbaum, T. E., Amer. J. Clin. Path. (Tech. Sec.) 10, 40 (1946).
10. Gundlach, H. G., Moore, S., and Stein, W. H., J. Biol. Chem. 234, 1754 (1959).
11. Cannan, R. K., Chem. Reviews 30, 395 (1942).
12. Alberty, R. A., in "The Proteins" (H. Neurath and K. Bailey, eds.) Vol. 1, Part A, p. 484, Academic Press, New York, 1953.
13. Determann, H., "Gelchromatographie," Springer-Verlag, Berlin-Heidelberg-New York 1967.
14. Maurer, H. R., "Disk-Elektrophorese," Walter de Gruyter & Co., Berlin, 1968.
15. Dose, K., and Risi, S., unpublished results.
16. Rohlfing, D. L., Ph. D. dissertation, Florida State University, 1964.

Received 2 August, 1971

STEREO-ENRICHED POLY-α-AMINO ACIDS: SYNTHESIS

UNDER POSTULATED PREBIOTIC CONDITIONS

Duane L. Rohlfing and Clarence E. Fouche, Jr.

Department of Biology
University of South Carolina, Columbia, S. C. 29208

INTRODUCTION

In 1954, Fox and Middlebrook reported (1) that some amino acids common to protein copolymerize thermally, under simulated prebiotic conditions. Many subsequent studies (reviewed in references 2 and 3) have extended the thermal method to produce oligo- or hetero-tonic (4) polyamino acids [termed <u>proteinoid</u> when some of each of the proteinous (5) amino acids are incorporated]. The thermal polymers exhibit many properties in common with present-day protein, including molecular weight range, catalytic activity, non-randomness, selective associations, and morphogenecity. They are regarded as models for prebiotic protein (2 - 4).

The extent to which L-amino acids may be racemized during thermal polymerization has not been systematically studied. The few pertinent reports have shown that aspartic acid and lysine are extensively racemized (6, 7), that glutamic acid and isoleucine are racemized to a lesser extent (6, 8), and that the polymers themselves can be optically active (6, 9, 10). Additional information on this subject would provide another basis for comparison of the polymers with contemporary proteins, which of course are almost exclusively comprised of L-amino acid residues. Also, inferences concerning the origins of stereo-enriched primitive protein could result. In the current study, which amplifies preliminary reports (11, 12), a systematic survey was made of the extent of racemization of individual L-amino acids during thermal copolymerization with L-glutamic acid and (separately) with L-lysine.

MATERIALS AND METHODS

L-Lysine free base was from Sigma Chemical Corporation; all other L-amino acids were from Calbiochem.   DNP-L-amino acids were from Nutritional Biochemicals Corporation.

Two series of polymers were prepared by the general method of Fox (8); two-component polymers were chosen for analytical and interpretive simplicity.   In one series, termed "Glu:X series," each of sixteen L-amino acids was heated separately under a stream of nitrogen gas with L-glutamic acid (1:1 molar ratio, 170°C, 4 hours).   In the other series, termed "Lys:X series" each amino acid was heated separately with L-lysine free base (4:7 molar ratio, 195°C, 1.5 hours).   The crude products were dissolved in dilute NaOH (Glu:X series; a few preparations were not completely soluble) or dilute acetic acid plus NaOH (Lys:X series).   They were then dialyzed against running distilled water and lyophilized after the removal of any insoluble material.   Color intensities of the polymers of the Glu:X series were determined at 450 nm in 1.3 $\underline{N}$ NH$_4$OH with a Beckman DB-G recording spectrophotometer.

Polymers were hydrolyzed for amino acid analysis (13; Beckman Model 120-C analyzer) in 6 $\underline{N}$ HCl, under N$_2$ in sealed tubes, 105 - 110°C, 2 days (3 days for the Lys:X polymers).   Basic hydrolysis (14; tubes were nitrogen-flushed, instead of evacuated) was used for determinations involving tryptophan.   Nascent amino acids (present in hydrolyzates but not in initial reactants) were tentatively identified by their elution times on the analyzer.

All polarimetric measurements were made at room temperature (ca. 25°C) at the Na$_D$ line with a Rudolph Model 80 polarimeter equipped with an oscillating polarizer.   At least five measurements, in 1.0 or 2.0 dcm tubes, were made for each intact polymer or each constituent of the Lys:X series, and at least eight for each sample from the Glu:X series (six and seven measurements in two cases).   (The standard deviations (15) in Tables II and III indicate the precision of the polarimetric technique, and do not reflect possible errors in the determinations of concentration.)

Specific rotations of intact polymers were determined in 1.3 $\underline{N}$ NH$_4$OH (Glu:X series, conc. = 0.2 - 1.4 wt. % except in three cases where the high color intensity dictated lower concentrations), or in 0.1 $\underline{M}$ phosphate buffer, pH 6.8 (Lys:X series, conc. = 0.5%; distilled water was used for the Lys:His and Lys:Tyr polymers).   Polymer concentrations

were determined gravimetrically. Preparative quantities of constituent
amino acids of the Glu:X series were isolated from hydrolyzates of both
the undialyzed and dialyzed products, by using ca. 0.8 x 30 cm^3 Dowex-1
columns (16) and eluting with 0.5 N acetic acid. After being evaporated
to dryness, the isolated amino acids were dissolved in a solvent suitable
for polarimetry (either distilled water, 5 N HC1, or glacial acetic acid).
Aliquots of each solution were used after suitable dilution to determine
the amino acid concentrations (17), which ranged from 0.1 to 2.8 wt. %.
For the Lys:X series, hydrolyzates of the dialyzed polymers were dini-
trophenylated (18) and the resulting DNP-amino acids were separated on
silica gel G plates (0.75 mm thick), using chloroform:methanol:glacial
acetic acid (95:5:1, v:v:v) as the solvent. The specific rotations of the
DNP-amino acids, after being eluted from the gel, were determined in
1.0 N NaOH (glacial acetic acid for DNP-glutamic acid and DNP-cystine).
Aliquots of these solutions were used after dilution to spectrophotometrically determine the concentration of DNP-amino acid, authentic DNP-
amino acids being used as standards. Concentrations ranged from 0.014
to 0.13 wt. %. Authentic glutamic acid and lysine, when dinitrophenyla-
ted and processed as above, showed specific rotations within 2% of the
literature values.

     The percentage of optical purity (the inverse of the percentage of
racemization) of amino acids was calculated from the ratio of specific
activity of the sample to that of published values (19, 20) for authentic
L-amino acids or DNP-L-amino acids.   The optical purities of lysine
from four samples were determined by Glenn Pollock by a gas-chroma-
tographic technique (21). Whether the hydrolytic procedures contributed
to racemization (or decomposition) of amino acids was not determined;
optical purities of the amino acids in polymeric form thus could be somewhat higher (22) than those indicated in the tables.

                          RESULTS AND DISCUSSION

     Yields and Color Intensities of Polymers. Yields (Table I) of water-
soluble, nondiffusible polymers of the Glu:X series were with three exceptions 2% or less, negligible amounts being obtained in five cases.
More than half of the lysine-containing polymers, however, were obtained
in greater than 10% yield, the Lys:His polymer being the highest at   24%
The tendency of lysine but not glutamic acid to homopolymerize (4) may
be pertinent in this context. These relative yields are in general accord
with those reported for other thermal polyamino acids that contain pre-

TABLE I. Yields, Color Intensities, and Compositions of Polymers

Copolymer of Glu or Lys and: (X)	Glu:X Series				Lys:X Series		
	Yield, wt %[a]	Color Intensity[b]	Amino Acid content[c]		Yield, wt %[a]	Amino Acid content[c]	
			Glu	X		Lys	X
Lys	0.3	3	48	47	Not Prepared		
His	tr[d]	–	–	–	24	72	22
Arg	tr	–	–	–	14	71	26
Asp	16	1	15	84	1	74	16
Thr	1.5	80	59	3	8	62	2
Ser	3.6	470	63	2	4	61	2
Glu	tr[e]	–	–	–	15	58	35
Pro	tr	–	–	–	1	16	70
Ala	tr	–	–	–	16	68	25
(Cys)½	6.0	387	52	tr	3	19	73
Val	0.6	14	40	57	13	87	10
Met	0.7	22	39	42	19	77	18
Ile	0.6	5	42	32	14	85	11
Leu	1.4	7	28	69	15	82	15
Tyr	1.6	13	25	71	12	83	13
Phe	1.5	3	37	61	3	64	30
Trp	2.0	12	34	64	3	72	3

[a]Soluble, dialyzed material. Percentages of insoluble material after dialysis for the Glu:X series were: Glu:Thr, 0.5; Glu:Tyr, 11.0: Glu:Phe, 5.5; and Glu:Trp, 0.2.

[b]Absorbance at 450 nm per gram of soluble polymer per liter.

[c]Moles per 100 moles of amino acid. Minor amounts of nascent Ala and Gly were common (1 - 2% for the Glu:X series, up to 7% for the Lys:X series). Major nascent constituents were: Glu:Thr, 28% α-amino-n-butyric acid; Glu:Ser, 32% Ala; Glu:Cys, 35% Ala, 8% Gly; Glu:Met, 11% methionine sulfoxide, 7% Asp; Glu:Ile, 24% alloIle; Lys:Thr, 16% α-amino-n-butyric acid, 16% Gly; Lys:Ser, 30% Ala; Lys:Ile, 5% alloIle; Lys:Trp, 8% Gly, 9% Glu.

[d]tr = trace; dashes = insufficient sample for measurement.

[e]Only Glu was present in the reactants.

dominantly dicarboxylic amino acids or lysine (8, 23).   The very low yields of several of the polymers prohibited subsequent characterization.

The absorption of the Glu:X polymers increased with decrease in wavelength from 600 to 350 nm, with distinct peaks or shoulders being noted (at ca. 580 nm) only with the polymers containing phenylalanine, leucine, and isoleucine. Color intensities at 450 nm (Table I) were very large for the Glu:Cys and Glu:Ser preparations, and moderately large for the Glu:Thr and Glu:Met samples ; this   coloration was accompanied by extensive decomposition of the "second" amino acid; see below.

Amino Acid Compositions. Amino acid analyses were conducted on hydrolyzates of all undialyzed products (which had received no purification or fractionation other than loss of volatile material during synthesis) of the Glu:X series. Except for five cases mentioned below, each crude preparation contained nearly identical amounts (moles per 100 moles of amino acid) of glutamic acid and the "second" amino acid, the greatest deviation from equality being the 53% glutamic acid, 47% leucine noted for that preparation. Because the reactants contained equimolar proportions of the two amino acids, the near-equality of values in the crude product indicates that either little or no decomposition occurred during the heating of these amino acids, or that glutamic acid and each of the other amino acids (the five exceptions below excluded) decomposed to the same extent. The former possibility is regarded as the more probable . However, extensive decomposition of serine, threonine, cystine, and tryptophan is indicated by the finding in the undialyzed products of only 1, 1, < 1, and 10 mole %, respectively, of these amino acids. Methionine was somewhat decomposed, 39% being found. The serine, threonine, and cystine undialyzed preparations contained 18% alanine, 18% α-amino-n-butyric acid, and 17% alanine, respectively, indicative of reductive processes. [These amino acids were identified only by their elution times. However, other amino acids that elute in similar times (e.g., cystine and α-aminoisobutyric acid near α-amino-n-butyric acid) are regarded as less probable derivatives of the parent amino acid s].

A different pattern was noted (Table I) with hydrolyzates of dialyzed polymers.   Only with the Glu:Lys and Glu:Cys polymers was the percentage of glutamic acid nearly identical to that in the reactants (50%). Glutamic acid was a minor constituent in all cases except for the polymers prepared with serine, cystine, or threonine.   (Reduction was extensive in these cases; hydrolyzates contained 37% alanine, 35% alanine, and 28% α-amino-n-butyric acid, respectively.) The high proportion of aspartic acid in the Glu:Asp sample is typical of one class (4) of thermal polymers.

That the compositions of the products did not closely reflect those of the reactants (or crude products) indicates a selective, or non-random (3), incorporation of amino acids during thermal polymerization.

About 40% of the total isoleucine was of the allo form (stereoisomer undetermined), which indicates partial racemization of only one of the two asymmetric carbon atoms. All polymer hydrolyzates contained a few per cent of nascent amino acids (Table I, footnote c), glycine and alanine being the most prevalent.

Lysine, which constituted 63.5% of the reactants of the Lys:X series, was the major component of all but two of the dialyzed polymers (the cystine and proline preparation; these were the only cases in this series in which the proportion of "second" amino acid was greater in the product than in the reactants). In ten of sixteen cases, the percentage of lysine in the product was greater than that in the reactants, selective incorporation again being indicated. Sixteen per cent each of glycine and α-amino-n-butyric acid was found in the threonine preparation, and 30% alanine in the serine polymer. Nearly half of the isoleucine was in the allo form, connoting partial racemization as in the Glu:Ile preparation. A compound eluting in the glutamic acid region (9% of total amino acids, calculated using the ninhydrin color value for glutamic acid) was noted in the Lys: Trp preparation. Small amounts of nascent alanine and especially glycine were found in nearly all polymers, along with trace amounts of other amino acids.

Optical Properties. The specific rotations (Table II) of intact polymers of the Glu:X series ranged from ca. 0 to -10°. Values up to -17.6° have been obtained for other thermal polyamino acids (6, 10); globular proteins show much higher values, e.g., -63° for albumin (24). No apparent correlation between rotation of polymer and optical purity of constituent amino acids was evident, although the Glu:Asp polymer, largely without optical activity, was comprised largely of racemized aspartic acid. Values for the polymers prepared with cystine, serine, or threonine are unprecise, because of the large color intensity of these polymers.

The optical purities of glutamic acid from undialyzed products (Table II, underlined values) ranged from 16 to 70%. The lowest values resulted when histidine, threonine, serine, proline, alanine, or cystine was present in the reactants. The extent to which glutamic acid was racemized thus depended on the particular amino acid present during thermal copolymerization. (No correlation was evident, however, be-

TABLE II. Rotatory Data Pertaining to Glu:X Copolymers[a]

Copolymer of Glu and:	$[\alpha]_D^{25}$								
	of intact polymer	of glutamic acid from:				of 2nd amino acid from:			
		crude product		dialyzed product		crude product		dialyzed product	
Lys	-7.2 ± 0.8	+23.6 ± 0.1[b]	74[c]	+26.8 ± 1.0[b]	84[c]	+3.7 ± 0.2[b]	14[c]	+4.4 ± 0.5[b]	17[c]
His	-	+9.0 ± 0.1	28	-	-	-1.5 ± 0.1	4	-	-
Arg	-	+11.8 ± 0.2	37	-	-	+3.7 ± 0.2	12	-	-
Asp	-1.0 ± 0.5	+18.8 ± 0.1	59	+7.0 ± 0.3	22	+0.5 ± 0.2	2	-0.1 ± 0.1	~0[d]
Thr	-9.6 ± 4.5	+5.1 ± 0.2	16	+5.5 ± 0.8	17	+2.9 ± 0.7	8[e]	+10.0 ± 5.4	24[e]
Ser	0 ± 27	+6.7 ± 0.1	21	+7.9 ± 0.2	25	-7.2 ± 2.6	?[d,f]	+17.1 ± 0.9	52[f]
Pro	-	+5.7 ± 0.7	18	-	-	-2.2 ± 0.2	3	-	-
Ala	-	+8.1 ± 0.5	25	-	-	+1.6 ± 0.5	5	-	-
(Cys)½	-24.4 ± 44	+10.0 ± 0.3	31	+0.9 ± 1.1	3	-0.6 ± 2.1	~0[d,f]	+5.6 ± 1.5	17[f]
Val	-11.4 ± 1.8	+21.5 ± 0.1	68	+10.3 ± 0.8	32	+13.4 ± 0.2	22	+20.0 ± 0.8	32
Met	-5.5 ± 1.9	+14.1 ± 0.1	44	+16.4 ± 1.0	52	+0.3 ± 0.1	1	+1.0 ± 0.3	4
Ile	-7.9 ± 1.0	+15.5 ± 0.1	49	+11.7 ± 1.0	37	+18.6 ± 0.1	38	+14.0 ± 0.2	29
Leu	-6.6 ± 0.6	+14.2 ± 0.1	45	-5.2 ± 4.8	~0[d]	+1.4 ± 0.1	6	+3.1 ± 1.5	14
Tyr	-9.6 ± 1.1	+16.3 ± 0.2	51	+11.3 ± 0.6	36	-2.8 ± 2.3	28	-6.8 ± 0.3	69
Phe	-3.1 ± 0.6	+17.0 ± 0.1	55	+5.0 ± 1.4	16	-0.9 ± 0.2	2	-4.9 ± 0.2	14
Trp	-6.4 ± 0.7	-1.0 ± 0.4	~0[d]	Not Measured		-1.4 ± 1.4	4	-1.4 ± 1.1	4

aSolvents were: 1.3 N NH₄OH for intact polymers; 5 N HCl for Glu, Lys, Arg, Met, and Tyr; water for His and Phe; glacial acetic acid for the others. Dashes indicate insufficient sample for measurement.

bStandard deviation (15); see Methods section.

cUnderlined values are percentages of optical purity; see Methods section.

dObserved rotation of wrong sign.

eBased on nascent α-amino-n-butyric acid.

fBased on nascent α-alanine.

tween the optical purities of glutamic acid and the pertinent second amino acid.) The range of optical purities of glutamic acid from the dialyzed products was greater [cf. 0 to 84%; cf. reports (6, 8) of 47 and 52%]. In seven of eleven comparable cases in Table II, the glutamic acid from the dialyzed product was appreciably less optically pure than that from the undialyzed product. Because the undialyzed products contained predominantly low molecular weight material (the yields of dialyzed materials were very low), it is concluded that in most cases glutamic acid was more vulnerable to racemization in polymeric form rather than in monomeric or oligomeric form.   Consistent with this is the finding that glutamic acid when heated alone was 72% optically pure.   Glutamic acid forms pyroglutamic acid (pyrrolidone-5-carboxylic acid) on heating , rather than a polymer; the ring structure of pyroglutamic acid, which is opened during peptide-bond formation in copolymerizations (2), may stabilize the α-carbon atom.

Optical purities of the "second amino acid" of undialyzed products from the glutamic acid series were in general less (range, ca. 0 to 35%) than those of glutamic acid.   With the dialyzed products, the range was ca. 0 to 70%.   The optical purity of the "second" amino acid from the dialyzed product was greater than that from the undialyzed product in nine of twelve comparable cases, only two of them being marginal. Thus, unlike glutamic acid, many of the "second" amino acids apparently were less vulnerable to racemization in polymeric form than in monomeric or oligomeric form.

The values of optical purity for isoleucine here and in Table III are probably incorrect, because of the contribution to rotation of the stereoisomer alloisoleucine; the L isomers of the two compounds have nearly identical specific rotations (19). With the threonine, serine, and cystine preparations, optical purities refer to those of nascent α-amino-n-butyric acid, α-alanine, and α-alanine, respectively. That these amino acids were not completely racemic indicates that, in their formation from the hydroxyamino acids, the α-carbon atom did not assume a planar configuration.

Specific rotations of intact lysine-containing polymers ranged from -4.6 to +4.8°, some being near zero. The optical purities of the "second amino acid" from dialyzed products ranged from 0 to 47%. These values may be compared with those of the Glu:X series in eight cases (including the optical purities of glutamic acid from the Glu:Lys and Lys:Glu polymers).   Except for tryptophan, those amino acids with relatively high values in one series also tended to show relatively high values in the other

TABLE III. Rotatory Data Pertaining to Lys:X Copolymers[a]

Copolymer of Lys and: (X)	$[\alpha]_D^{25}$			
	Intact Polymer		DNP Derivative of X From Dialyzed Product	
His	-1.2 ± 0.8[b]		-8.6 ± 6.1[b]	39[c]
Arg	+3.2 ± 0.7		-	-	-
Thr	+4.8 ± 0.6		-	-	-
Glu	-4.6 ± 0.7		-32.4 ± 1.1	43
Ala	+2.4 ± 0.7		+67.8 ± 23	47
(Cys)½	-	-	-45.9 ± 0.6	12[d]
Val	-2.8 ± 0.8		+31.2 ± 7.4	29
Met	+0.6 ± 0.7		0.0 ± 0.0	0
Ile	+1.8 ± 0.8		+36.3 ± 10	43
Leu	+0.8 ± 0.7		+9.0 ± 4.6	15
Tyr	+3.4 ± 0.6		-	-	24[e]
Trp	-	-	-159.7 ± 0.8	46

[a]See Methods section for solvents; dashes indicate no measurement.

[b]Standard deviation (15); see Methods section.

[c]Underlined values are percentages of optical purity; see Methods section.

[d]Based on cystine.

[e]Determined by Glenn Pollock by a gas chromatographic method (21); values of 31, 5, and 14% were found by this technique for Glu, Ala, and Val, respectively, and lysine from four samples was extensively racemized.

series, although sizable differences in actual percentages occurred in some cases. In four tested cases (the preparations containing glutamic acid, tyrosine, valine, or alanine), lysine was found by Glenn Pollock via a different technique (21) to be completely racemized, which is in accord with the report (7) that lysine racemizes rapidly and extensively during thermal homopolymerization.

   Conclusions and Interpretations. This study has shown that several amino acids are not extensively racemized during copolymerization under hypohydrous conditions at 170-195°C. The results, augmented by earlier reports (6, 8), contraindicate the assumption (cf. 25) that amino acids lose their optical activity during thermal polymerization. In no case in this study, however, did a preparation result that was comprised solely of optically pure residues.

   The effect of conditions of synthesis on racemization was not explored. However, one comparison concerning the temperature of heating is available from these data. The optical purity of glutamic acid from the Glu:Lys polymer (prepared at 170°, 4 hours), was appreciably greater (84%), in spite of the longer heating period, than that (43%) of the glutamic acid from the Lys:Glu polymer (195°, 1.5 hours). Although the proportions of amino acids in the reactants was also a variable, these results do suggest that the use of lower temperatures during polymerizations [e.g., < 100°C (26)] could further reduce the extent of racemization.

   The polyamino acids of this study were prepared under postulated prebiotic conditions. That several of the amino acids were not extensively racemized is interpreted to indicate how stereo-enriched (and optically active) primitive polyamino acids could have been formed. Although the technique presupposes an environment enriched with L (or D) amino acids, several methods, e.g., asymmetric forces, selective seeding, selective absorption on asymmetric surfaces, have been proposed (27 - 30) to account for such stereo-enrichment. Incidentally, this study would indicate that optical activity, of proteins or polyamino acids for example, be applied as a criterion for extraterrestrial life (31) only with much caution.

   Stereo-enriched primitive protein might be expected to have had properties different from those of "racemic protein." For example, helicity (which has not yet been demonstrated for thermal polyamino acids) in part depends upon sequences of amino acid residues of the same stereo-configuration (32, 33). The results of this study suggest that helicity should be sought in thermal polymers of the "right" composition (preferably not enriched with lysyl or aspartyl residues because they are extensively racemized).

One difference in behavior between kinds of thermal polyamino acids has been reported that perhaps is explicable in terms of stereo-enrichment.   The thermal polymers show a variety of catalytic activities (4); one catalyzed reaction, demonstrated by Krampitz and coworkers (34), is the amination of 2-oxoglutaric acid, in the presence of an NH_2 donor, to form glutamic acid.   When the catalyst was thermal polylysine, the product was completely racemic. When, however, the catalyst was lysine proteinoid, containing many kinds of residues, the glutamic acid was 74% of the L-form (48% optically pure). In the latter case, an asymmetric catalyst is connoted from thermodynamic considerations.   The results of the current study suggest that several of the amino acids in the lysine proteinoid used by Krampitz et al., may not have been completely racemized, whereas the homopolymer probably was totally comprised of racemized lysine (cf. 7).

In theory, stereo-enriched and catalytically active primitive protein could not only participate in asymmetric syntheses as above, but could also "select" between D and L substrates, yielding, e.g., L product (and leaving behind D substrate).   Alternatively, an L (or D) substrate could exert a selective effect among D, L, or "racemic protein," by being acted upon only by, e.g., "L-protein." These processes would lead to relatively high concentrations of particular enantiomorphic forms that in turn would enhance the opportunities for additional stereoselective interactions. The processes would thus tend to be self-propagating, and would result in an accelerated stereo-enrichment of the environment. The results of this study suggest how the stereo-enriched polyamino acids necessary for such processes may have originated.

ACKNOWLEDGMENTS

We thank Glenn Pollock for gas chromatographic analyses, Robert Bly for use of the polarimeter, and Richard Boan for conducting some of the amino acid analyses. Taken in part from the M.S. thesis of C. E. Fouche, Jr., University of South Carolina, 1971. Supported in part by NASA Grant No. NGR 41-002-025.

REFERENCES

1.   Fox, S. W. , and Middlebrook, M. , Federation Proc.13, 221 (1954).
2.   Fox, S. W. , Nature 205, 328 (1965).
3.   Fox, S. W. , Naturwissenschaften 56, 1 (1969).
4.   Rohlfing, D. L., and Fox, S. W. , Advances Catal. 20, 373 (1969).
5.   Hayakawa, T. , Windsor, C. R. , and Fox, S. W. , Arch. Biochem. Biophys.118, 265 (1968).
6.   Rohlfing, D. L. , Ph.D. Dissertation, Florida State University, 1964.
7.   Heinrich, M. R. , Rohlfing, D. L. , and Bugna, E. Arch. Biochem. Biophys. 130, 441 (1969).
8.   Fox, S. W. , and Harada, K. , J. Am. Chem. Soc. 82, 3745 (1960).
9.   Fox, S. W. , Harada, K. , and Rohlfing, D. L. , in "Polyamino Acids, Polypeptides, and Proteins" (M. Stahmann, ed.), p. 47. University of Wisconsin Press, Madison, 1962.
10. Rohlfing, D. L. , Nature 216, 657 (1967).
11. Fouche, C. E. ; Rohlfing, D. L. , Bull. S. C. Acad. Sci. 33, 56-57 (1971).
12. Rohlfing, D. L. , and Fouche, C. E. , Federation Proc. 30, 1068 Abs. (1971).
13. "Model 120C Amino Acid Analyzer" Instruction manual, Spinco Division, Beckman Instrument Co. , Palo Alto, 1965.
14. Blackburn, S. , "Amino Acid Determination; Methods and Techniques," pp. 23-24. Dekker, New York, 1968.

15. Van Norman, R. W. , "Experimental Biology," pp. 244-246. Prentiss-Hall, Englewood Cliffs, N. J. , 1971.
16. Hirs, C. H. W. , Moore, S. , and Stein, W. H. , J. Am. Chem. Soc. 76, 6063 (1954).
17. Moore, S. , and Stein, W. H. , J. Biol. Chem.211, 907 (1954).
18. Levy, A. L. , and Chung, D. , J. Am Chem. Soc. 77, 2899 (1955).
19. Greenstein, J. P. , Birnbaum, S. M. , and Otey, M. C. , J. Biol. Chem. 204, 307 (1953).
20. Rao, K. R. , and Sober, H. A. , J. Am. Chem. Soc. 76, 1328 (1954).
21. Pollock, G. E. , and Oyama, V. I. , J. Gas. Chrom. 4, 126 (1966).
22. Blackburn, S. , op. cit. , p. 15.
23. Harada, K. , Bull. Chem. Soc. Japan 32,1007 (1959).
24. Jirgensens, B. , "Optical Rotatory Dispersion of Proteins and Other Macromolecules," p. 53. Springer-Verlag, New York, 1969.
25. Lemmon, R. M. , Chem. Revs. 70, 95 (1970).
26. Harada, K. , and Fox, S. W. , in "The Origins of Prebiological Systems" (S. W. Fox, ed.), p. 289. Academic Press, New York, 1965.
27. Hanafusa, H. , and Akabori, S. , Bull. Chem. Soc. Japan 32, 626 (1959).

28. Harada, K., <u>Nature</u> <u>205</u>, 590 (1965).
29. Harada, K., <u>Nature</u> <u>206</u>, 1354 (1965).
30. Harada, K., <u>Nature</u> <u>194</u>, 768 (1962).
31. Stryer, L., in "Biology and the Exploration of Mars" (C. S. Pitten-drigh, W. Vishniac, and J. P. T. Pearman, eds.), p. 141. National Academy of Sciences National Research Council, publication 1296, Washington, 1966.
32. Bresler, S. E., "Introduction to Molecular Biology," p. 47ff. Academic Press, New York, 1971.
33. Shecter, I., Benderly, H., Berger, A., Lotan, N., and Scheraga, H. A., Abstracts of Paper, Div. of Biol. Chem. No. 171A, 154th Meeting, <u>Am. Chem. Soc.</u>, Chicago, 1967.
34. Krampitz, G., Diehl, S., and Nakashima, T. <u>Naturwissenschaften</u> <u>54</u>, 516 (1967).

Received 1 March, 1972

LIGHT ENHANCED DECARBOXYLATIONS BY PROTEINOIDS

A. Wood and H. G. Hardebeck

Waite Agricultural Research Institute
University of Adelaide, Glen Osmond, South Australia
and
Institut fur Anatomie und Physiologie der Haustiere
Universitat Bonn, Bonn, Germany

INTRODUCTION

The work reported here was carried out during the period 1966-67 when the authors were privileged to work at the Institute of Molecular Evolution under Professor Fox.   His enthusiasm for the thermally-prepared proteinoid as a model for the origin of prebiotic protein led us to investigate whether, in addition to its catalytic and structural properties, it could function as an energy absorbant.   This stemmed from the observation that, despite prolonged dialysis, the proteinoid remained pigmented.

The preparation and some of the characteristics of thermal proteinoids have been described in considerable detail (1-3).   The intrinsic properties of these polyamino acids have suggested a model of the assembly of a self-ordered structure preceding the cell (4).   Pursuing further the analogy between the polypeptide nature of proteinoid and natural proteins, synthetic thermal polyamino acids have been shown to possess catalytic activity for a variety of substrates. Rohlfing and Fox (5) found that certain proteinoids accelerated the hydrolysis of p-nitrophenyl acetate and were able to ascribe the catalytic action to the presence of aspartoylimide and imidazole groups in the polymer.   Rohlfing (6) was later able to study the catalytic breakdown of oxalacetate on a manometric scale.   Fox and Krampitz (7) described a two-stage reaction for the breakdown of glucose via the formation of glucuronic acid.   Hardebeck et al. (8) studied the decarboxylation of pyruvate in considerable detail and have established kinetic data for the reaction.   Aminotransferase activity (9, 10) has been shown to be more particularly facilitated by the more basic, lysine-rich, proteinoids.

However, while simple energy-yielding reactions were feasible in the context of the reducing atmosphere of the earth of earlier times (given a continued synthesis of substrate by electrical discharge, ultra-violet light, heat or thunder), the more complex syntheses required in the living cell have needed, in addition to enzymes, the direct harnessing of energy and the provision of general 'energy-donor' molecules such as ATP and $NADH_2$. We have therefore investigated whether the pigmented thermal proteinoids could function as an energy-transferring matrix. The transfer of absorbed light energy to the breakdown of a substrate molecule would constitute a simple test, and such energy-transferring systems have been described using dyes (11). We initially investigated the complex formed between acidic proteinoid and crystal violet, but the addition of the photodynamic dye was later found to be unnecessary.

METHODS

The preparation and purification of proteinoids has been fully des-cribed previously (1-3). The designation "2:2:1" in this paper refers to the polyamino acid product prepared from two parts by weight glutamic acid to two parts aspartic acid to one part of an equimolar mixture of sixteen other amino acids. (Specific numerals after the designation re-fer to particular preparation, e.g., 2:2:1/7.) The end product contained at least 50% of material of molecular weight exceeding 1500 as estimated by gel filtration procedures. The polymers differed considerably de-pending on temperature and length of heating. Other polymers of differ-ing amino acid content were donated by Dr. T. Nakashima and Dr. T. V. Waehneldt (12). A product, prepared by Mr. A. Weber (13) from glycine alone, was used in some experiments; others were prepared from amino acid mixtures of the proportions obtained by Harada and Fox (14) from the passage of gases over heated sand.

Experimental solutions of proteinoids plus ^{14}C-labelled substrates (approx. 5-8 μCi), usually 3.0 ml in volume, were maintained in a water bath under anaerobic conditions by a slow flush of water-saturated nitrogen. The stream swept evolved ^{14}C-carbon dioxide (from decar-boxylation of the substrate) to a trap of methanolic hyamine hydroxide (Packard) or 0.1 N sodium hydroxide. A Millipore filter (0.22 micron) was placed in the gas stream before the sample tubes, and all solutions were autoclaved or Millipore filtered.

The samples were contained in test tubes of Corning 7910 glass sealed with serum caps pierced by hypodermic needles for the passage of nitrogen.   A 300 watt tungsten filament bulb irradiated the samples from a distance of 10 cm; heat was removed by a filter 5 mm thick of 10% aq. $CuSO_4$. The samples themselves were immersed in a water bath circulating tap water.

The hyamine trap was changed periodically and its contents were immediately transferred to standard glass spectrometer vials and mixed with 10 ml of a solution prepared from 22.8g POP and 2.28g POPOP per gallon of analytical grade toleune.   The vials were counted in a Packard model 314 E scintillation spectrometer and the efficiencies calculated by channel ratios method from a calibrated quenching curve.

To permit a correction to be made for volatile substrates being carried to the trap, the identity of $^{14}CO_2$ was checked by using a barium carbonate precipitation procedure (8).   In this case the trap contained 0.1 N sodium hydroxide, which was stirred with an excess of 5% barium chloride and boiled before the slow addition of carrier sodium carbonate. The precipitate was washed with hot water and with methanol before carbon dioxide was liberated by the addition of excess hydrochloric acid. The evolved gas was trapped in the usual manner in hyamine hydroxide.

Chromatograms were run with aliquots of experimental solutions to identify intermediate and end products.   Whatman No. 3MM paper was used with ascending 1-butanol, ethanol, water (70:10:20 v/v/v).   Other solvents were used to confirm identifications against authentic compounds. Kodak Medical Blue Brand X-ray film was used to locate radioactivity on the chromatograms.

Radioisotopes were obtained from New England Nuclear or Nuclear Chicago and used directly.   After the radioactive substrate was added to the proteinoid solutions or controls, the whole system was flushed with nitrogen for 18-24 h in darkness to remove any accumulated products of radiolysis prior to the start of the experiment.

Proteinoid spectra were traced on a Beckman DK-2 photometer before and after the irradiation period.   At the conclusion of the experiments samples were tested for contamination by both aerobic and anaerobic microorganisms.

Since pyruvic acid is of relatively low molecular weight, a correction was required for its carryover into the trap. This was obtained by the comparison of $2\text{-}^{14}C$ pyruvate and $3\text{-}^{14}C$ pyruvate.

The background rate of pyruvate transfer in the gas stream at pH 7.1 was found to be of the order of $1.0 \times 10^{-4} \mu M/h$. In a simultaneous assay less than 1% of these counts in darkness were found to be due to CO_2, as shown in Table I. The level of counts in experimental samples was frequently many times the 'background' rate, but a correction was applied to all samples of $1 \times 10^{-4} \mu M/h$.

TABLE I

Volatility of Pyruvate

Substrate	Specific Activity	Possible Radioactive Products	Total dpm h^{-1}	μM h^{-1} $x\ 10^4$	$\mu M h^{-1}$ $x\ 10^4$ as CO_2
$2\text{-}^{14}C$ pyruvate	10.0mC/mM	pyruvate, acetate	2443	1.10	0.01
$3\text{-}^{14}C$ pyruvate	6.5mC/mM	pyruvate, acetate	1270	0.88	0.01
$1\text{-}^{14}C$ pyruvate	various	pyruvate, CO_2			

RESULTS

A comparison of different polymers for catalytic activity was desirable on the basis of equal optical densities at a given wavelength, but the control solution of unheated amino acids showed no absorption of visible radiation. Comparison was therefore made on an equal weight basis. Solutions with the highest optical densities did not, in any case, necessarily give the highest rates of decarboxylation (15).

Table II shows that the nature of the proteinoid is the largest variable affecting the decarboxylation of citrate. The dialyzed 2:2:1 proteinoid is more optically dense at 400 $m\mu$ than an equal weight of its parent crude

TABLE II

Decarboxylation of Citrate -1, 5 -^{14}C. Proteinoid solutions of 1.5 mg/ml irradiated by 300 watt tungsten at 10 cm in Britton-Robinson buffer pH 7.1. Control: equimolar mixture of 18 monomeric amino acids. Rate 760 dpm/h. illuminated.

Proteinoid	OD 400 mu	Excess over Control dpm/h	Excess over Control per OD$_{400}$ unit
2:2:1 dialysed	.730	623	854
2:2:1 crude	.220	446	2030
2:0:1	.530	291	549
1:0:1	.930	304	327
1:0:2	.950	181	190
0:0:1	.970	130	134

product, suggesting a firm binding of pigment to the high molecular weight portion of the polyamino acid product.

Since tryptophan has high absorption in the near UV and has been indicated as the amino acid most likely to be involved in energy transfer, either as a monomer or as a constituent of protein (16, 17), its contribution to the decarboxylation of pyruvate was investigated. Table III presents results for the activity of various solutions (pH 7.1) containing equal weights of tryptophan in solution, assuming that approximately 5% of a tryptophan-tyrosine-phenylalanine-cysteine-glycine-16 amino acid mix (20:20:10:10:30:10 molar percent) was incorporated as tryptophan (number 4 in Table III) as indicated by analysis of hydrolyzates. Sample 5 contains a much smaller amount of tryptophan than the others as its solubility was too low to bring it to the required concentration. It was still almost 10 times as active as a saturated solution than the next most active sample,

TABLE III

Contribution of Tryptophan to the Decarboxylation of $1\text{-}^{14}C$ Pyruvate during Irradiation.   Approximately equal weight basis for tryptophan, except 5 & 7. Irradiation as in methods.

Proteinoid	Corrected Rates			
	mg/ml	dpm/h	mµM/h	Ratio
1 Control, monomeric amino acids	1.68	2770	.034	1
2 Free tryptophan	0.073	2990	.045	1.32
3 Free tryptophan + phenylalanine + tyrosine	0.255	2900	.040	1.18
4 tryptophan-rich polymer	0.33	7100	.244	7.18
5 2:2:1 proteinoid, crude	1.47	45940	2.123	62.44
6 *TW1 lysine rich proteinoid	1.147	5570	.170	5.0
7 Control *TW3 proteinoid	1.47	6770	.228	6.71

* Prepared by Dr. Waehneldt.   TW1 contains 50% lysine, the remaining 50% from 17 other amino acids. TW3 contains 50% lysine and 50% other amino acids, omitting benzenoid acids.

which was itself tryptophan free. The polymer containing augmented tryptophan was also six times more active than the same weight of free acid. We concluded that at the wavelengths of light used in these experiments, tryptophan is not a primary requirement for decarboxylating activity.

The result of carrying out the irradiations of $1\text{-}^{14}C$-pyruvate containing solutions at two different pH values is shown in Fig. 1.   The pH optimum for spontaneous and proteinoid catalyzed decarboxylation of py-

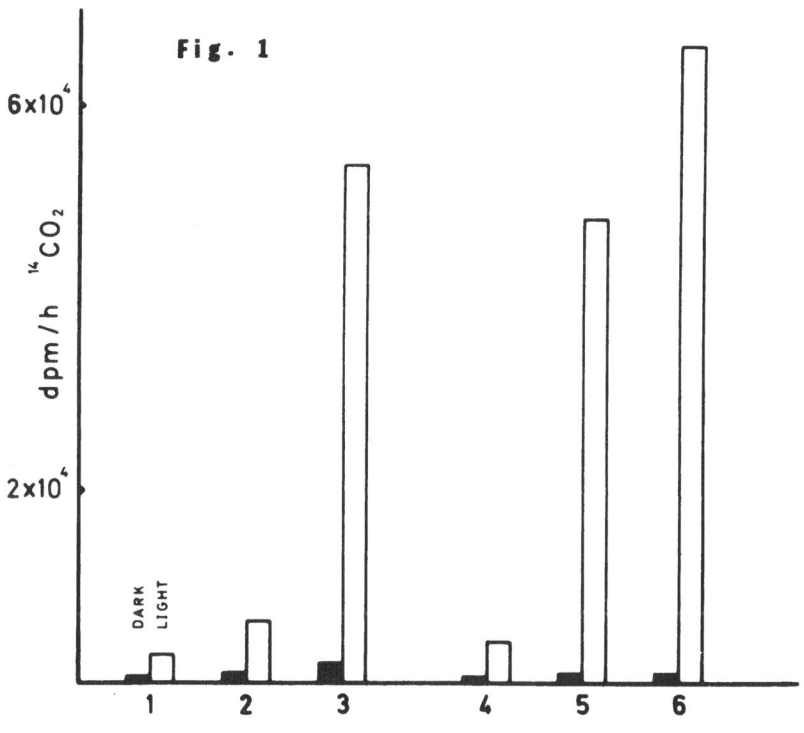

Photochemically Enhanced Decarboxylation
of Na-Pyruvate-1-^{14}C.

Key

1. 2:2:1-amino acid mixture,   pH 3.5
2. Lysine-rich proteinoid (TW 1), pH 3.5
3. 2:2:1-proteinoid (AW 7),   pH 3.5

4. 2:2:1-amino acid mixture,   pH 7.1
5. Glycine polymer (ALW),   pH 7.1
6. 2:2:1-proteinoid (AW 7),   pH 7.1

Data corrected to spec. act. of 10 mc/mM.

ruvate (8) is near 8.3, which accounts for the control at pH 7.1 being more
reactive than that at pH 3.5 (samples 1 and 4). Although the difference of
the controls between darkness and light is not readily explained, it can be
seen that there is both a small effect of proteinoid in darkness and a very
large enhancement of decarboxylation in the light.   There is also indication of some specificity of proteinoid, e.g., compare samples 2 and 3.
The effect of switching the light on and off was very marked.   Although
the system was rather slow to equilibrate (3-4 h) from light to dark, the
result of switching on the light was immediate.   Over several alternating

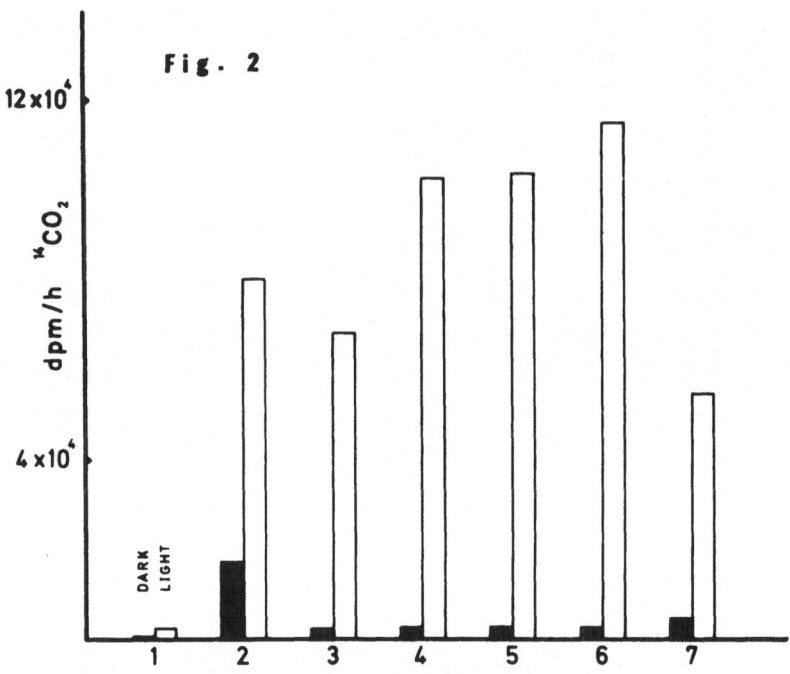

Photochemically Enhanced Decarboxylation
of Na-Glyoxalate-1-^{14}C.

Key

1. Glyoxalate in buffer solution only

2. 2:2:1-proteinoid, prepared at 130° for 24 h
3. 2:2:1-proteinoid, prepared at 150° for 24 h
4. 2:2:1-proteinoid, prepared at 170° for 24 h
5. 2:2:1-proteinoid, prepared at 170° for 48 h
6. 2:2:1-proteinoid, prepared at 200° for 7 h (AW7)
7. 1:1:1-proteinoid, prepared at 170° for 11 h (TN)*

   *Prepared by Dr. T. Nakashima.

Data corrected to spec. act. of 10 mc/mM.
Incubation solutions buffered at pH 7.1.

periods of light and dark steady linear rates could be obtained for each
state which were reproducible in subsequent periods.   It should be noted
that the pH is less of a critical factor for the photochemical process as
comparison with the curve in ref. 8 for a simple catalytic reaction shows.

     Data for the decarboxylation of glyoxalate at pH 7.1 is presented in
Fig. 2. The figure is notable for the high dark rate recorded for sample 2,
which may indicate that the particular proteinoid used had a higher than

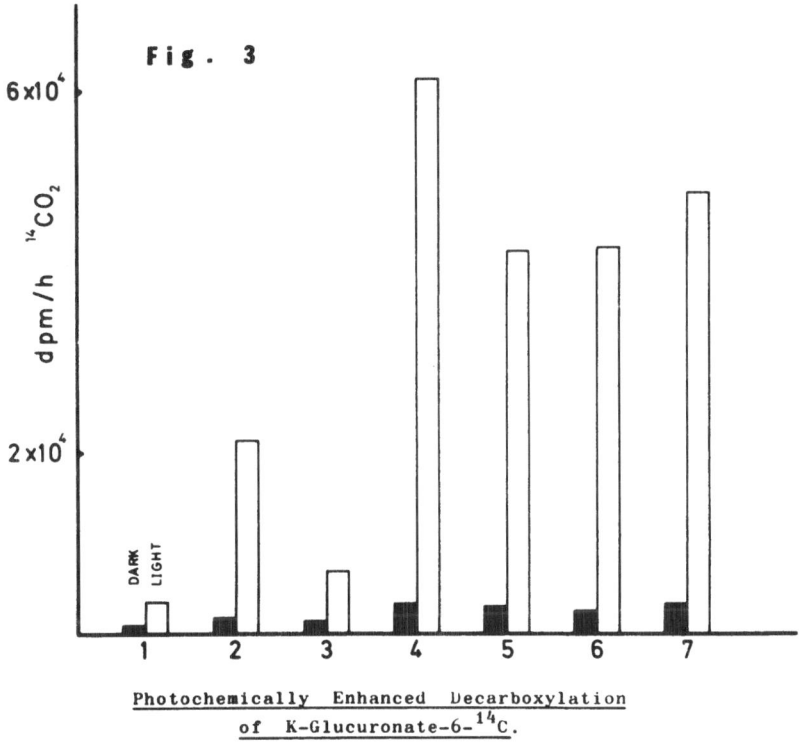

Photochemically Enhanced Decarboxylation
of K-Glucuronate-6-^{14}C.

Key

1. Equimolar mixture of 18 amino acids
2. Glycine polymer (ALW)
3. Lysine-rich ptd., prepared at 180° for 5 h (DHD)
4. Lysine-rich ptd., prepared at 185° for 6 h (HH1)
5. 2:2:1-proteinoid, prepared at 135° for 72 h (AW11)
6. 2:2:1-proteinoid, prepared at 185° for 6 h (HH12)
7. 2:2:1-proteinoid, prepared at 200° for 7 h (AW7)

Data corrected to spec. act. of 10mc/mM.
Incubation solutions buffered at pH 8.0.

usual affinity for the substrate.   Fig. 3 presents a similar experiment,
this time for a third substrate, K-D-glucuronate-6-^{14}C.   This substrate
was investigated, as glucuronate has been suggested as a rate-limiting
intermediate in the proteinoid-catalyzed breakdown of glucose (7).   The
data here suggests that glucuronate decarboxylation responds to irradiation to approximately the same extent as the other substrates tested; however, the most effective proteinoid was lysine-rich, unlike the situation
for pyruvate.

TABLE IV

Decarboxylation of 1-^{14}C Pyruvate by 2:2:1 Proteinoid. Proteinoid 2:2:1/7; 3.90 mg in 3.0 ml total volume, Britton Robinson buffer pH 7.10, pyruvate 0.565 to 582 μM.

Substrate Concentration (S) μM	Rate in Light (V_1) mμ M/h	$\dfrac{S}{V_1}$	% Exhaustion of Substrate
0.565	2.857	197.7	16
6.378	6.95	917.7	3.6
19.945	14.14	1410.5	2.2
58.710	25.58	2295.2	1.6
194.380	45.47	4274.9	0.98
582.020	74.37	7826.0	0.74

A plot of S against S/V obtained from samples at various concentrations of pyruvate did not yield a straight line, suggesting that a more complex reaction than catalysis is occurring in the light (see Table IV). There is in addition the possibility that light of the required wavelength was limiting, and that there may be a requirement for a degree of molecular proximity before energy transfer becomes feasible. Hardebeck et al. (8) and Durant and Fox (18) have previously been able to obtain straight line plots when using proteinoid as a catalyst without external energy sources. Note that the spontaneous decay rate for pyruvate plus free amino acids in light was 3.2×10^{-6} μmole/min compared to the lowest value here 50.5×10^{-6}.

In a comparison of proteinoids with coloured proteins or protein products, both ferritin and casein hydrolysate gave a sparing effect on the pyruvate substrate, while 2:2:1 and a glycine thermal product again gave rates substantially greater than those of the control.

DISCUSSION

Natural enzymes of present-day organisms, the product of millions of years of selective evolution, accelerate reactions 10^6 to 10^{11} times (19).   The polyamino acids studied here have produced accelerations of the order of 10^1 to 10^2 with possibly 10^4 if optimal conditions could be specified.   The catalytic action of proteinoids, therefore, despite the restriction of product variability imposed upon their nature by our choice of the proportions of the starting material, appears to be intermediate between the spontaneous decay rate of a substance and the same reaction assisted by an enzyme.

The phenomenon of sensitization of a reaction or a living organism by a dye has been well documented.   That photodynamic action in biological systems is fundamentally similar to photosensitized oxidation in vitro was suggested as early as 1904, to be developed later by Gaffron (20) and by Blum (21), the simplified mechanism being given as:

$$D + h\nu \rightarrow D^1$$

$$D^1 + X \rightarrow X^1 + D$$

$X^1 + O_2 \rightarrow X_{ox}$, where D and D^1 represent the inactive and activated dye forms respectively and X and X^1 the inactive and active substance being oxidized.   Blum concluded a review "The dye is taken up by a cell so that it is in intimate association with oxidizable substances upon which the cell is dependent, probably protein.   When the photosensitive molecule is activated by capture of a quantum of radiant energy, it transfers its activation to the substrate which reacts with oxygen, a reaction which would not occur in the normal course of cell structure."

The results presented in this paper suggest that the association of a pigment, substrate and proteinaceous matrix (the proteinoid microsphere (3, 4), whose stability and particulate nature provide the necessary longevity) may have occurred during the evolution of a primitive organized unit.

The contribution of glycine, the simplest amino acid in this scheme, is interesting.   Unpublished work has suggested that glycine is required for the production of the unidentified pigment(s).   The pyrolytic decomposition (22) of glycine has been shown to give a wide range of products, including other amino acids, carboxylic acids (fumaric, oxalic, succinic) plus hydrogen cyanide.   HCN itself has been shown to be active in spontaneous polymerization (23), and, although the presence of peptide bonds is not certain (24), microsphere-like bodies can be formed by the products and they have catalytic properties (25, 26). The HCN polymers (azulmic acids) do, however, yield glycine on hydrolysis. It is, therefore, supposed that the pigment(s) in proteinoid preparations could be of azulmic acid nature originating themselves by spontaneous polymerization of pyrolytic decomposition products of glycine and other suitable amino acids.

The products of polymerization of glycine alone were not, however, superior to other proteinoids as sensitizers in our system. Either the other amino acids are required to produce polypeptides of a considerable molecular weight as an essential reaction surface, or the conditions of synthesis in the presence of other acids may be important. During the thermal process on the dry amino acid powder, glycine and certain other amino acid monomers have been shown to disappear almost completely even prior to the appearance of a liquid phase (15).

Glycine rich polymers, however, and HCN products are deeply coloured, even after exhaustive dialysis, unlike the 2:2:1 polymer which is pale yellow. It is apparent that even very simple compounds such as glycine and HCN, whose origin can be traced directly from a simple gas mixture, can give rise to a wide range of products, some of which are of considerable complexity. The relevance of the findings presented here to prebiological chemistry lies in the coupling, as products of one simple plausible reaction (the heating of dry amino acids), of morphological polypeptide units and an energy-trapping pigment. This could have a bearing on whether the first living organisms were heterotrophs or autotrophs. The metabolism envisaged would be occurring in the second and third evolutionary eras proposed by Gaffron (20), covering the loss of hydrogen from the earth by diffusion to the loss of ultraviolet radiation by the appearance of an ozone barrier in the atmosphere. By era IV the loss of free 'food' is presumed and energy is postulated as being derived by a more complex photochemical reaction involving the splitting of water by a 2-quanta process. Life would probably exist at this point.

REFERENCES

1. Fox, S. W., and Harada, K., J. Am. Chem. Soc. 82, 3745 (1960).
2. Harada, K., and Fox, S. W., Arch. Biochem. Biophys. 109, 49 (1965).
3. Fox, S. W., Harada, K., Woods, K. R., and Windsor, C. R., Arch. Biochem. Biophys. 102, 439 (1963).

4. Fox, S. W., McCauley, R. J., and Wood, A., Comp. Biochem. Physiol. 20, 773 (1967).
5. Rohlfing, D. L., and Fox, S. W., Arch. Biochem. Biophys. 118, 122 (1967).
6. Rohlfing, D. L., Arch. Biochem. Biophys. 118, 468 (1967).
7. Fox, S. W., and Krampitz, G., Nature 203, 1362 (1964).
8. Hardebeck, H. G., Krampitz, G., and Wulf, L., Arch. Biochem. Biophys. 123, 72 (1968).
9. Krampitz, G., Diehl, S., and Nakashima, T., Naturwissenschaften 54, 516 (1967).
10. Krampitz, G., Baars-Diehl, S., Haas, W., and Nakashima, T., Experientia 24, 140 (1968).
11. Posthuma, J., and Berends, W., Biochim. Biophys. Acta 112, 422 (1966).
12. Waehneldt, T. V., and Fox, S. W., Biochim. Biophys. Acta 160, 239 (1968).
13. Weber, A. L., Wood, A., Hardebeck, H. G., and Fox, S. W., Federation Proc. 27, 830 (1968).
14. Harada, K., and Fox, S. W., Nature 201, 335 (1964).
15. Wood, A., and Hardebeck, H. G., A.S.P.P. Proc., Canberra, 1968.
16. Karreman, G., Steele, R. N., and Szent-Gyorgi, A., Proc. Nat. Acad. Sci. U. S. 44, 140 (1958).
17. Chen, R. F., Vurek, G. G., and Alexander, N., Science 156, 949 (1967).
18. Durant, D. H., and Fox, S. W., Federation Proc. 25, 342 (1966).
19. Koshland, D. E., J. Cell. Comp. Physiol. (Sup. 1) 47, 217 (1956).
20. Gaffron, H., in "Origins of Prebiological Systems" (S. W. Fox, ed.), p. 437. Academic Press, New York, 1965.
21. Blum, H. F., "Photodynamic Action of Light and Diseases Caused by Light," Reinhold, New York, 1941.
22. Heyns, K., and Pavel, K., Z. Naturforsch. 126, 97 (1957).
23. Matthews, C. N., and Moser, R. E., Nature 215, 1230 (1967).
24. Labadie, M., Jensen, R., and Neuzil, E., Biochim. Biophys. Acta 165, 525 (1968).
25. Labadie, M., Ducastaing, W., and Breton, J. C., Bull. Soc. Pharm. Bordeaux 107, 61 (1968).
26. Hardebeck, H. G., and Haas, W., Jahresber. Kernforschungsanlage Juelich, 1968, p. 55.

Received 26 October, 1971

SYNTHESES AND CONFORMATIONAL STUDIES OF POLYACIDIC

AMINO ACIDS CONTAINING OPTICAL ACTIVE SIDE CHAINS

Tadao Hayakawa and Hiroyuki Yamamoto

Institute of High Polymer Research, Faculty of Textile Science
and Technology, Shinshu University, Ueda, Japan

INTRODUCTION

The syntheses and physicochemical properties of poly (β-alkyl aspartate) and poly(γ-alkyl glutamate) have been widely investigated (1 - 11). Optical rotatory dispersion (ORD) and circular dichroism (CD) have led to the conclusion that many L-polypeptides have a right-handed helical sense. Poly(γ-alkyl L-glutamate) assumes a right-handed helical conformation in a solvent such as chloroform, while it assumes a disordered conformation in a solvent such as dichloroacetic acid (DCA) or trifluoroacetic acid (TFA). Polymers of β-methyl and β-benzyl L-aspartates represent exceptions to the general rule and exist as left-handed helices in chloroform and in methylene dichloride solutions. The introduction of a nitro, methyl, chloro, or cyano group into the <u>para</u> position of the aromatic ring in the side chain of poly(β-benzyl L-aspartate) causes a reversal of the left-handed helix of the poly- α-amino acid in chloroform (12, 13). In addition, poly(β-benzyl L-aspartate) is known to undergo a polymorphic transition from the α-helix to the ω-helix when a solid film is heated <u>in vacuo</u> (14).

Poly(β , <u>N</u>-benzyl DL-asparagine) (15) and poly(β, <u>N</u>-diphenylmethyl L-asparagine) (16) have been synthesized via the corresponding <u>N</u>-carboxyanhydrides. The resulting polyasparagine and its derivatives had an average molecular weight of 5000, as determined from their sedimentation and diffusion rates and by an analysis of the amino-nitrogen. Because of their low molecular weights and solubilities, the secondary structures of polyasparagine derivatives have not, however, been reported.

In this paper, poly[β -(l)-menthyl D- and L-aspartates], poly[γ -(l)-menthyl D- and L-glutamates], poly(β , N-benzyl L-asparagine), and poly-[β, N-(d), (l) and (dl)-(1)-phenethyl L-asparagines] were synthesized, and the secondary structures of these polymers have been studied in order to investigate the effect of the contribution of side chains with optically active protective groups to the polypeptide structure (17-20). The syntheses of these polymers were carried out by the polymerization of the corresponding amino acid N-carboxyanhydride (NCA) such as β -(l)-menthyl D-aspartate, β-(l)-menthyl L-aspartate, γ-(l)-menthyl D-glutamate, γ-(l)-menthyl L-glutamate, β, N-benzyl L-asparagine, β , N-(d)-(1)-phenethyl L-asparagine, and β, N-(l)-(1)-phenethyl L-asparagine. Some properties of the polymers are listed in Table I.

The number-average molecular weight was measured by (a) the titration of the amino endgroup with 0.02 N perchloric acid in dioxane and in m-cresol, crystal violet being used as the indicator (21), and (b) estimation from viscosity studies by use of the empirical equation of Doty et al. (22). The solubilities of poly(β-menthyl aspartate) and poly(γ -menthyl glutamate) were similar in many solvents. However, their solubilities were different from those of the usual polypeptides. Poly(menthyl aspartates and glutamates) were soluble in n-hexane, diethyl ether, chloroform, tetrahydrofuran, dioxane, DCA, and TFA. They were slightly soluble in isopropyl ether and petroleum ether, and were insoluble in water. The solubility of poly(β, N-alkyl L-asparagine) was rather low. Poly(β, N-benzyl L-asparagine) was soluble in DCA, TFA, and methanesulfonic acid, slightly soluble in m-cresol, and insoluble in chloroform, dioxane, tetrahydrofuran (THF), benzene, dimethylformamide (DMF), alcohol, and water. Poly[β, N-(1)-phenethyl L-asparagine] was soluble in only DCA, TFA, and methanesulfonic acid.

EXPERIMENTAL

Materials

β -(l)-Menthyl D- and L-Aspartates. β -(l)-Menthyl D- and L-aspartates were prepared by hydrazinolysis of the corresponding N-phthalyl derivatives (18). β-(l)-Menthyl D-aspartate; mp, 170°C. $[\alpha]_D^{27} = -61.2°$ (c=1.02, acetic acid). β-(l)-Menthyl L-aspartate; mp, 173°C. $[\alpha]_D^{20} = -63.1°$ (c=1.04, acetic acid).

γ-(l)-Menthyl D- and L-Glutamates. γ -(l)-Menthyl D- and L-glutamates were also prepared by hydrazinolysis of the corresponding N-phthalyl derivatives (17). γ-(l)-Menthyl D-glutamate; mp, 196°C. $[\alpha]_D^{15} =$ -73.6° (c=1.06, acetic acid). γ-(l)-Menthyl L-glutamate; mp, 194-195° C. $[\alpha]_D^{15} = -49.3°$ (c=1.00, acetic acid).

β, N-Benzyl L-Asparagine, β, N-(d) and (l)-(1)-Phenethyl L-Asparagines. β, N-Alkyl L-asparagines were prepared from the corresponding phthalyl and benzyloxycarbonyl-β, N-alkyl L-asparagines (20). β, N-Benzyl L-asparagine; mp, 253°C. $[\alpha]^{22} = +44.4°$ (c=0.50, 2 N hydrochloric acid). β, N-(d)-(1)-Phenethyl L-asparagine; mp, 215°C $[\alpha]_D^{24}=$ +153.0° (c=0.52, 2 N hydrochloric acid). β, N-(l)-(1)-Phenethyl L-asparagine hydrobromide; mp, 184°C, $[\alpha]_D^{24} = -59.8°$ (c=0.50, water).

Preparation of Amino Acid NCAs. β -(l)-Menthyl D-aspartate NCA (mp, 76°C), β -(l)-menthyl L-aspartate NCA (mp, 138°C), γ -(l)-menthyl D-glutamate NCA (mp, 108°C), γ -(l)-menthyl L-glutamate NCA (mp, 103°C), β, N-benzyl L-asparagine NCA (mp, 130°C), β, N-(d)-(1)-phenethyl L-asparagine NCA (mp, 75°C), and β, N-(l)-(1)-phenethyl L-asparagine NCA (mp, above 220°C) were prepared from the corresponding amino acids and phosgene in dry dioxane.

Preparation of Polypeptides.   The above NCAs were polymerized in dioxane with triethylamine as an initiator. Poly[β, N-(dl)-(1)-phenethyl L-asparagine] was prepared by the copolymerization of the same amount of β, N-(d) and (l)-(1)-phenethyl L-asparagine NCAs. The results are listed in Table I.

Methods

Optical rotatory dispersion, circular dichroism, and infrared absorption spectra measurements were made on the ORD/UV 5 instrument with the CD attachment and the IR DS-301 instrument, both made by the Japan Spectroscopic Co., Ltd. The optical rotatory dispersion and circular dichroism were measured with 0.1-10 mm cells for the 200-600 nm wavelengths. The optical rotations were expressed as a reduced molar residue rotation. As for CD, the measured values of $\varepsilon_L - \varepsilon_R$ were converted to the molar ellipticity. The value of the parameter b_0, derived from the Moffitt-Yang equation (23, 24), was determined from the ORD curve of the solution. A linear plot of $[R']$ $[(\lambda^2 - \lambda_0^2)/\lambda_0^2]$ versus $\lambda_0^2/(\lambda^2 - \lambda_0^2)$ was obtained over the wavelength range of 300 ~600 nm by using a λ_0^2 value of 212 nm.

TABLE I. Molecular Weight of Synthesized Polypeptides.[a]

Polymer	Yield %	Mn[b]	$[\eta]^c$ dl/g	Calcd.[e] Mn
β-Menthyl D-Asp	87	10700	0.069	7900
β-Menthyl L-Asp	96	13300	0.125	15900
γ-Menthyl D-Glu	73	29500	0.28^d	
γ-Menthyl L-Glu	100	18300	0.22^d	
β,N-Benzyl L-Asn	94	10700	0.138	11400
β,N-(d)-Phenethyl L-Asn	87		0.182	21200
β,N-(l)-Phenethyl L-Asn	91		0.218	26700
β,N-(dl)-Phenethyl L-Asn	65		0.135	11800

[a]Polymerized in dioxane, with molar ratio of anhydride to initiator, A/I, of 100.

[b]Determined from the amino endgroup titration.

[c]In DCA at 25°C.

[d]$c=0.5\%$ in DCA at 25°C.

[e]Molecular weight estimated from the empirical equation of Doty et al.

RESULTS AND DISCUSSION

Infrared Spectra

The infrared absorption spectra of the synthesized polypeptides showed absorption typical of polypeptides. In the case of poly(menthyl aspartate and glutamate), the position of the amide I is shifted to $1682 \sim 1690$ cm^{-1}, which is different from those of the α-helix, β-form, and random coil. This could be interpreted as indicating that the amide I of these polypeptides is shifted to $1682 \sim 1690$ cm^{-1} because of the stacking of the menthyl groups. It is difficult to determine the relationship between the frequencies and the backbone conformations of the poly(β, N-alkyl L-asparagines), because the monomer has the absorption bands of the amide linkage in its side chain.

Optical Rotatory Dispersion and Circular Dichroism

The conformation of the synthesized polypeptides in several solvents and in a DCA-chloroform mixed solvent has been studied by means of ORD and CD measurements.

Poly[β-(l)-menthyl D- and L-aspartates]. The ORD curves of poly-[β-(l)-menthyl D- and L-aspartates] in THF and in TFA are shown in Fig. 1. In THF, the D-polymer exhibits a peak at 230 nm. The electronic transition at 230 nm may be due to the n-π* transition. This behavior is essentially identical to that of the β-form of the other polyamino acids. In THF, the ORD behavior of the L-polymer is identical to that of the random-coil structure. In diethyl ether, n-hexane, and dioxane, both the D- and L-polymers gave similar curves in all region. In TFA, both the D- and L-polymers exhibit ORD behaviors characteristic of random coil.

The CD curves of the poly(β-menthyl aspartates) were recorded in order to resolve any possible overlapping of the Cotton effect in this region. The CD curves are shown in Fig. 2. In THF, diethyl ether, n-hexane, and dioxane, a positive band at 215 nm, with $[\theta]_{215} = 3000$–5000, is observed for poly[β-(l)-menthyl D-aspartate]. The ellipticity band at 215 nm appears to be n-π* peptide electronic transition associated with the β-form of polypeptide.

In THF and in diethyl ether, a negative dichroism band below 204 nm, with $[\theta]_{204} = -4200$, is observed for poly[β-(l)-menthyl L-aspartate].

This dichroism band of poly[β-(l)-menthyl L-aspartate] is found in approximately the same position and is found to have the same sign as that observed for the random coil structures of L-polyamino acid.

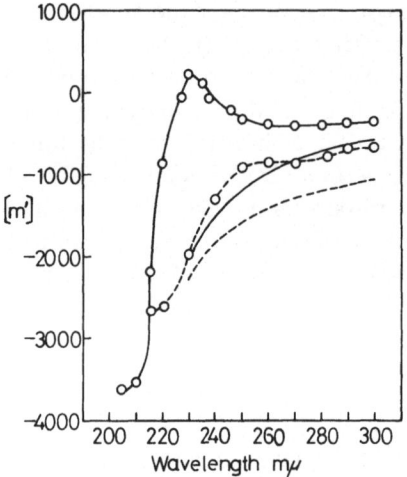

Fig. 1. Optical rotatory dispersions of poly[β -(l)-menthyl (———) d- and (-----) L-aspartates] at 25°C: o, in THF; and ———, in TFA.

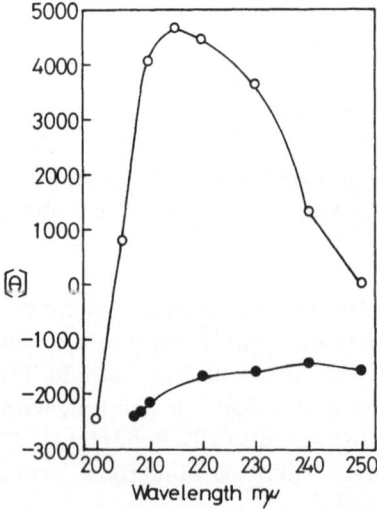

Fig. 2. Circular dichroisms of poly[β-(l)-menthyl (—o—) d- and (—●—) L-aspartates] in THF at 25°C.

Poly[γ-(l)-menthyl D- and L-glutamates]. The ORD and CD curves of poly[γ-(l)-menthyl D- and L-glutamates] in THF, in diethyl ether, in chloroform, in n-hexane (α-helix), in DCA, and in TFA (random coil) were measured. The ORD curves of the polymers in THF and in TFA are shown in Fig. 3. In THF, the L-polymer exhibits a deep trough at 233 nm, a zero value of [m'] near 225 nm, a shoulder at 212 nm, and a strong peak near 198 nm. This ORD behavior is essentially identical to that of the right-handed α-helical form of the other polyamino acid (25, 26). The [m'] value of 233 nm is about -11000 deg-cm^2/dm. The D-polymer exhibits the same behavior with a deep trough of about +10100 deg-cm^2/dm at the same wavelength (233 nm); it was found to have the left-handed α-helical structure.

The CD curves of the polypeptides in THF are also shown in Fig. 3. Two negative dichroism bands at 222 nm and 206 nm, with [θ]$_{222}$ = -28000 and [θ]$_{206}$ = -25000, are observed for poly[γ-(l)-menthyl L-glutamate] in THF, and a positive dichroism band can be expected below 200 nm.

The negative ellipticity band at 222 nm appears to be a n-π* peptide electronic transition associated with the α-helical conformation of poly-peptide, and the negative band at 206 nm and the positive band below 200 nm are to be assigned to the parallel-polarized and perpendicular-polar-

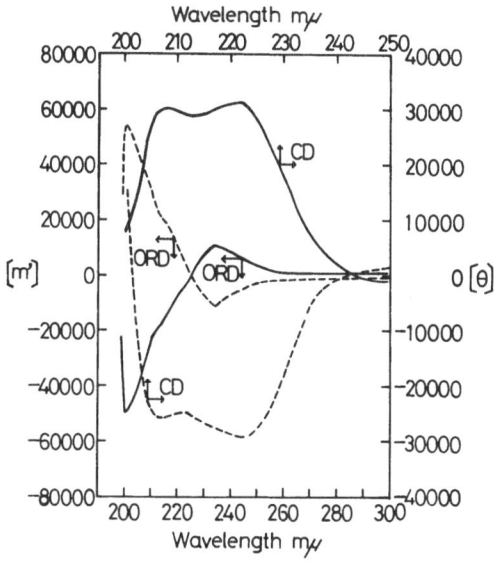

Fig. 3. Optical rotatory dispersions and circular dichroisms of poly[γ -(l)-menthyl (——) D- and (-----) L-glutamates] in THF at 25°C.

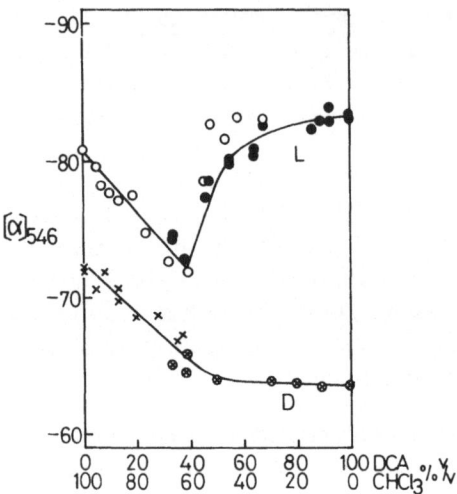

Fig. 4. Specific rotation [α]$_{546}$ values of poly[γ-(l)-menthyl D- and L-glutamates] of varying solvent composition: x and o values for D- and L-polymers dissolved in chloroform (solution diluted with DCA); ⊗ and ● for D- and L-polymers dissolved in DCA (solution diluted with chloroform) at room temperature.

ized π - π* exciton transitions of the peptide groups, respectively (27). These dichroism bands of poly[γ-(l)-menthyl L-glutamate] are found in approximately the same positions and have the same signs and magnitudes as has been observed in right-handed α -helical polypeptides. The polymer gave similar curves in diethyl ether, in chloroform, and in n-hexane above the 205 nm region.

It has been known that many α -helical polypeptides undergo a helix-coil transition in a DCA-chloroform (or TFA-chloroform) mixed solvent (28). In the case of poly(α-benzyl L-glutamate), the helix-coil transition occured at 68% DCA-32% chloroform. Plots of the [α]$_{546}$ values versus the solvent composition of poly[γ -(l)-menthyl D- and L-glutamates] are shown in Fig. 4. These polymers cause a gradual helix-coil transition at about 40% DCA. Poly[γ -(l)-menthyl L-glutamate] gradually loses its helical structure upon an increase in the DCA, whereupon the specific rotation decreases to a minimal value of [α]$_{546}$ = -72° at 40% DCA, finally reaching [α]$_{546}$ = -83° at above 40% DCA. On the other hand, poly[γ-(l)-menthyl D-glutamate] decreases its optical rotation at below 40% DCA and reaches a constant value of [α]$_{546}$ = -64°. In comparison

TABLE II. b_0 and $[\alpha]_{546}$ Values of Poly(β, $\underline{N}$-Benzyl L-Asparagine)

	CHCl$_3$ %	Polymer concentration (g/100ml)	$[\alpha]_{546}$	b_0
1	0	1.009	+ 4.0	- 33
2	33.3	0.668	+ 6.7	- 33
3	50.0	0.504	+ 5.7	- 21
4	66.7	0.334	- 3.9	- 8
5	75.0	0.252	- 45.9	+ 51
6	80.0	0.201	- 77.1	+102
7	83.3	0.168	-128.5	+335
8	87.5	0.126	-152.6	+417
9	90.0	0.101	-153.1	+463
10	93.3	0.067	-164.7	+467

with poly(benzyl glutamate), the helical structure of poly(menthyl glutamates) are made rather unstable by the introduction of the γ-menthyl group.

These poly[γ-($\underline{l}$)-menthyl D- and L-glutamates] are thought to have the helical and the random coil conformations as their secondary structures in a DCA-chloroform mixed solvent. In the random coil region, the $[\alpha]_{546}$ values of D- and L-polypeptides are different.

Poly(β, $\underline{N}$-alkyl L-asparagines).   The ORD curves of poly(β , $\underline{N}$-benzyl L-asparagine), poly[β , N-($\underline{d}$), ($\underline{l}$), ($\underline{dl}$), and ($\underline{d+1}$, 1:1)-(1)-phenethyl L-asparagines] in DCA and in a DCA-chloroform mixed solvent were

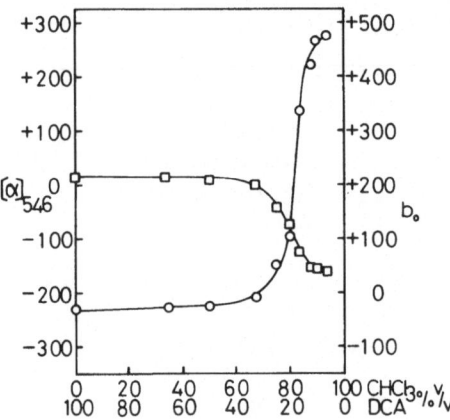

Fig. 5. Conformational transition of poly (β ,
N-benzyl L-asparagine) in the DCA-chloroform
mixtures at 22°C: (o), b_0; (□), [α]$_{546}$.

measured. Poly(β,N-benzyl L-asparagine) is dissolved initially in DCA
and diluted with chloroform successively until it is 93.3% chloroform.
The results of ORD measurements in various solvent compositions are
shown in Table II.   The specific rotation reverses from dextro to levo
rotation at about 60% chloroform.   The parameters calculated are the
specific rotation [α]$_{546}$ and the b_0 value; they are shown in Fig. 5. The
transition from a random coil to a left-handed helical conformation is
apparent at about 30% DCA-70% chloroform. The tendency of the [α]$_{546}$
is consistent with the b_0 value. Bradbury (2), Goodman (29), and Hashimoto (13) have reported that the b_0 value of the left-handed α-helix of
poly(L-aspartate) is +630 ∿ +680. This b_0 value of +475 indicates about
a 75% helical form at 6.7% DCA-93.3% chloroform. It is very interesting that the helix sense of poly(β ,N-benzyl L-asparagine) is the same
as that of poly(β -benzyl L-aspartate), i.e., left-handed, as was found
by Goodman, et al. (29). Blout et al. reported that the helix-random coil
transition for poly(β-benzyl L-aspartate) occurs at 5∿ 8% DCA in a DCA-
chloroform mixed solvent (5).   In comparison with poly(β-benzyl L-
aspartate), the helical structure of poly(β ,N-benzyl L-asparagine) is
made rather stable by the amide linkage in the side chain.

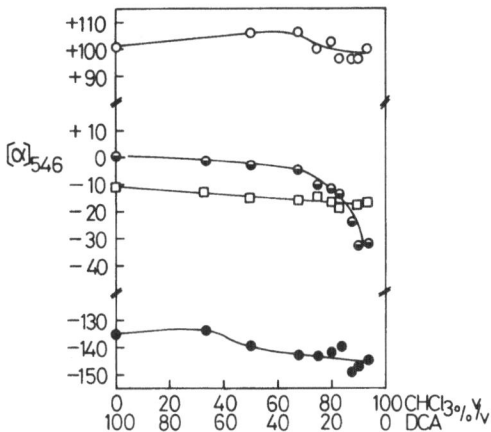

Fig. 6. Specific rotation $[\alpha]_{546}$ values of poly[β , $\underline{N}$-(1)-phenethyl L-asparagines] of varying solvent composition at 22ºC: (o), $\underline{d}$-polymer; (●), $\underline{l}$-polymer; (◕), $\underline{dl}$-polymer; and (□), $\underline{d} + \underline{l}$, 1:1-polymer.

The solubilities of poly[β , $\underline{N}$-($\underline{d}$) and ($\underline{l}$)-(1)-phenethyl L-asparagines] containing an optical active alkylamide linkage in their side chains were also poor. Moreover, it is impossible to calculate the b_0 values from the Moffitt-Yang plots because of their optically active side chains. Therefore, the $[\alpha]_{546}$ value was calculated from the ORD curves in order to clarify the conformational change of the polymers in the same way as that used for poly(β , $\underline{N}$-benzyl L-asparagine). The $[\alpha]_{546}$ value is plotted against the solvent composition in Fig. 6. The ($\underline{d}$), ($\underline{l}$), and ($\underline{d}+\underline{l}$, 1:1)* polymers show a nearly constant specific rotation in all the DCA-chloroform solvent systems and do not cause any apparent transition. On the other hand, the $[\alpha]_{546}$ value of poly[β, $\underline{N}$-($\underline{dl}$)-(1)-phenethyl L-asparagine] with an optically inactive side chain is greatly increased the levo rotation at a DCA content of less than 30%, as in the case of poly(β , N-benzyl L-asparagine). This may suggest the polymer causes a gradual folding of the helix.

---

* ($\underline{d}+\underline{l}$, 1:1): mixed polymer containing the same weighted poly-β , $\underline{N}$-($\underline{d}$) and ($\underline{l}$)-(1)-phenethyl L-asparagines.

CONCLUSION

The effects of the side chain on the secondary structure of the syn-thesized poly [D- and L-acidic amino acid ω-(l)-menthyl esters] and poly-(L-aspartic acid ω, N-optically active alkylamides) were shown by the interesting behavior in solutions.   Because of the introduction of optically active groups in polyacidic amino acids, poly[β-(l)-menthyl D-aspartate] exhibited a β-form and poly[β-(l)-menthyl L-aspartate] was a random coil structure.   The helical stabilities of poly[γ-(l)-menthyl D- and L-glutamates] are made rather unstable in comparison with that of poly(γ-benzyl glutamates). Also, poly[β, N-(1)-phenethyl L-asparagines) showed a nearly constant specific rotation in the DCA-chloroform solvent system. However, poly[β, N-benzyl and (dl)-(1)-phenethyl L-asparagines] in DCA, which have optically inactive alkylamide groups, are changed gradually into the left-handed α-helix structure when chloroform was added to the solution. In this case, the helical stabilities of these polymers are made rather stable compared to that of poly(β-benzyl L-aspartate).

These results can be considered to prove that the asymmetry of the l-menthyl and (d) and (l)-(1)-phenethyl chromophores in the side chain causes an extraordinary optical rotation; that is, this phenomenon arises from interaction between the chromophore in the side chain and the optical active centers in the polypeptide main chain, as has been described by Goodman and Sarker (30-33) for polypeptides with aromatic and azoaromatic side chains, and by Kubota et al. for poly(O-acetylthreonine) and poly(O-acetylallothreonine) (34).

SUMMARY

Poly[β-(l)-menthyl D- and L-aspartates], poly[γ-(l)-menthyl D- and L-glutamates], and poly[β, N-benzyl- and (1)-phenethyl L-asparagines] were prepared by the polymerization of the corresponding amino acid N carboxyanhydrides. From the results obtained by a study of the infrared absorption spectra, the optical rotatory dispersions, and the circular dichroisms, poly[β-(l)-menthyl D-aspartate] was found to be a β-form structure, poly[β-(l)-menthyl L-aspartate] was a random coil structure, and poly[γ-(l)-menthyl D- and L-glutamates] were a α-helix structure in solvents such as chloroform, tetrahydrofuran, dioxane, and diethyl ether. On the other hand, poly(β, N-benzyl L-asparagine) was a random coil structure in dichloroacetic acid, and the optical rotatory dispersion curves gradually changed into the left-handed α -helix structure when chloroform was added to the solution.   The helix to coil transition of poly[γ-(l)-menthyl glutamates] and the coil to helix transition of poly(β, N-benzyl L-asparagine) were observed in the vicinity of 40% dichloroacetic acid for the former, while the later in the vicinity of 20% dichloroacetic acid in a dichloroacetic acid-chloroform mixture.

Poly[β, N-(d), (l), and (d+l, 1:1)-(1)-phenethyl L-asparagines] showed a nearly constant specific rotation in the dichloroacetic acid-chloroform solvent system.   Poly[β, N-(dl)-(1)-phenethyl L-asparagine] caused a gradual folding of the helix at dichloroacetic acid content of less than 20%.

REFERENCES

1. Coleman, D., J. Chem. Soc. 1951, 2294.
2. Bradbury, E. M., Carpenter, B. G., and Goldman, H., Biopolymers 6, 837 (1968).
3. Frankel, M., and Berger, A., J. Org. Chem. 16, 1513 (1951).
4. Berger, A., and Katchalski, E., J. Am. Chem. Soc. 73, 4084 (1951).
5. Karlson, R. H., Norland, K. S., Fasman, G. D., and Blout, E. R., J. Am. Chem. Soc. 82, 2268 (1960).
6. Blout, E. R., and Karlson, R. H., J. Am. Chem. Soc. 80, 1259 (1958).
7. Bamford, C. H., Elliott, A., and Hanby, W. E., "Synthetic Polypeptides," Academic Press, New York, 1956.
8. Katchalski, E., and Sela, M., Advan. Protein Chem. 13, 243 (1958).
9. Sugai, S., Kamashima, K., Makino, S., and Noguchi, J., J. Polym. Sci. A-2, 4, 183 (1966).
10. Adler, A. J., Fasman, G. D., and Blout, E. R., J. Am. Chem. Soc. 85, 90 (1963).
11. Blout, E. R., and Asadourian, A., J. Am. Chem. Soc., 78, 955 (1956).
12. Goodman, M., Deber, C. M., and Felix, M., J. Am. Chem. Soc. 84, 3773 (1962).
13. Hashimoto, M., and Aritomi, J., Bull. Chem. Soc. Japan, 39, 2707 (1966).
14. Bradbury, E. M., Carpenter, B. G., and Stephans, R. M., Biopolymers 6, 905 (1968).
15. Frankel, M., Liwschitz, Y., and Zilkla, A., J. Am. Chem. Soc. 75, 3270 (1953).
16. Ariely, S., Fridkin, M., and Patchornic, A., Biopolymers 7, 417 (1969).

17. Yamamoto, H., Kondo, Y., and Hayakawa, T., Biopolymers 9, 41 (1970).
18. Yamamoto, H., and Hayakawa, T., Biopolymers 10, 309 (1971).
19. Yamamoto, H., and Hayakawa, T., Bull. Chem. Soc. Japan 44, 1990 (1971).
20. Hayakawa, T., Yamamoto, H., and Aoto, N., Biopolymers 11, 185 (1972).
21. Sela, M., and Berger, A., J. Am. Chem. Soc. 77, 1893 (1955).
22. Doty, P., Bradbury, T. H., and Holtzer, A. M., J. Am. Chem. Soc. 78, 947 (1956).
23. Moffitt, W., and Yang, J. T., Proc. Natl. Acad. Sci. U. S. 42, 596 (1956).
24. Yang, J. T., in "Poly-α-Amino Acids," (G. D. Fasman, ed.), p. 239. Marcel Dekker Inc., New York, 1967.
25. Carver, J. P., Shechter, E., and Blout, E. R., J. Am. Chem. Soc. 88, 2550 (1966).
26. Ingwall, R. T., Scheraga, H. A., Lotan, N., Berger, A., and Katchalski, E., Biopolymers 6, 331 (1968).
27. Holzwarth, G., and Doty, P., J. Am. Chem. Soc. 87, 218 (1965).
28. Fasman, G. D., in "Polyamino acids, Polypeptides and Proteins" (M. A. Stahman, ed.), p. 221. University of Wisconsin Press, Madison, 1962.
29. Goodman, M., Boardman, F., and Litowsky, L., J. Am. Chem. Soc. 85, 2491 (1963).
30. Goodman, M., and Benedetti, E., Biochemistry 7, 4226 (1968).
31. Goodman, M., and Kossay, A., J. Am. Chem. Soc. 88, 5010 (1966).
32. Toniolo, C., Falxa, M. L., and Goodman, M., Biopolymers 6, 1579 (1968).
33. Sarker, P. K., and Doty, P., Proc. Natl. Acad. Sci. U. S. 45, 1601 (1959).
34. Kubota, S., Sugai, S., and Noguchi, J., Biopolymers 6, 1311 (1968).

Received 17 June, 1971.

A MECHANISM FOR POLYPEPTIDE SYNTHESIS

ON A PROTEIN TEMPLATE*

Fritz Lipmann

The Rockefeller University
New York, N. Y. 10021

INTRODUCTION

With this contribution, I am happy to take part in celebrating Dr. Fox's anniversary. It was his invitation to the memorable Wakulla Springs Conference which became, for me, the beginning of an infatuation with origin of life problematics.

The protein structure has a vast capacity for specific binding and catalysis. This is due, obviously, to the arrangement of twenty amino acids in a variety of sequences which produce folding of a specified kind. The number of amino acids and their chemical structure have remained invariant for about two billion years, if Barghoorn's suggestion is accepted that his electronmicrographs represent fossilized bacteria related to those now living (1). In spite of the difficulty of imagining a prebiotic evolution without the genetic information transfer machinery, I feel, prompted by common sense, that the twenty now-invariant components of protein structure, to judge from their amazing capacity to perform, should have been the result of a selection process.

With proteins the main target of genetic information transfer, one is inclined to assume that prior to this choice, the unusual capacity of poly-

---

* All of the work reported in this study has been supported by a grant from the U. S. Public Health Service, GM-13972.

peptide structures for catalytic function had to become apparent before the information transfer machinery to produce it did develop.   It was from such premises that I decided to look in present-day organisms for a process that produced polypeptide chains independently of messenger RNA, presumably on protein templates.   Such was suggested to exist in the mechanism for biosynthesis of certain antibiotic polypeptides produced by spore-forming bacteria.   The analysis of this process has yielded now a quite well defined process, which in itself deserves attention because of biochemical uniqueness.

The experimental background for this mechanism has been published extensively, and this permits its discussion without going into details. The cell-free synthesis of the antibiotics gramicidin S (GS) (2–5) and ty-rocidine (Ty) (5–7) are very similar, and information about GS, which has been worked out more thoroughly, may be transferred to Ty.   Ty contains a sequence of ten amino acids:

$$
\begin{array}{ccccc}
 & & (\text{Trp}) & (\text{D-Trp}) & \\
\text{D-Phe} & \rightarrow \text{Pro} \rightarrow & \text{Phe} \rightarrow & \text{D-Phe} \rightarrow & \text{Asn} \\
\uparrow & & & & \downarrow \\
\text{Leu} & \leftarrow \text{Orn} & \leftarrow \text{Val} & \leftarrow \text{Tyr} & \leftarrow \text{Gln} \\
 & & & (\text{Phe}) &
\end{array}
$$

GS synthesis sequences only five amino acids, which cyclize by anti-parallel condensation of two of them:

$$
\begin{array}{ccccc}
\text{D-Phe} & \rightarrow & \text{Pro} \rightarrow & \text{Val} \rightarrow & \text{Orn} \rightarrow & \text{Leu} \\
\uparrow & & & & & \downarrow \\
\text{Leu} & \leftarrow & \text{Orn} & \leftarrow \text{Val} & \leftarrow \text{Pro} & \leftarrow \text{D-Phe}
\end{array}
$$

Ty synthesis being more complex, will serve as the example and will be discussed here.

Three enzymes take part in the reaction: two small ones that start from the N-terminal and activate and carry separately the first and second amino acids in the sequence;   the eight following amino acids in the deca-peptide are all activated and carried by a larger enzyme with the rela-tively low molecular weight of 460,000 daltons. Additional information may be gained from Table I.   It shows the sequence which is also indicated in the formula by arrows, the dotted arrows showing the release of enzyme-bound polypeptide by cyclization.

TABLE I

Enzyme activities in the biosynthesis of GS and TY

Decapeptide synthesized	No.	Mol. wt. of enzymes	Amino acid activated and fixed in sequence read downwards	Mol. wt. per aa.	Pantetheine content (mole/enz.)
GS*	1	100,000	D-Phe	–	None
	2	280,000	Pro, Val, Orn, Leu	70	1
Ty	1	100,000	D-Phe	–	None
	2	230,000	Pro	–	None
	3	460,000	Phe, D-Phe, Asn, Gln, Phe, Val, Orn, Leu	57.5	1

*by antiparallel cyclization of two pentapeptides

The selective positioning of the amino acids is due to their fixation to enzyme-linked sulfhydryls after primary activation with ATP:

$$ATP + aa \rightleftharpoons AMP{\sim}aa \tag{1}$$

$$AMP{\sim}aa + E\ SH \rightleftharpoons AMP + E \cdot S{\sim}aa \tag{2}$$

Ty biosynthesis begins with the fixation of amino acids by the enzymes. When separated, the small enzyme binds and racemizes phenylalanine, an enzyme of intermediate size binds proline, and a large enzyme binds in equimolar proportion the eight following amino acids (Table I); it also racemizes the third phenylalanine in the overall sequence. Polymerization is initiated by a selective transfer from the N-terminal D-phenylalanine on enzyme 1 to proline on enzyme 2 (Fig. 1). The transfer of D-Phe-Pro from enzyme 2 to the first phenylalanine on the large enzyme triggers the sequential addition of the rest of the amino acids to form thioester-linked polypeptides until the last, leucine, is reached (Table II). There polymerization terminates by cyclization through reaction of the thioester-linked carboxyl of leucine with the N-terminal amino group of D-phenylalanine.

TABLE II

Formation of protein-bound nascent peptide chains by sequential addition
of amino acids with ^{14}C-marker in the proline

	Peptide chains formed	Bound radioactivity (pmol)	Increment
1	$\cdots - ^{14}$C-Pro	6	6
2	D-*Phe*–^{14}C-Pro		
3	D-*Phe*–^{14}C-Pro – *Phe*	20.5	14.5
4	D-*Phe*–^{14}C-Pro – *Phe* – D-*Phe*		
5	D-Phe–^{14}C-Pro–Phe – D-Phe–*Asn*	27.6	7.1
6	D-Phe–^{14}C-Pro–Phe– D-Phe–Asn – *Gln*	39	11.4
7	D-Phe–^{14}C-Pro–Phe– D-Phe–Asn–*Gln*–Phe*		
8	D-Phe–^{14}C-Pro–Phe– D-Phe–Asn–Gln–Phe–*Val*	43.5	4.5
9	D-Phe–^{14}C-Pro–Phe– D-Phe–Asn–Gln–Phe–Val–*Orn*	49.3	5.8
10	D-Phe–^{14}C-Pro–Phe–D-Phe–Asn–Gln–Phe–Val–Orn–*Leu*	54.7	5.4
11	$\cdots -^{14}$C-Pro$\cdots$ (Asn,Gln,Val,Orn,Leu)	7.0	
12	D-Phe–^{14}C-Pro–Phe– D-Phe$\cdots$ (Gln,Phe,Val,Orn,Leu)	21	
13	D-Phe–^{14}C-Pro–Phe– D- Phe–Asn–Gln–Phe–Val–$\cdots$(Leu)	43.7	

*This Phe waiting in position 7 does not cause an increment before Asn, 5, and Gln, 6, are added, indicating that binding out of sequence does not cause elongation.

In 11, 12, and 13, omission of an amino acid, indicated by dots, stops further incorporation of ^{14}C-Pro posterior to omission; the amino acids in parenthesis are present on the enzyme in peripheral thioester links but are not polymerized due to the omission.   For experimental details consult Ruskoski et al. (6).

Initiation

$$E_{II}\text{–S} \sim \text{Phe NH}_2 + E_I\text{–S} \sim \text{Pro : NH}$$

$$\downarrow$$

$$E_{II}\text{–SH} + E_I\text{–S} \sim \text{Pro–Phe NH}_2$$

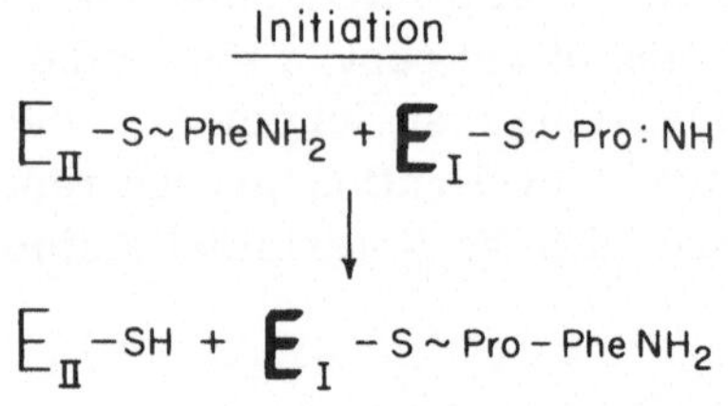

Fig. 1. Initiation reaction for biosynthesis of GS (5).

The most interesting part of the reaction is the sequential elongation on the large enzyme.   The large enzyme may be charged with all eight amino acids, one mole of each, without causing polymerization, before transfer of the dipeptide from enzyme 2 to form the tripeptide that transthiolates from peripheral to the swinging -SH of pantetheine (Fig. 2). Vectorial addition in the right order is proved by the stoppage of elongation on omission of one of the amino acids in the sequence (Lines 11-13, Table 2). We visualize the large enzyme as an assemblage of eight amino acid-specific activating enzyme subunits arranged vectorially.   Dividing the molecular weight of 460,000 by 8, one obtains 57,000 as an average weight for the individual activating units. So far, however, we have been unable to break up the large enzyme into subunits.

In contrast to the small enzymes, the large enzyme contains one mole of 4'-phosphopantetheine (8) linked to the enzyme protein by a phosphate bridge to an hydroxyl group, presumably of a serine.   Our data indicate that, beginning with the transthiolation of D-Phe-Pro-Phe to the pantetheine-SH, the latter then becomes the peptidyl donor to the amino group of the next amino acid in sequence. Transpeptidation to the enzyme-bound amino acids liberates the -SH of pantetheine, which then appears to accept the elongated peptide by transthiolation (9).   This corresponds rather strikingly to translocation following transpeptidation in the ribo-

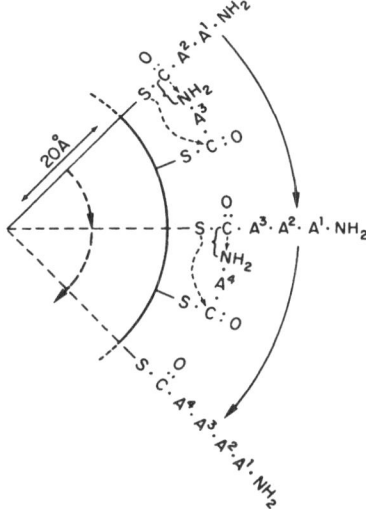

Fig. 2. A schematic representation of transpeptidation and transthiolation, presumably mediated by enzyme-bound pantetheine in antibiotic peptide biosynthesis (12).

somal process.   The sequence of reactions is schematically illustrated
in Fig. 2.

The template function of the poly-activating enzyme is thus accomplished by the arrangement in the right order of the amino acid-specific
fixation points from which the sequence is constructed, and assembling
them through the intermediacy of pantetheine.   The same pantetheine
functions similarly as carrier in the fatty acid synthesis polyenzyme
(Fig. 3).   Here, the initial condensation product, a β-keto acid of increasing chain length, is carried through a three-step transformation,
catalyzed by a polyenzyme assemblage and succeeded by repeated elongation through condensation with a two-carbon fragment.   The crucial feature of the system is to use the recognition of the different amino acids
by the activating enzyme for positioning, assemble them in the demanded
sequence on a polyenzyme, and polymerize the fixed amino acids from
the assemblage.   It is a compact process that combines activation and
transpeptidation into one polyenzyme.   The use of the recognition site of
an activating enzyme as a basis for a polymerization template is a plausible device.   It fuses activation and polymerization into one system of
reasonably small size.   Work presently proceeding on the biosynthesis of

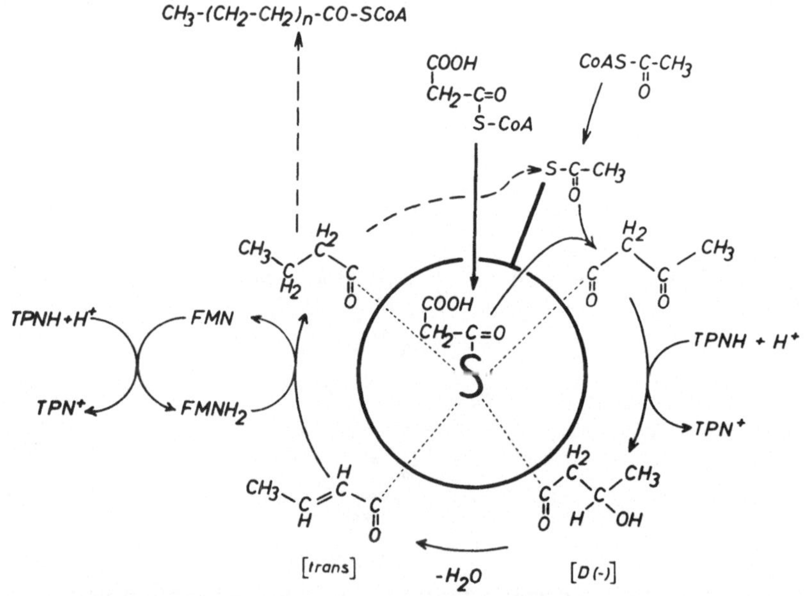

Fig. 3. Lynen's scheme for the mechanism of fatty acid biosynthesis (10).

gramicidin A appears to extend the limit of polypeptide size derivable from the process to a chain of 15 amino acids, or, including N-terminal formate and carboxyl terminal ethanolamine, to 17 units (Fig. 4).

DISCUSSION

The studies described here were initiated in the hope of finding a mechanism from which it might be possible to project backwards towards prebiotic evolution (11). The process that we have analyzed is produced by enzyme proteins made, of course, from coding transmitted through mRNA's and using the ribosomal machinery. Yet, the study has revealed a mechanism that uses the amino acid selection mechanism of the activating enzymes for positioning and fixing the amino acid in energy-rich thioester linkage on a polyenzyme that can be triggered into sequentially ordered polymerization. The transfer of ATP-activated amino acids to protein-linked -SH groups was an unexpected feature. This, however, seems to present the missing link which completes the long-realized parallels between acetate and amino acid activation as well as between β-keto acid and peptide condensations. Thereby, a process evolution is suggested from fatty acid over antibiotic polypeptide to protein synthesis. This has been extensively discussed in a recent paper (12).

The fact that in the present-day organism production of a process is taken over by the ribosomal system seems not to obviate the possibility that, prebiotically, amino acids may have been fixed in sequence to an -SH-carrying polypeptide and then induced to polymerize. Recent experiments have shown that at least in the case of D-phenylalanine the amino acid may be fixed to the enzyme from D-Phe-thiophenol (13), circumvent-

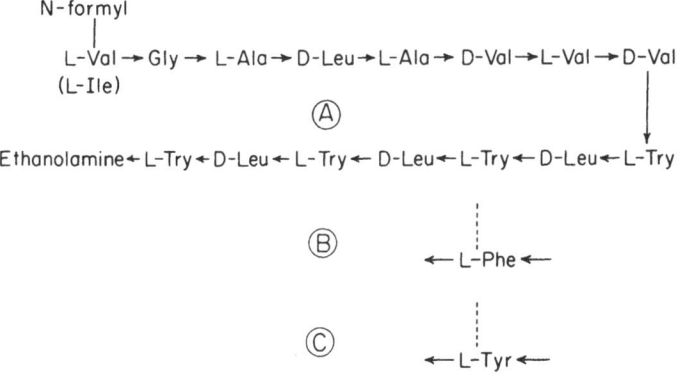

Fig. 4.   Formulae of gramicidin A, B, and C.

ing the ATP-linked initiation.   We hope to continue attempts to reconstruct polymerization of amino acids after transfer to protein-linked
-SH groups.

SUMMARY

    The mechanism of biosynthesis of the two cyclic decapeptides grami-
cidin S and tyrocidine has been described in detail.   The present paper
presents a synopsis of this mechanism of polypeptide synthesis on a protein template.   Specific ATP-linked amino acid activation followed by
transfer to enzyme-bound -SH positions the amino acids in sequence.
This is succeeded by their polymerization to enzyme-bound peptidyl thio-
esters with participation of 4'-phosphopantetheine as cofactor.   The process terminates with release of the decapeptide through cyclization.
Omission of an amino acid in the sequence   results in cessation of polymerization at the point of omission, which indicates a vectorial arrangement of amino acid activating subunits on a polyenzyme.

REFERENCES

1. Schopf, J. W., Barghoorn, E. S., Maser, M. D., and Gordon, R. O.
    Science 149, 1365 (1965).
2. Gevers, W., Kleinkauf, H., and Lipmann, F., Proc. Nat. Acad.
    Sci. 60, 269 (1968).
3. Kleinkauf, H., Gevers, W., and Lipmann, F., Proc. Nat. Acad.
    Sci. 62, 226 (1969).
4. Gevers, W., Kleinkauf, H., and Lipmann, F., Proc. Nat. Acad.
    Sci. 63, 1335 (1969).
5. Kleinkauf, H., and Gevers, W., Cold Spring Harbor Symp. Quant.
    Biol. 34, 805 (1969).
6. Roskoski, R., Jr., Gevers, W., Kleinkauf, H., and Lipmann, F.,
    Biochemistry 9, 4839, 4846 (1970).
7. Lipmann, F., Gevers, W., Kleinkauf, H., and Roskoski, R. Jr.,
    Adv. Enzymol. 35, 1 (1971).
8. Kleinkauf, H., Gevers, W., Roskoski, R. Jr., and Lipmann, F.,
    Biochem. Biophys. Res. Commun. 41, 1218 (1970).
9. Kleinkauf, H., Roskoski, R. Jr., and Lipmann, F., Proc. Nat.
    Acad. Sci. 68, 2069 (1971).
10. Lynen, F., Oesterhelt, D., Schweizer, E., and Willecke, K., in
    "Cellular Compartmentalization and Control of Fatty Acid Metabo-
    lism" (F. C. Gran, ed.), p. 1. Universitetsforlaget, Oslo, 1968.

11. Lipmann, E., in "Origins of Prebiological Systems" (S. W. Fox, ed.), p. 259, Academic Press, New York, 1965.
12. Lipmann, F., Science 173, 875 (1971).
13. Roskoski, R., Jr., Ryan, G., Kleinkauf, H., Gevers, W., and Lipmann, F., Arch. Biochem. Biophys. 143, 485 (1971).

NOTE ADDED IN PROOF

In the papers published so far on Ty synthesis, impure preparations were used. We already had some misgivings about the disproportion between the size of 230,000 Daltons of the intermediate enzyme (IE) and its apparently activating only one amino acid, proline. Dr. Sung G. Lee in this laboratory has now purified to near homogeneity the IE as well as the large enzyme (LE), and the assignment of only proline for activation by IE has to be revised. In confirmation of a recent report by Kambe, Sakamoto and Kurahashi [J. Biochem. (Tokyo) 69, 1131 (1971)], our purified IE activates and fixes the three amino acids following the initiating D-phenylalanine, namely, proline, L-phenylalanine, and D-phenylalanine, and the LE activates and fixes the six remaining amino acids, asparagine, glutamine, phenylalanine, valine, ornithine, and leucine. This now corrects the proportion between molecular weights and number of amino acids as follows: for IE 230,000 ÷ 3 = 76,000 and for LE 450,000 ÷ 6 = 75,000. Experiments in progress indicate that fragments of this size can indeed be obtained from IE and LE. Furthermore, IE and LE each contain 1 mole of pantetheine. Our molecular weights using purified polyenzymes for approximate determination by sucrose gradient centrifugation compare well with the data obtained with impure preparations.

Received 14 September, 1971

MYELOPEROXIDASE, THE PEROXIDASE OF A PRIMITIVE

CELL; ITS REACTION WITH FE AND H_2O_2*

J. Schultz** and S. Rosenthal***

Papanicolaou Cancer Research Institute, Miami, Florida

INTRODUCTION

Evidence that mitochondria appeared sometime during the evolution of the cell as a result of a symbiotic bacteria which eventually became an integral part of the cellular structure has been reviewed by Sagan (1). If one stops to think about the system of terminal oxidation in the pre-mitochondrial age, one might look to the peroxidases. Peroxidases require hydrogen peroxide, with which they combine to oxidize a large variety of substrates which serve as donors. It is the diversity and widespread chemical structures which can be peroxidized which makes them excellent candidates for the role of terminal oxidases. The fact that flavins, DPNH, and oxygen, which were early cellular constituents, react with the formation of hydrogen peroxide serves well for this concept.

---

* This work was supported in part by Grant CA-03715 from the National Cancer Institute of the National Institutes of Health when both investigators were at Hahnemann Medical College, where part of the work was done; and by the Papanicolaou Cancer Research Institute at Miami under Grant CA 10904.

** All inquiries concerning reprints, etc., address is: Dr. Julius Schultz, Papanicolaou Cancer Research Institute, 1155 N.W. 14th Street, Miami, Florida 33136.

*** Present Address: Institute for Experimental Therapeutics, University of Pennsylvania School of Medicine, Philadelphia, Pennsylvania.

In his book on peroxidases Saunders (2) reviews the hundreds of kinds of hydrogen donors which react with peroxidase-hydrogen peroxide complex. More recently, studies by Hager (3) on halogenation by chloroperoxidase stimulated interest in the extent to which other halogens and other peroxidases carry out this reaction. In addition to the substrates already reported by Saunders and the halogenation reaction, a recent report by Zgliczynski, et al. (4) extended these peroxidation reactions to decarboxylation and deamination of amino acids forming aldehydes.

It is quite possible than that as mitochondrial electron transport systems took over in the course of evolution, the flavin peroxidase system became less important. However, in the normal neutrophil this system is still retained. When the neutrophil, a primitive cell, undergoes phagocytosis, a flavin enzyme reacts with DPNH and oxygen to form hydrogen peroxide. Myeloperoxidase released from the lysosome combines with the H_2O_2 and attacks foreign bodies through chlorination, iodination, decarboxylation, and deamination among other reactions (5, 6).

Finally, myeloperoxidase has certain structural features of cytochrome oxidase of mitochondrial electron transport (7).

From the above, one can accept the possibility that peroxidase represents one of the terminal electron transport systems in pre-mitochondrial cells. It might be relevant to report some of our investigations of the reaction of the native myeloperoxidase with two substances of evolutionary interest, that is, H_2O_2 and ferric ion.

When myeloperoxidase is treated with ferrous ion in the presence of oxygen, it undergoes an irreversible inactivation (8). In the absence of knowledge of the nature of the prosthetic group and its apoprotein, speculation as to the mechanism of this reaction is limited; however, a series of experiments from this laboratory based on studies of factors such as phosphate (9), ATP and ADP (10), and imidazole (11), suggested H_2O_2 as an intermediate in the reaction, inasmuch as these factors seemed to affect the conversion of ferrous to ferric ion which was shown to take place during the inactivation process (9). In the present report, inactivation of the enzyme by both ferric ion and hydrogen peroxide was therefore examined. When hydrogen peroxide was added before ferric ion, the effect was additive; when added after ferric ion, the effect was synergistic-- that is, approximately 50% inactivation of the peroxidase took place in the presence of ferric ion itself, or after peroxide addition, but greater than 90% of the activity was lost when Fe (III) was added prior to the addition of H_2O_2.

Data obtained from experiments on the kinetics and the pH optima of the reaction of the peroxidase with ferric ion and hydrogen peroxide when reacting together or independently can be explained by postulating two sensitive sites on the enzyme, not necessarily on the same molecule but part of at least two components. The latter is supported by alterations in the light absorption in the Soret region of the native enzyme treated with Fe (III) and H_2O_2 and by isolation of the two peptide fractions found in tryptic digests of the peroxidase, one of which behaves like a typical heme on reduction with dithionite, and the other is atypical in that it is bleached by dithionite.

METHODS AND PROCEDURES

The myeloperoxidase used in these experiments was prepared according to the method of Schultz and Shmukler (7), from the leucocytes of normal human blood. Enzyme, free of dialyzable phosphate and having an A_{430}/A_{280} of greater than 0.8 was diluted to the appropriate concentration with 10% ammonium sulfate and the pH adjusted to 6.0 - 6.4 with ammonium hydroxide. Hydrogen peroxide solutions were prepared by dilution of Superoxol 30% shortly before each experiment.

Iron was determined by Drabkin's procedure (12); nitrogen, as ammonia, by micro-Kjeldahl distillation with the elimination of the digestion step; and phosphate by the Fiske and Subbarow method as described by Gomori (13).

Experiments with ferrous ion were carried out as previously described (8); that is, the amount of iron, enzyme, and phosphate when present were added as indicated in each experiment so that the final volume of the solution was 0.5 or 1.0 ml. Dried air was bubbled through the solution at room temperature and the enzyme activity measured (14, 15) at various intervals, usually 2, 5, and 15 minutes. For the kinetic studies the amount of ferric ion formed was determined spectrophotometrically by adding samples of reaction mixtures to one of two balanced cuvettes containing 1 N sulfuric acid and scanning the range of 400 - 304 nm. An $E_M(A_{304} - A_{400})$ of 2.23 x 10^3 x $M^{-1}cm^{-1}$ was used to determine the amount of ferric ion (16).

In the experiments in which ferric ion and hydrogen peroxide were used, the final volume of the test solution after addition of all reagents was 0.5 ml. Anaerobic conditions, where indicated, were established by equilibrating the enzyme solution with nitrogen before adding the reagents and then bubbling nitrogen through the solution during incubation.

Tryptic Digestion. Partial digestion of myeloperoxidase by trypsin was carried out according to the following procedure. Eight mg of myeloperoxidase were treated with 0.3 ml of 0.67 $\underline{N}$ hydrochloric acid for 45 minutes at 0° and neutralized with 0.1 ml of $2\,\underline{N}$ sodium hydroxide and 1.6 ml of potassium phosphate buffer (pH 7.8, 0.1 $\underline{M}$). For digestion 0.14 ml of crystalline trypsin (2.5 mg/ml) was added and the solution incubated for two hours at 37° and centrifuged to yield a clear greenish-yellow solution (S-1) and a dark green-brown precipitate (R-1). The precipitate (R-1) was resuspended in 1.7 ml of phosphate buffer containing 0.1 ml of the trypsin solution, digested for two hours at 37°, and again centrifuged to yield a second clear greenish-yellow solution (S-2) and more dark greenish precipitate (R-2). A third two-hour digestion of the precipitate (R-2) with 0.05 ml of trypsin solution in 1 ml of phosphate buffer yielded a 3 ml soluble fraction (S-3) and a small amount of residue (R-3).

Spectrophotometric measurements were made on a Process and Instrument recording spectrophotometer, which has a scale expander so that the absorbancies of 0 to 0.100 can be read over an 11 inch scale within accuracy of $\pm$ 0.002.

RESULTS

The Requirement for Phosphate in the Ferrous Ion Inactivation of Myeloperoxidase. Fig. 1 shows that when ferrous sulfate is added to a solution of myeloperoxidase under aerobic conditions, both the appearance of ferric ion and loss of enzyme activity increase linearly with increasing phosphate concentration. The proportional increase in ferric ion formation and myeloperoxidase inactivation continues to completion after 15 minutes; at the highest phosphate concentration, essentially all of the enzyme activity is destroyed and all the iron converted to the ferric state.

The Analysis of Precipitate Formed in the Presence of Phosphate. The reaction mixtures which contain phosphate become turbid in the presence of iron. The precipitate obtained was analyzed for inorganic ions. Table I shows the results of analyses of the washed precipitates obtained from several reaction mixtures. Iron, ammonia, and phosphate were found in a molar ratio of 1:1.12:1.93. These three ions accounted for 76% of the weight of the dried solid. Acidic hydrolysis and chromatography indicated that some ninhydrin material was present. Sulfate was not found, contrary to our earlier report (8) based on qualitative tests which erroneously suggested its presence.

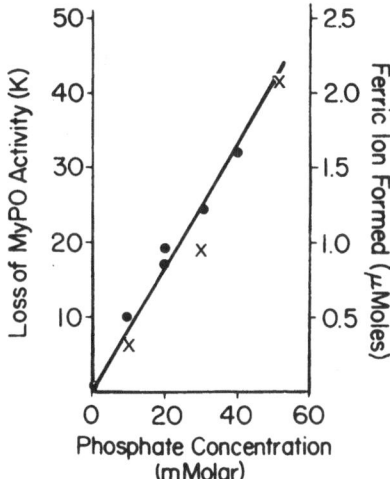

Fig. 1.   Phosphate dependence of the inactivation of myeloperoxidase by iron (II) and oxygen. Dried air was bubbled through a solution containing 100 K of peroxidase in 1 ml of 10% ammonium sulfate containing 0.5 μmoles of ferrous sulfate.   Ferric ion formation is indicated by (x) and the right hand ordinate; loss of peroxidase activity (•) is indicated by the left hand ordinate.

TABLE I.   Analysis of Precipitate.[a]

Element analyzed	Aliquot weight	Found	Calculated per mg.	Atomic ratio to Fe
	mg.	μg.	μ atoms	
Iron	0.02	3.3	2.93	1.0
	0.04	6.3	2.81	
Phosphorous	0.08	13.7	5.53	1.93
	0.08	13.7	5.55	
Ammonia	9.60	440.0	3.20	1.12
	4.16	190.0	3.26	

[a]The precipitate obtained when myeloperoxidase was inactivated by ferrous sulfate in the presence of air was washed twice with 10% ammonium sulfate and twice with distilled water.   After drying with methanol and ethyl ether, it was further dried in a vacuum desiccator.   For iron and phosphorous analyses, a sample was prepared by dissolving 2.0 mg in 100 ml of 0.08 N hydrochloric acid, and treated as described in the text under Methods.

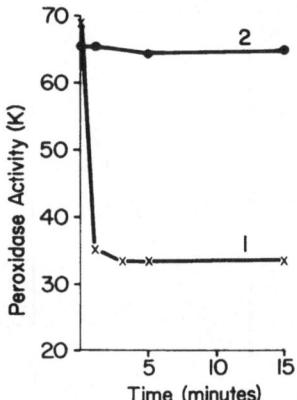

Fig. 2.   Effect of phosphate on inactivation of myeloperoxidase by iron
(III).   Curve 1 represents the peroxidase activity at the times indicated
when a solution containing 0.5 µ mole   of ferric sulfate is added to 0.5
ml of 10% ammonium sulfate (pH 6.2) containing the peroxidase and a
stream of nitrogen bubbled through the reaction vessel. Curve 2, conditions were the same as Curve 1 except for the presence of 50 µ moles of
potassium phosphate.

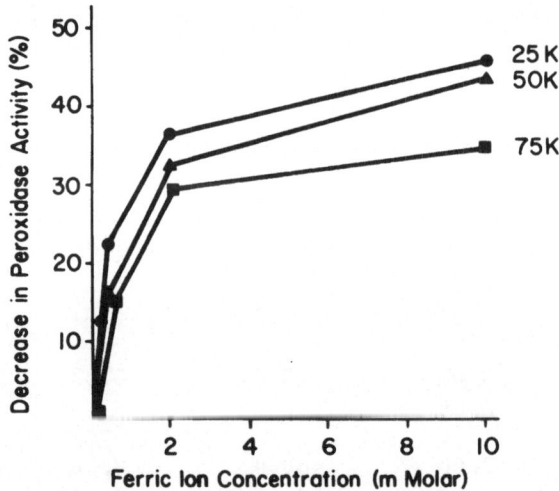

Fig. 3.   Effect of various concentrations of ferric ion on the inactivation
of myeloperoxidase at three levels of enzyme activity. Each point on the
curve represents the loss of enzyme activity two minutes after the addition of the indicated quantities of ferric sulfate to 0.5 ml of 10% ammonium
sulfate solution of the enzyme (pH 6.2) where activity in Curve 1 was 25 K,
Curve 2, 50 K, and Curve 3, 75 K. Note that less than 50% of the activity was lost in each case.

Inactivation of Myeloperoxidase by Ferric Ion.   Fig. 2, Curve 1 shows that the addition of 1 mM ferric sulfate to a myeloperoxidase solution in 10% ammonium sulfate inactivates approximately 50% of the enzyme under anaerobic conditions. The reaction is rapid and is essentially complete in one minute.   Curve 2, Fig. 2 demonstrates that in the presence of 100 mM phosphate there is no loss of peroxidase activity with the ferric ion. At lower concentrations of phosphate only a partial loss of enzyme activity is observed. These observations explain why no inactivation was obtained with ferric ion in our earlier reports (8) which were carried out with enzyme solutions contaminated with relatively high phosphate concentrations accumulated during the purification procedure (14). The dependence of myeloperoxidase inactivation on ferric ion concentration at three enzyme concentrations is seen in Fig. 3.   In Curves 1 and 2, the amount of inactivation appears to level off as it approaches 50%.   At the highest myeloperoxidase concentration the amount of enzyme inactivation also appears to be levelling off, although the percentage of inactivation observed (36%) is significantly less than 50%.   In similar experiments, which were carried out aerobically, the same percentage of inactivation took place as was obtained anaerobically (Curve 1 of Fig. 2).   In an experiment not shown, 10 mM ferric ion added to 50 K of myeloperoxidase produced a decrease of 22 K, which is the same decrease obtained at 5 mM.

Effect of Hydrogen Peroxide on the Ferric Ion Inactivation of Myeloperoxidase. When 48.5 K of myeloperoxidase were treated with hydrogen peroxide (0.18 mM) in the absence of substrate, a 9.3% loss in peroxidase activity was observed (Table II). Hydrogen peroxide in combination with ferric ions produced an extensive inactivation of myeloperoxidase in a two-step reaction sequence - (experiments 5 and 6 of Table II). Ferric ion produced a 44% decrease in peroxidase activity (experiment 5), and the addition of hydrogen peroxide resulted in additional loss of 48% (experiment 6).   In contrast to this the presence of ferrous ions did not cause enzyme inactivation, nor did it enhance the effect of hydrogen peroxide on the enzyme under these conditions.

Effect of Order of Addition on the Ferric Ion Hydrogen Peroxide System.   Throughout these studies we observed that when ferric ion was added to a solution of the enzyme before the addition of hydrogen peroxide, a greater total inactivation occurred after the addition of the peroxide. This is illustrated in Table III, Experiments 1 and 2, where the final activity is decreased to 14.5 K when iron was added first but only decreased to 26 K when hydrogen peroxide was added before the iron. To determine whether or not this difference resulted from an enzymatic decomposition of hydrogen peroxide which effectively reduces the hydrogen peroxide

TABLE II.   The effect of Ferrous and Ferric Ions on the Inactivation of Myeloperoxidase by H_2O_2.[a]

Experiment no.	Additions	Peroxidase activity, K/ml	Loss of activity, %
1	None	48.5	--
2	+H_2O_2	44	9.3
3	+Fe^{+2}	48.5	0
4	+Fe^{+2}+H_2O_2	42	12.4
5	+Fe^{+3}	27	44.4
6	+Fe^{+3}+H_2O_2	4	92

[a]In each experiment the amount of reactants used per 0.5 ml test solution were: H_2O_2, 0.09 µmole; ferrous sulfate, 0.5 µmole; ferric sulfate, 0.5 µmole. Aliquots were analyzed for peroxidase activity after 2 - 5 minutes. Loss of activity is the percent decrease taking the control as 100.

concentration before the iron is added, the time which lapsed between the addition of hydrogen peroxide and ferric ion was reduced from two minutes (used in most experiments) to 45 seconds. Under these conditions the total inactivation (iron plus peroxide) was independent of the order of addition (Experiment 3, Table III). A second addition of hydrogen peroxide, however, in the presence of iron did enhance the inactivation.

pH Optima of the Ferric Ion and Ferric Ion-Hydrogen Peroxide Systems. The optimum for the inactivation of myeloperoxidase by ferric ion was found to be at pH 8 (Figure 1, Curve 1). The fact that ferric ion has its optimum effect at least 1.5 pH units above the ferrous ion phosphate-oxygen system suggests that its attack on the enzyme differs from that of the ferrous ion system. After addition of hydrogen peroxide (0.18 mM) to the above reaction mixture, a further inactivation of the peroxidase results; but this system has (Figure 4, Curve 2) the same pH optimum as the ferrous ion oxygen system (8). Curve 3 shows that at a higher hydrogen peroxide concentration (0.18 mM) the maximal inactivation occurs at pH 8. The higher hydrogen peroxide concentration was necessary in order to obtain significant inactivation. The significance of the observation that the pH optimum for the independent inacti-

TABLE III. Effect of Order of Addition of Fe^{+3} and H_2O_2 on Inactivation of Myeloperoxidase.[a]

Experiment no.	Additions	Time of additions, seconds,	Peroxidase activity, K
1	None	--	53
	$+Fe^{+3}$	0	30.5
	$+H_2O_2$	120	14.5
2	None	--	53
	$+H_2O_2$	0	44
	$+Fe^{+3}$	120	26
3	None	--	44
	$+H_2O_2$	0	43
	$+Fe^{+3}$	45	11
4	None	--	45
	$+Fe^{+3}$	0	27.5
	$+H_2O_2$	120	16
	$+H_2O_2$	120	7.5

[a]In each experiment myeloperoxidase was incubated for two minutes in 0.5 ml of 10% ammonium sulfate at pH 6.2 in the presence of ferric sulfate (0.5 μmole) or hydrogen peroxide (0.09 μmole). At the time indicated additions were made, and the solution assayed for peroxidase activity after two minutes, except in case of Experiment 3 where an aliquot was taken for assay 25 seconds after the addition of hydrogen peroxide.

vation of myeloperoxidase by hydrogen peroxide or ferric ion is different from that of the reaction with hydrogen peroxide in the presence of ferric ion will be discussed in another section of this report.

Effect of Ferric Ion and Hydrogen Peroxide on the Absorption Spectrum of Myeloperoxidase. The effect of ferric ion on the absorption spectrum of myeloperoxidase under aerobic conditions was determined by the use of iron difference spectra. Curve 1 in Figure 5.A is a typical spectrum of myeloperoxidase from 500 nm to 260 nm. The addition of 0.2 mM ferric sulfate to both the sample and reference cuvettes produced a disappearance of the shoulder at 360 nm, a marked absorption increase

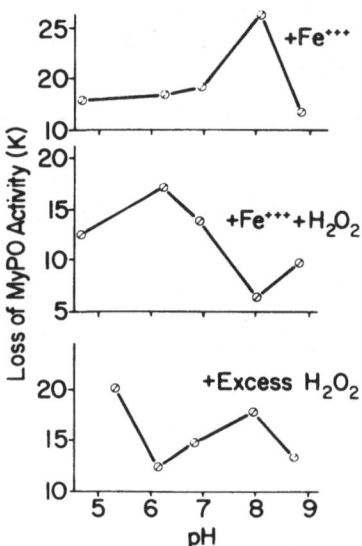

Fig. 4.   Effect of pH on the inactivation of myeloperoxidase by ferric ion and hydrogen peroxide. The top curve shows the loss of activity as a function of pH when 0.5 μmoles of ferric sulfate is added to 0.5 ml of a solution containing 50 K of myeloperoxidate in 10% ammonium sulfate. The middle curve illustrates the additional loss that is obtained when 0.09 μ moles of hydrogen peroxide is added to reaction mixtures in Curve 1.   The bottom curve represents the loss of peroxidase activity after addition of 0.45 μ moles of hydrogen peroxide to 0.4 ml of solution containing 50 K of myeloperoxidase.

in the 260–280 nm region, and an apparent decrease in the soret peak at 428 nm (Curve 2).   When hydrogen peroxide (0.18 mM) was added to the latter, the peak at 428 nm shifted to 454–458 nm without any significant change in the ultra violet region (Curve 3, Figure 5.A). Similar experiments were carried out by adding iron and hydrogen peroxide to myelo perocidase in the reverse order (Figure 5.B).   In presence of hydrogen peroxide the 428 nm shifted to 454 nm without changing the absorption at 360 nm (Curve 2, Figure 5.B) to give the typical myeloperoxidase-H_2O_2 complex II spectrum (17). Ferric ion, as shown in Figure 5.B and Curve 3, caused a marked reduction in the 360 nm region and an increase in the 260–280 absorption.   The latter increase is significantly less than that observed in Figure 5.A, Curve 3 and probably is related to hydrogen peroxide decomposition which occurs during the time interval (about 5 minutes) which lapsed between the addition of hydrogen peroxide and iron as shown in an earlier section of this paper.

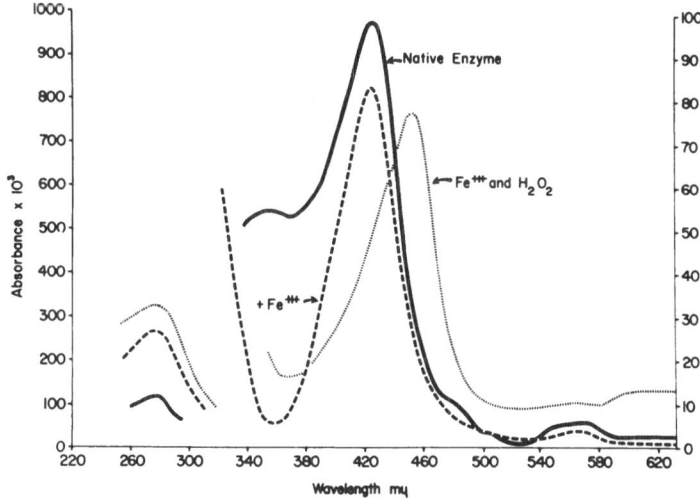

Fig. 5.A. Effect of order of addition of ferric ion and H_2O_2 on the absorption spectrum of myeloperoxidase in 10% ammonium sulfate at pH 6.2   (1) Absorption spectrum of native enzyme, (————). (2)   When both reference and sample cuvettes were made 0.2 mM to ferric sulfate, (--------).   (3) When the cuvettes in case of 2 were made 0.18 mM to hydrogen peroxide (------).   Right-hand ordinate refers to the visible spectrum, and left-hand ordinate refers to the ultraviolet.

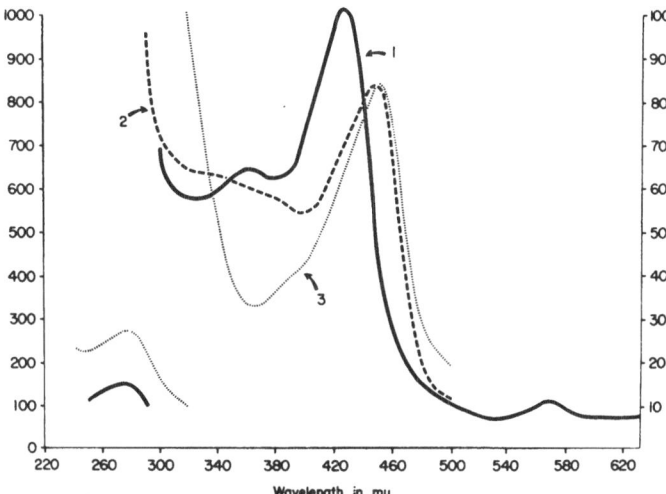

Fig. 5.B.   (1) Native enzyme (————); Curve 2, both cuvettes are 0.18 mM to H_2O_2 (-------); and Curve 3, (-------) ferric sulfate added to both cuvettes of Curve 2, to 0.2 mM.   Ordinates are the same as in 5.A.

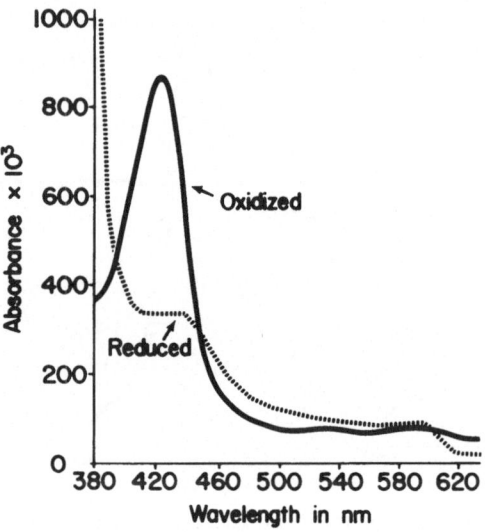

Fig. 6. The oxidized and dithionate-reduced spectra of the tryptic peptide fraction appearing within two hours after digestion of acid denatured myeloperoxidase. See S-1 under Methods.

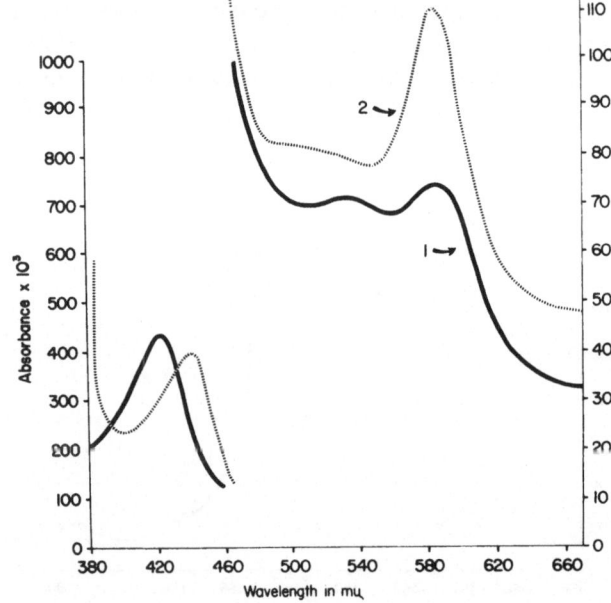

Fig. 7. Curve 1 is the oxidized and Curve 2 is the dithionite-reduced spectrum of the peptide fraction obtained after four hours digestion of acid denatured myeloperoxidase (Product S-2 under Methods). Right-hand ordinate refers to visible spectrum and left-hand ordinate refers to the Soret region.

Resolution of Myeloperoxidase Into Two Fractions of Distinctly Different Spectrophotometric Properties. From the spectrophotometric evidence in the previous section, one might speculate that myeloperoxidase contains two prosthetic groups which are affected independently by ferric ion and hydrogen peroxide. In this connection, two components of myeloperoxidase were separated by partial digestion of acid denatured enzyme with trypsin at pH 7.8 as described in the methods section. The material solubilized after two hours, fraction (S-1) has a major absorption maximum at 424 nm and minor peaks at 540 nm and 590 nm (Fig. 6, oxidized). The peak at 424 nm of this fraction on reduction (Fig. 6, reduced) is bleached, and a small increase in the absorption of the 450 nm to 600 nm region appears. The material which was solubilized after 4 to 6 hours of digestion with trypsin, fractions S-2, also had a major absorption maximum at 424 nm and minor absorption peaks at 540 nm and 590 nm as shown in Fig. 7, Curve 1. On reduction, the 424 nm peak shifted to 440 nm, the absorption at 590 nm increased, and the peak at 540 nm disappeared, Curve 2.

The absorption spectra of the oxidized and the reduced forms of the latter fraction are typical of a hemeprotein, while those of the former fraction have characteristics of a flavoprotein, or a linear tetrapyrrol resembling ring cleavage of a porphyrin. Careful inspection of the spectra in Figures 6 and 7 suggest that fraction S-1 is contaminated with a small amount of fraction S-3 and that the absorption at 540 nm and 590 nm is not related to the major component of S-1. This is because in the reduced state the ratios of the absorption at 440 nm and 590 nm are the same in both fractions while the absolute absorption in the 440 nm region differs by a factor of 3 (Figures 6 and 7).

Effect of Flavin Mononucleotide and Atabrine on the Inactivation of Myeloperoxidase by Ferric Ion and Hydrogen Peroxide. Flavin mononucleotide (FMN) has little effect on the inactivation of myeloperoxidase when compared to the control values (Table IV) without FMN. On addition of hydrogen peroxide to the control or to the reaction mixture containing FMN, essentially complete inactivation of the peroxidase takes place. In contrast to this, in Experiment 3 atabrine has no protective effect against the ferric ion but in the presence of added atabrine affords almost complete protection to the remaining peroxidase activity, which is rapidly lost without it.

Effect of Ferric Ion and Hydrogen Peroxide on Horseradish Peroxidase. Fig. 8 shows the effect of three concentrations of ferric sulfate on the activity of horseradish peroxidase over a ten-minute period. At

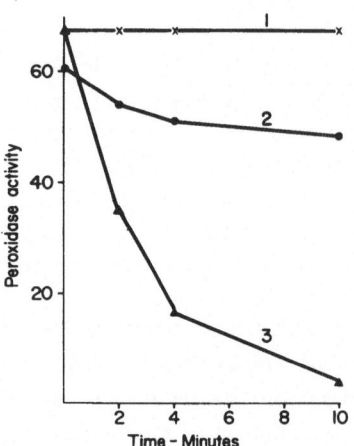

Fig. 8. Effect of ferric ion on horseradish peroxidase. Peroxidase activity expressed in terms of K was determined at the indicated times after addition of ferric sulfate to a 10% ammonium sulfate solution (pH 6.4) of horseradish peroxidase. Curve 1 was obtained at iron (III) concentration of 0.04 mM, Curve 2, at 1 mM and Curve 3, at 5 mM, under conditions described in the section on Methods for myeloperoxidase. Note that at low iron concentration, a more rapid rate of inactivation takes place in the case of myeloperoxidase (Fig. 2) than with horseradish peroxidase as seen above.

TABLE IV.   Effect of FMN and Atabrine on the Inactivation of Myeloperoxidase by Ferric Ion and Hydrogen Peroxide.[a]

Experiment No.	Additions	Activity, K
1	None	47.5
	+Fe^{+3} (0.5 mole)	26.5
	+H_2O_2 (0.09 mole)	4
2	FMN (1.0 mole)	46
	+Fe^{+3} (0.5 mole)	31.5
	+H_2O_2 (0.09 mole)	2
3	Atabrine (1.0 mole)	46
	+Fe^{+3} (0.5 mole)	26.5
	+H_2O_2 (0.09 mole)	21.5

[a]In each experiment the indicated additions of ferric sulfate and hydrogen peroxide were made in consecutive order to 0.5 ml of myeloperoxidase in 10% ammonium sulfate at pH 6.2.

0.04 mM ferric sulfate no effect on the peroxidase activity is observed; at 1 mM, 19% inactivation is observed. In contrast to this, ferric ion inactivates myeloperoxidase in a rapid reaction (complete in one minute) at a lower concentration of iron, but never produces more than a 50% loss in activity in this time interval.

DISCUSSION

The effect of ferric ion and hydrogen peroxide on myeloperoxidase illuminates properties of the enzyme that suggest the presence of two prosthetic groups of different chemical reactivity, contrary to one report indicating that the reactivity of two proposed hemes toward hydrogen peroxide were equal (17) and in agreement with a statement by K. Paul (18) that they were unequal.

Of particular significance in this light, and in relation to the mechanism of inactivation, is the fact that the alteration of the absorption spectrum of myeloperoxidase at 280 nm in the presence of ferric ion is highly suggestive of a change in conformation of the protein. It is suggested that this change exposes the site of a second component protein which becomes more susceptible to attack by oxygen reduction products, or hydrogen peroxide, and subsequently, causes complete inactivation. This reflects the implication that these products may react with one prosthetic group before iron treatment, and the second only after reaction with iron.

If ferric ion alone was the active agent, then in its presence the inactivation of myeloperoxidase should proceed anaerobically and in the absence of phosphate. Experimentally, ferric ions alone produce a partial inactivation of myeloperoxidase in a reaction which is complete within one minute, at a time when 50% or less of the enzymatic activity is destroyed. This partial inactivation of myeloperoxidase by ferric ions occurred under aerobic or anaerobic conditions, in contrast to the ferrous-ion-phosphate-oxygen system which slowly (15 minutes) produces a complete inactivation of the enzyme and has an absolute requirement for oxygen. Since ferric ion alone does not completely inactivate myeloperoxidase even under aerobic conditions, it seems likely that a reduced oxygen product formed during the oxidation of the ferrous ion may also participate in the inactivation process.

The search for the particular species of reduced oxygen between O_2^- and H_2O_2 led to the testing of H_2O_2, for the latter is known to be formed in the autoxidation of ferrous ion (19, 20). Following the reaction with ferric ion, hydrogen peroxide was found to completely inactivate the enzyme at concentrations of the peroxide which have little or no effect on the enzyme in the absence of ferric ion. It is thus very likely that the ferric ion and hydrogen peroxide participate in the ferrous ion-oxygen-phosphate system which slowly inactivates myeloperoxidase. Factors such as ATP, ADP, and imidazole, which were shown by Baker and Schultz (10, 11) to enhance the rate of inactivation of this system, may act through complexing ferric ion, as in the case of phosphate as suggested by Taube (21), "The $H_2PO_4^-$ associates much more strongly with ferric than with ferrous ion, thus, the reducing strength of ferrous ion is increased and apparently the reduction of O_2 to O_2^- becomes possible."

The relationship of the two systems [Fe (II)-$O_2^-PO_4$ and FE III - H_2O_2] is brought even closer together when one considers the effect of pH on the inactivation. Schultz and Rosenthal have reported (8) that the ferrous ion-phosphate oxygen system produces maximum myeloperoxidase inactivation at pH 6.0 to 6.5. Results reported here show that the partial inactivation of myeloperoxidase by ferric ion has a pH optimum of 8, while the action of hydrogen peroxide in the presence of ferric ion has a pH optimum of 6.0 to 6.5. Thus, the similarity of the pH optima for enzyme inactivation by hydrogen peroxide-ferric ion system and the ferrous ion system gives further support to the identification of the reaction intermediates (inasmuch as ferric ion when acting alone exhibits optimal activity at pH 8, rather than pH 6.0 to 6.5) and suggests that the total inactivation of myeloperoxidase proceeds in 2 steps and that the pH 6 to 6.5 step is the rate limiting step.

Spectrophotometric analyses of the enzyme after treatment with ferric ion and hydrogen peroxide have provided evidence for the existence of two distinct prosthetic groups. O e is a heme component which in the presence of hydrogen peroxide forms a 454 nm peak characteristic of myeloperoxidase complex II (17). The other component which accounts for the shoulder at 360 nm in the native enzyme undergoes two changes in the presence of ferric ion: first, the shoulder at 360 nm is bleached and secondly, the absorption in the 260-280 nm region increases. When ferric ion and hydrogen peroxide are both added to the native enzyme, the spectrum obtained is similar to, but not identical with, that which would be obtained by graphically adding the individual changes which occur when ferric ion and hydrogen peroxide are added to separate enzyme solutions. In our earlier report (8) on the ferrous ion inactivation of myeloperoxidase we demonstrated that a shift in the absorption maximum from 428 nm to 457 nm accompanied the inactivation. Because of the difficulty in obtaining ultraviolet absorption spectra in the presence of phosphate and a ferrous ion-ferric ion mixture, no information is available on the changes in the ultraviolet region with the ferrous ion system. That the 457 nm peak obtained with the ferrous ion system is identical with the peak obtained with ferric ion and hydrogen peroxide is supported by data obtained with kinetics and pH optima showing the probable identity of the two reactions.   Whether or not the absorption maximum obtained with hydrogen peroxide and ferric ion is identical to peroxide complex II is still unresolved.

By the use of partial trypsin digestion, a physical separation of the two components was obtained.   One fraction shows the properties of a heme-protein, in that on reduction the peak at 420 nm shifts to 440 nm, and the peak at 590 nm becomes intensified.   The other fraction shows the properties other than a heme, in that on reduction, the peak at 420 nm becomes bleached and a slight increase in absorption occurs from 450 nm to 600 nm.

There is no evidence in the literature wherein tryptic digestion, per se, results in modification of the covalent bonds found in iron-cyclic-tetra-pyrrole compounds. It is quite possible that hydrolysis of peptide bonds could result in exposing the prosthetic groups of components of myeloperoxidase, otherwise tightly encased in the globular molecule. This is evident by the alteration in the spectrum as seen here when one compares the native enzyme (reduced maxima at 475 nm and 637 nm) and the tryptic hemepeptides, one of whose reduced maxima are at 440 and 590 nm [similar to the pyridine hemochromogen (?)]; the other must represent a second kind of group which in the native enzyme is protected but in the digested enzyme is free of protection; the absorption in the soret suggests a porphyrin, but the bleaching on reduction is not consistent with heme-protein chemistry of typical iron-porphyrins.

If one must consider the presence of a non-heme prosthetic group, the evidence based on light absorption itself would lead to two other possibilities amongst the major naturally occurring compounds, the flavines and the ferrodoxins, for each of these bleach on reduction in the near ultraviolet.   Attempts to isolate a flavine component by classical techniques (22, 23) have failed, and further conjecture in this regard must rest on related compounds covalently bound to the protein with unique properties yet to be discovered. For example, atabrine, which inhibits flavine enzymes, has recently been shown to inhibit DNA-containing catalysts (24).   Evidence for a ferridoxin-like structure may depend on the nature of the labile sulfur shown to be present (7) and the current finding of a red fluorescent component formed on heating the native peroxidase for 15 to 20 minutes at 75 degrees during which a loss of only 15-21% of the initial activity takes place (25). Continuing studies on the preparation of the components of myeloperoxidase may clarify the reaction of Fe and perhaps reveal some of the factors involved in the recent interest in the reaction of Fe and proteins (26).

Finally, at least ten components of myeloperoxidase have been separated by electrophoresis on the Brinkman Free-Flow in 6 $\underline{M}$ urea, and these can be divided into two groups--those with Soret absorption that bleach on reduction and those that shift to longer wave lengths as seen in the case of the native meyloperoxidase (27, 28).

Current work in this respect showing the multiple nature of myeloperoxidase has recently been reported by Felberg and Schultz (29) on the isozymic nature of myeloperoxidase. Summaries of the relationship of the present report and other studies of the structure of myeloperoxidase will appear in Miami Winter Symposia Volume 4.

REFERENCES

1. Sagan, L., J. Theor. Bio., 14, 225 (1967).
2. Saunders, B. C., "Peroxidase," Butterworth, Washington, 1964.
3. Hager, L. P., in "Biochemistry of the Phagocytic Process," (J. Schultz, ed.), North Holland Publ., Amsterdam, 1970.
4. Zgliczynski, J. M., Stelmaszynska, T., Ostrowski, W., Naskalski, J., and Sznajd, J., Eur. J. Bioch. 4, 540 (1968).
5. Klebanoff, S. J., In "Biochemistry of the Phagocytic Process," (J. Schultz, ed.), North Holland Publ., Amsterdam, 1970.
6. Sbarra, A. J., Paul, B. B., Strauss, R. R., Jacobs, A. A., and Mitchell, G. W., Advances in Exper. Med. & Biol. 15, 209 (1960).
7. Schultz, J., and Shmukler, H., Biochem. 3, 1234 (1964).
8. Schultz, J., and Rosenthal, S., J. Biol. Chem. 234, 2486 (1959).
9. Schultz, J., and Rosenthal, S., ABSTR. Amer. Chem. Soc. 137, 35 c (1960).
10. Baker, A., and Schultz, J., Federation Proc. 22, 587 (1963).
11. Baker, A., and Schultz, J., ABSTR. Amer. Chem. Soc. 145, 49 c (1963).
12. Drabkin, D. L., J. Biol. Chem. 140, 387 (1940).
13. Gomori, G., J. Lab. Clin. Med. 27, 955 (1942).
14. Schultz, J., Gordon, A., and Shay, H., J. Am. Chem. Soc. 79, 1632 (1957).

15. Schultz, J., Shay, H., and Gruenstein, M., Cancer Res. 14, 157 (1954).
16. Rigg, T., Taylor, W., and Weiss, J., J. Chem. Phys. 22, 575 (1954).
17. Agner, K., Acta. Chem. Scan. 17, 332 (1963).
18. Paul, K. G., The Enzymes 8, 262 (1963).
19. Weiss, J., Experientia 9, 61 (1953).
20. King, J., and Davidson, N., J. Am. Chem. Soc. 80, 1542 (1958).
21. Taube, H., in "Oxygen" (Symposium, N. H. Heart Association), p. 29. Little, Brown & Co., Boston, 1965.
22. Beinert, H., The Enzymes 2, 339 (1963).
23. Hiromi, K., and Sturtevant, J., J. Biol. Chem. 240, 4662 (1965).
24. Yates, M. G., and Nason, A., J. Biol. Chem. 241, 4861 (1966).
25. Bonner, M. J., and Schultz, J., ABSTR., Federation Proc., March–April, 26 (1967).
26. Taborsky, G., and Grant, C. T., Biochemistry 5, 544 (1966).
27. Schultz, J., Felberg, N., and John, S., Biochem. Biophys. Res. Comm. 28, 543 (1967).
28. Schultz, J., and Felberg, N., Anal. Biochem. 23, 241 (1968).
29. Felberg, N., and Schultz, J., Arch. Biochem. Biophys. 148, 407 (1972).

Received 11 February, 1972

DEPENDENCE OF POLY U-DIRECTED CELL-FREE SYSTEM ON

RATIOS OF DIVALENT AND MONOVALENT CATIONS

A. S. Spirin, S. A. Bogatyreva, and E. N. Trifonov

A. N. Bakh Institute of Biochemistry
Academy of Sciences of the USSR, Moscow, USSR

An obligatory component of a protein-synthesizing cell-free system are inorganic cations: divalent ones, primarily Mg^{++}, and monovalent ones, NH_4^+ or K^+ (1). Up to the present, the optimal concentration of these cations for use in cell-free systems was chosen by varying the concentration of one cation on the background of an invariable concentration of the other. However, an absolute optimum for both cations cannot be determined in such a manner, as the concentration of the second cation itself affects the position of the optimum for the first cation. The aim of the present communication is to determine the dependence of the cell-free system activity on the ratios of Mg^{++} and NH_4^+ cations in the medium and to find the optimum of the system for the two denoted factors. A polyU-directed cell-free system of polyphenylalanine synthesis on Escherichia coli ribosomes was taken as the object of the investigation.

"NH_4^+-salted out" E. coli ribosome preparations obtained by a technique described earlier (2) were used for the cell-free system. The enzyme fraction, free from ribosomes and nucleic acids, was prepared from the E. coli extract by prolonged centrifuging at 100,000 g with a following fractionation on DEAE-cellulose (3). The total tRNA preparation was isolated by phenol deproteinization of the ribosome-free 100,000 g E. coli supernatant with a following purification on a DEAE-cellulose column (4). Enzymatic aminoacylation of the total tRNA was performed with ^{14}C-L-phenylalanine, specific activity 125 μC/μmole, followed by phenol deproteinization of the mixture and Sephadex G25 column purification (5); a frozen-dried preparation of such a ^{14}C-phenylalanyl-tRNA had a specific

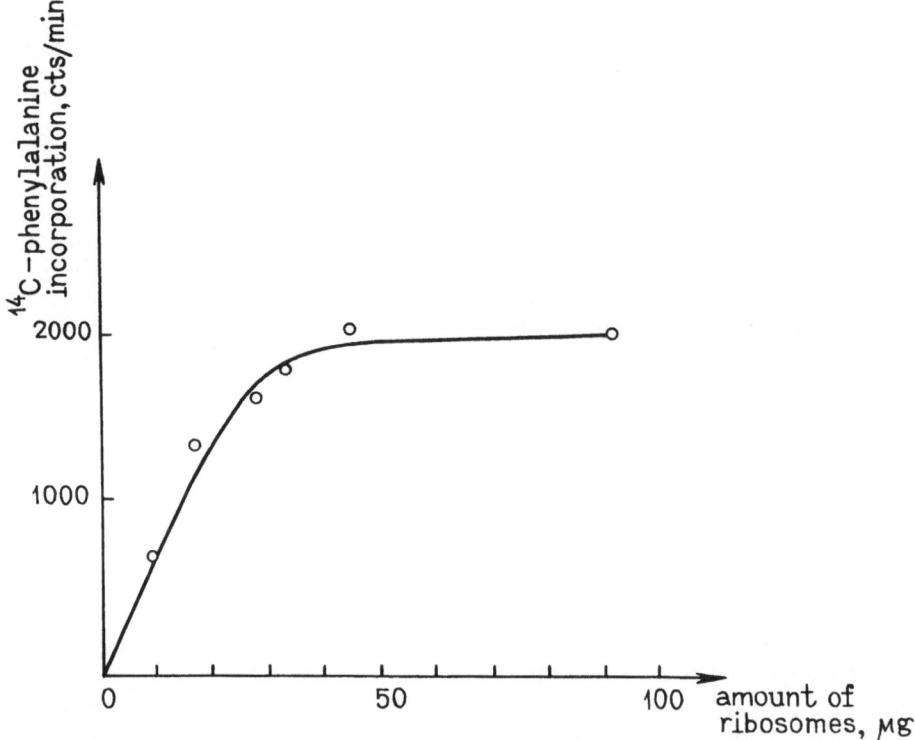

Fig. 1.   Dependence of ^{14}C-phenylalanine incorporation in the
cell-free system on the amount of ribosomes at the standard
amounts of other components (see text); 0.015 $\underline{M}$ MgCl$_2$, 0.07
$\underline{M}$ NH$_4$Cl, 37°, 15 min.

activity of 140,000 cts/min per mg of total tRNA (counting efficiency 40%).
ATP (Na$^+$-salt), GTP (Na$^+$-salt), phosphoenolpyruvate and polyU (K$^+$-
salt) were from the Reanal Co. , Hungary; the pyruvate kinase was purchased from Boehringer, Germany.

The cell-free system, in a volume of 0.36 ml, contained the follow-
ing components: $\underline{E. \ coli}$ ribosomes, 10 µg; enzyme fraction of $\underline{E. \ coli}$
70 to 100 µg (calculated as dry protein); ^{14}C-phenylalanyl-tRNA, 75 µ g
(calculated as total tRNA); polyU, 15 µg; GTP, 0.06 µmole; phosphoenolpyruvate, 1.25 µmole; pyruvate kinase (dialysed against NH$_4^+$), 12 µ g.
The Tris-buffer concentration in the system was 0.025 M; mercaptoetha-
nol was present in a concentration of 0.0025 $\underline{M}$; the final pH value was
7.3 (at 37°C); the MgCl$_2$ and NH$_4$Cl concentrations varied in different
samples. The samples were incubated for 15 min at 37°C. After incuba-

tion the samples were hydrolysed with 5% trichloroacetic acid at 90° for 15 min, then cooled, and the precipitates were deposited onto No. 4 nitro-cellulose filters (Experimental Ultrafilter Factory, Mytishchi, Moscow Region); the filters were washed with 5% trichloroacetic acid, dried at 70-90°, and their radioactivity counted with a gas-flow methane counter (counting efficiency about 40%).

It is important to note that the number of ribosomes in the fractions was small as compared to the number of all the other components of the system. It is evident that the dependence on changes in ionic conditions of the activity of the ribosomes themselves can be observed only under conditions where there is an excess of all other components. The curve of ^{14}C-phenylalanine incorporation into the polypeptide depending on the concentration of ribosomes at the above-denoted amounts of all the other components and at the standard concentration of $MgCl_2$ and NH_4Cl (see following page) is shown in Fig. 1. It is seen that the concentration of ribosomes in the fractions (10 μg per 0.36 ml) corresponds to the initial linear section of the curve, and there is a manifold reserve before emergence onto a plateau.

In all the series of experiments with varying concentrations of Mg^{++} and NH_4^+, assays were made of so-called standard samples where the concentration of $MgCl_2$ was 0.015 $\underline{M}$ and that of NH_4Cl 0.07 $\underline{M}$ ("standard conditions"). The absolute value of ^{14}C-phenylalanine incorporation in these standard assays varied somewhat from series to series, depending on the ribosome preparation or enzyme fraction; variations in the main were from 700 to 1100 cts/min per 10 μg of ribosomes. The activity of the standard samples in each series of experiments was taken as 100%. The activities of the rest of the samples were expressed as a percentage of the value of ^{14}C-phenylalanine incorporation in the standard sample.

Table I shows the results of the experiments with varying concentrations of Mg^{++} and NH_4^+ expressed in this manner. Each value is the average of two or three parallel assays. The figures shown in the squares of the Table are the results of independent series of experiments.

The presented results permit one to determine the position of the maximum of ^{14}C-phenylalanine incorporation (optimum of the system) and give a general view of the dependence of the incorporation (Q) on the concentration of Mg^{++} (C_1) and NH_4^+ (C_2). This dependence represents a surface in coordinates $[C_1, C_2, Q]$. The lines in Fig. 2 denote the levels corresponding to identical values of incorporation. The experimentally de-

TABLE I. Relative Values of ^{14}C-phenylalanine Incorporation into the Polypeptide in a Cell-free System at Different Concentrations of Mg^{++} and NH_4^+.

[NH₄⁺], mM \ [Mg], mM	2	3	4	5	5.5	6	7	8	9	10	11	12	13	13.5	14	15	16	16.5	17	18	20	22	23
10											21 32												
20							112 138		85 102														
25	5	5	26								56 76												
30							216																
40								182 235															
50						137 159					85 105												
70		11	55				115		310 210		158	150 136				100					37 39		
100										200,190 179,200 337	213 278		140										
110											200												
120										285 339		210				125							
130										240	320 420,200 290,300 324,290												
140				24								240									74	96 53	
150								110 150	110	210 280	210			160	140	150,200 153,160 130,200 175							
175											292 222	175 114				170 160							
180														123									
200						10	34		53 39	160 159 150	170,216 187,216	218,170 219			181,142 110	180,200 115,160 180			80 103 126		56 84 70		
230																		150 168					
250										75 102					152 207				137 151		71	63	
280																				117			
300										37,43 20						117 105			79		78 76		48
320																				45 80			
340																					42		
360																						32	

termined dependence of $Q(C_1)$ at fixed C_2 values (70, 100, 120, 150, 175, 200 and 250 mM NH_4Cl) and $Q(C_2)$ at fixed C_1 values (10, 11, 12, 15 and 20 mM $MgCl_2$) were used to plot these "isoactive lines." In plotting these dependences and the levels, the existing scattering of experimental data was taken into account (see Table I), so that the obtained surface $Q(C_1, C_2)$ (Fig. 2) deviated from the experimentally determined incorporation values Q by not more than by the root-mean-square estimate of error of these values (16% and less for the repeated independent series of experiments).

As a result of the conducted study the optimal ionic conditions for the synthesis of polyphenylalanine in a E. coli cell-free system was found to

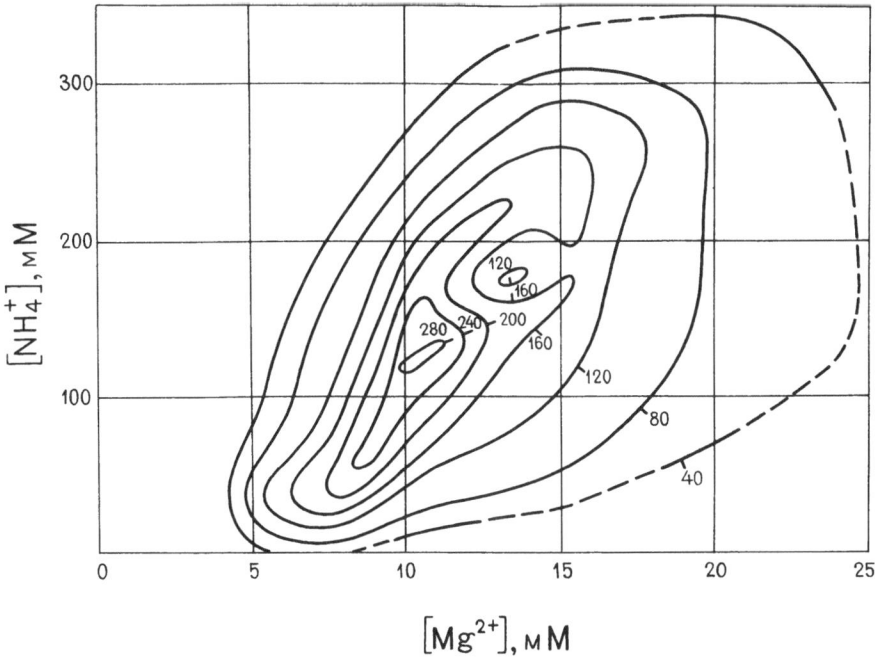

Fig. 2. Dependence of activity (^{14}C-phenylalanine incorporation) of the polyU-directed cell-free system on the concentration of Mg^{++} and NH$_4^+$. Lines correspond to levels of equal activity; figures on lines denote activity of the system expressed as a percentage to the activity of the standard system (15 m$\underline{M}$ Mg^{++}, 70 m$\underline{M}$ NH$_4^+$) taken as 100%.

be 10-11 mM MgCl$_2$ at 120-130 m$\underline{M}$ NH$_4$Cl (Fig. 2). In these conditions the activity of the cell-free system is higher in comparison with the standard conditions (15 mM MgCl$_2$, 70 m$\underline{M}$ NH$_4$Cl) by approximately three times.

Besides this distinctly expressed maximum a well-defined minimum "trough" can also be seen on one of the slopes of the obtained surface (Fig. 2) in the region of ∿13 m$\underline{M}$ Mg^{++}, 180 mM NH$_4^+$. It is seen from the presented dependence that the activity of the cell-free system is the most sensitive to variations of the MgCl$_2$ concentration in the region of about 9.5 m$\underline{M}$ Mg^{++}, 140 m$\underline{M}$ NH$_4^+$. The region with the maximum sensitivity to NH$_4$Cl fluctuations is about 9 m$\underline{M}$ Mg^{++}, 40 m$\underline{M}$ NH$_4^+$. The activity of the system is least sensitive to ion concentration fluctuations in the region of about 20 m$\underline{M}$ Mg^{++}, 200 m$\underline{M}$ NH$_4^+$.

In conclusion it should be noted that in conditions at which the maximum activity of a cell-free system is observed here (ca. 10 mM Mg^{++}, ca. 120 mM NH_4^+, 37°, pH 7.3) the association between ribosomal subparticles must be rather labile, so that at least the non-translating 70S ribosomes are somewhere near to the state of semi-dissociation (6).

REFERENCES

1. Spirin, A. S., and Gavrilova, L. P., "The Ribosome," Springer-Verlag, Berlin-Heidelberg-New York, 1969.
2. Spirin, A. S., Sofronova, M. Y., and Sabo, B., Molekularnaya Biologiya (USSR) 4, 618 (1970).
3. Traub, P., and Zillig, W., Z. Physiol. Chem. 343, 246 (1966).
4. Holley, R. W., Apgar, J., et al. J. Biol. Chem. 236, 200 (1961).
5. Levin, J. G., and Nirenberg, M., J. Mol. Biol. 34 467 (1968).
6. Sabo, B., and Spirin, A. S., Molekularnaya Biologiya (USSR) 4, 628 (1970).

Received 26 March, 1971

INFORMATIONAL BIOPOLYMER STRUCTURE IN

EARLY LIVING FORMS

M. O. Dayhoff, P. J. McLaughlin, W. C. Barker, and L. T. Hunt

National Biomedical Research Foundation
Georgetown University Medical Center
Washington, D. C. 20007

INTRODUCTION

The conservative nature of the evolutionary process has preserved in all living species "relics" of the biochemical nature of ancient organisms, including those which lived before the ordinary fossil record was formed. Protein and nucleic acid sequences provide evidence which is particularly amenable to quantitation because certain generally applicable and theoretically understandable principles have characterized their evolution; therefore, it is possible to apply logical and statistical methods to the data to extrapolate into the past. The biochemical evidence in protein and nucleic acid structures can provide information from which to construct a phylogenetic tree encompassing all the main lines of living organisms. Further, it can reveal the nature of ancient molecules in ancestral organisms and of the evolutionary process, and finally it can provide a time scale for measuring the relative antiquity of species divergences and gene duplications. By tracing the course of evolution of the more slowly evolving molecules, we may eventually be able to infer much about the protein and nucleic acid structures in what we have termed the "proto-organism"--the most recent common ancestor of all presently living species, which existed more than 3 billion years ago--and even in the more primitive organisms in the evolutionary line that preceded it. Potentially much information regarding the structure of informational biopolymers can be elicited; there is sufficient DNA to code for several thousand different kinds of proteins (of 500 residues) in the bacterium Escherichia coli and for more than a million kinds in a human being. Many related proteins and nucleic acids are found throughout a wide range of organisms--from bacteria to multicellular animals and plants. Chemical structures from five such families of informational molecules in very diverse organisms have now been examined: cytochrome c, ferredoxin, trypsin, transfer ribonucleic acid (RNA), and 5S ribosomal RNA (1). Even from these few, we can draw some interesting inferences about early living organisms.

The development of replicating genetic nucleic acids, the genetic coding apparatus, and the earliest coded proteins were particularly important in the evolution of life. Because very similar metabolic systems are used by all living things so far investigated, it is very likely that certain of these had evolved by the time of the proto-organism (2, 3). In all living organisms, genetic nucleic acid (DNA) is a polymer composed of only four different kinds of nucleotide monomers. The coded proteins are polymers composed of twenty different kinds of amino acids. Each successive triplet of DNA nucleotides ultimately directs the incorporation of a particular amino acid into the succession of amino acids in the protein chain. The coding process has a very complex chemistry, involving many nucleic acid and protein catalysts.

The earliest species divergences leading to lines which survive today is now thought by biologists to have given rise to various groups of prokaryotes (bacteria and blue-green algae) (4, 5). These very small organisms reproduce by direct division, and their genetic complement usually consists of one double helical loop of desoxyribonucleic acid (DNA) without any associated structural protein and without a nuclear membrane to separate it from the cytoplasm. From one of these prokaryote lines evolved the eukaryotes, the higher organisms which include the animals, the green plants, the other algae, and most protozoa and fungi (6). Eukaryote cells have a nuclear membrane surrounding the chromosomes and divide by mitosis or meiosis. The double helical strands of DNA in each chromosome are super-coiled, and structural protein is associated with the DNA.

It is not possible to decide, from the meager sequence evidence so far accumulated, which of the three large eukaryotic groups (plants, animals, and fungi) diverged first from the line leading to the other two. No proteins or nucleic acids from protozoa have been sequenced. Further, there is as yet little sequence evidence from which to derive the evolutionary history of the prokaryotes.

CYTOCHROME C

The variation in protein structure from one biological group to another is illustrated by the first half of an alignment of sequences of cytochrome c, a protein of fundamental importance which functions in mitochondrial electron transfer in higher organisms (see Fig. 1). Also shown is a clearly though very distantly related protein, cytochrome c_2, from a simple photosynthetic bacterium, Rhodospirillum rubrum, in which the protein is used in photosynthetic electron transfer.

All these sequences, even though from very diverse organisms, are so similar that there is no difficulty in aligning them. About 60% of the total number of amino acids at corresponding positions are identical in wheat and human chains while 30% are identical in human and R. rubrum. Certain positions contain the same amino acid in all of the sequences; others are filled by any one of several amino acids which are usually of similar shape or chemical properties.

Evidently the evolution of this protein has proceeded through a number of "point mutations," replacements of one amino acid for another. These changes have in turn been caused by the exchange of one nucleotide for another in the three-nucleotide codon of the DNA. Occasionally there have been insertions or deletions of one or a few amino acids. The mutations which we can observe in present-day sequences are those which have been accepted by natural selection and fixed in the populations in which they occurred. Many other changes have been eliminated.

The different positions in an alignment of sequences may be treated as separate traits from which a phylogeny of the species can be deduced in much the same way as from other biological traits (7). If a few of these traits are ultimately demonstrated to be linked or of varying value in establishing the order of past events, the reliability of the inferences will be improved by the incorporation of such details into the calculations. With the small amount of information presently available, it is not possible to gain significantly by such procedures. The mathematical techniques used reflect the premise that when identical changes in a trait are observed in two species, the most likely explanation is that the change occurred in a single organism which was a common ancestor to both. In general, to derive the topological configuration of the evolutionary tree we search for that tree requiring the minimum possible number of amino acid changes in the ancestral organisms. Because cytochrome c changes slowly and because there are few identical mutations occurring indepen-

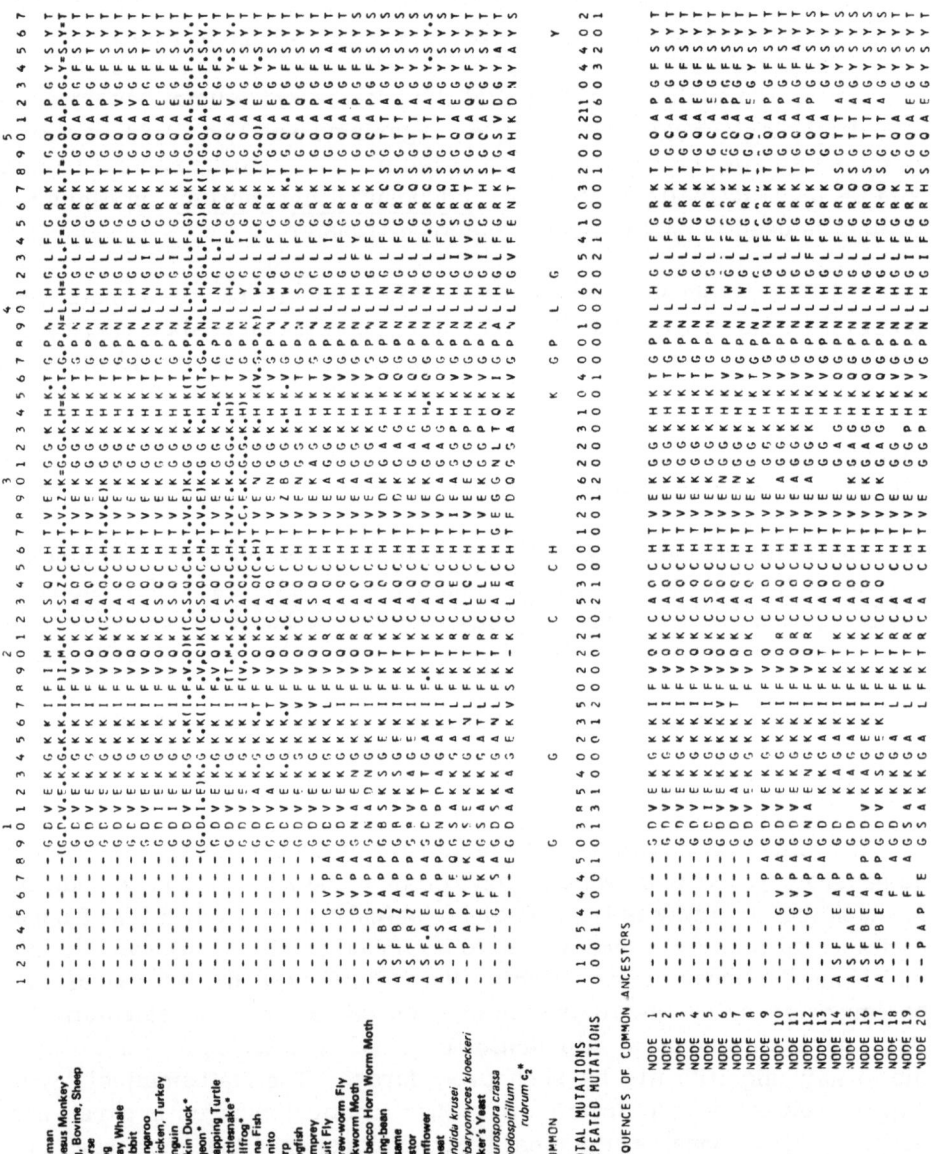

dently in different species, this tree would be expected to reflect quite closely the true historical course of events.

In principle, the computer tries every pattern of connected branches and all probable sequences for the ancestral structures at the branch points (nodes). For each pattern of branches we calculate the tree size, the minimum number of mutations which could have occurred during the history of the tree; this is derived from a count of the total number of changes between adjacent sequences at the nodes and at the ends of the branches. The pattern yielding the smallest tree size is chosen. The detailed procedure is described in Chapter 2 of the "Atlas of Protein Sequence and Structure 1972" (7). Fig. 2 shows the phylogenetic tree which we have derived from the cytochrome c data. This tree, within the limitations of the small quantity of data now available, does not contradict the known relationships shown by morphological evidence (8). Each point on the tree represents a definite time, a particular species, and a definite

Fig. 1. The first half of the cytochrome c proteins from 33 species of eukaryotes and the related cytochrome c₂ from a bacterium. Each amino acid residue is represented by a single letter. Sequences of closely related species are similar. Even though the proteins come from widely different biological groups, only a single amino acid is found in 11 of the 57 positions shown. Very few changes have been accepted in the other positions. Dashes have been inserted to maintain the alignment where insertions or deletions of genetic material must have occurred. Probable ancestral sequences at the nodes of Fig. 2, which were derived by the computer (7), are shown in the lower part of the figure. In the few places that are blank, two or more amino acids are equally likely to have been present. At each position, the total number of mutations and the number of repeated mutations are shown. These counts were made from the 28 eukaryote sequences which are well determined (those marked by * are excluded). It can readily be seen that some positions have not changed at all, while some have changed frequently, sometimes to be filled by any one of many amino acids as at position 52, and sometimes to be filled by first one and then the other of a single pair as at position 54. The abbreviations for the amino acids are: A, alanine; B, aspartic acid or asparagine; D, aspartic acid; E, glutamic acid; F, phenylalanine; G, glycine; H, histidine; I, isoleucine; K, lysine; L, leucine; M, methionine; N, asparagine; P, proline; Q, glutamine; R, arginine; S, serine; T, threonine; V, valine; W, tryptophan; Y, tyrosine; Z, glutamic acid or glutamine. (From the "Atlas of Protein Sequence and Structure 1972").

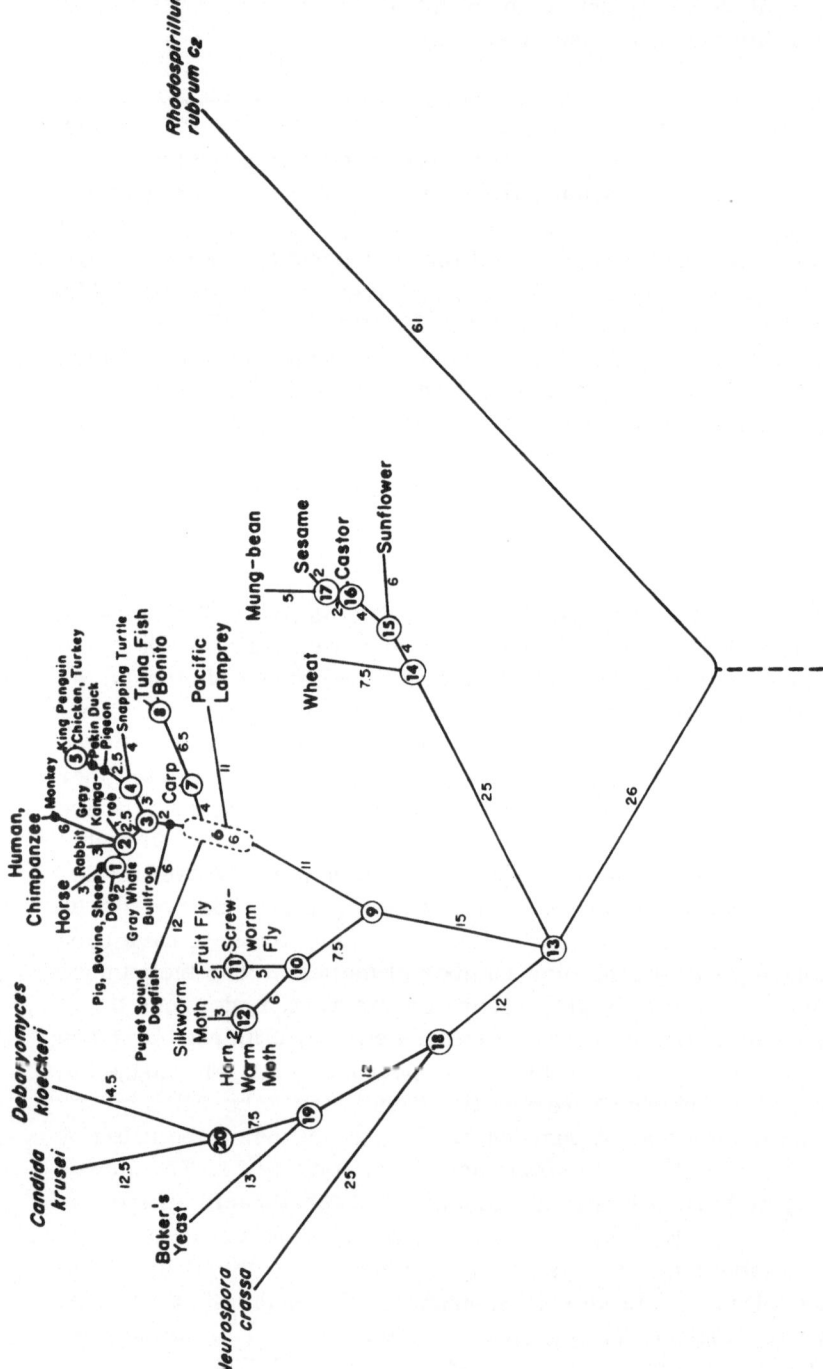

Fig. 2. Phylogenetic tree of cytochrome c. The topology has been derived from the sequences as explained elsewhere (7). The numbers of inferred amino acid changes per 100 links are shown on the branches of the tree. The point of earliest time cannot be determined directly from the sequences; we have placed it by assuming that, on the average, the cytochrome sequences in different species change at the same rate. (From the "Atlas of Protein Sequence and Structure 1972").

protein structure within the majority of individuals of this species. There is a "point of earliest time" on any such tree. Radiating from this point, time increases on all branches. Protein sequences from living organisms all lie at the ends of branches which represent the present time. The series of branchings in the tree then indicates the relative order in which the proteins (and the species containing them) became distinct from one another. The major groups fall clearly into the topological configuration shown. However, the more recent divergences cannot always be resolved because either there have been no mutations in cytochrome c or the evidence of these mutations has been obliterated by subsequent changes. The branch lengths are proportional to the number of changes which we infer to have occurred in each interval. The actual count described above has been corrected for the superimposed and parallel mutations which must have occurred (9). The location of the point of earliest time, that is, the connection of the trunk to the branching structure, cannot be inferred directly from the sequences, but must be estimated from other considerations. It can be seen that the sequences from all organisms have changed over the period of time represented by the tree, even though the morphology of some organisms may still appear primitive. Because there are insufficient data to calculate precise differential probabilities of mutation in different lines, we have usually taken a point equidistant from all of the present-day sequences as the point of earliest time.

The ancestral sequences generated by the computer are shown at the bottom of Fig. 1. If one amino acid is clearly most likely to have been ancestral, it is shown. Where two or more amino acids are nearly equal contenders, the position is left blank. Most positions of ancestral cytochrome c structures can thus be inferred because the evolution of this protein has been quite conservative. At node 13 is the sequence which is probably very close to that of the cytochrome c which functioned in the common ancestor of all eukaryotic organisms. With the advent of methods to synthesize proteins in the laboratory (10, 11), these ancestral structures take on added significance; they may one day be produced and their properties measured. Some of the chemical capabilities of ancient organisms may thus eventually be known.

At each position in the alignment of sequences, the total number of mutations, and the number of repeated mutations are shown. It can readily be seen that some positions have not changed at all, while some have changed frequently, sometimes to be filled by any one of many amino acids (see position 52) and sometimes to be filled by first one and then the other of a single pair (see position 54). From such data a more

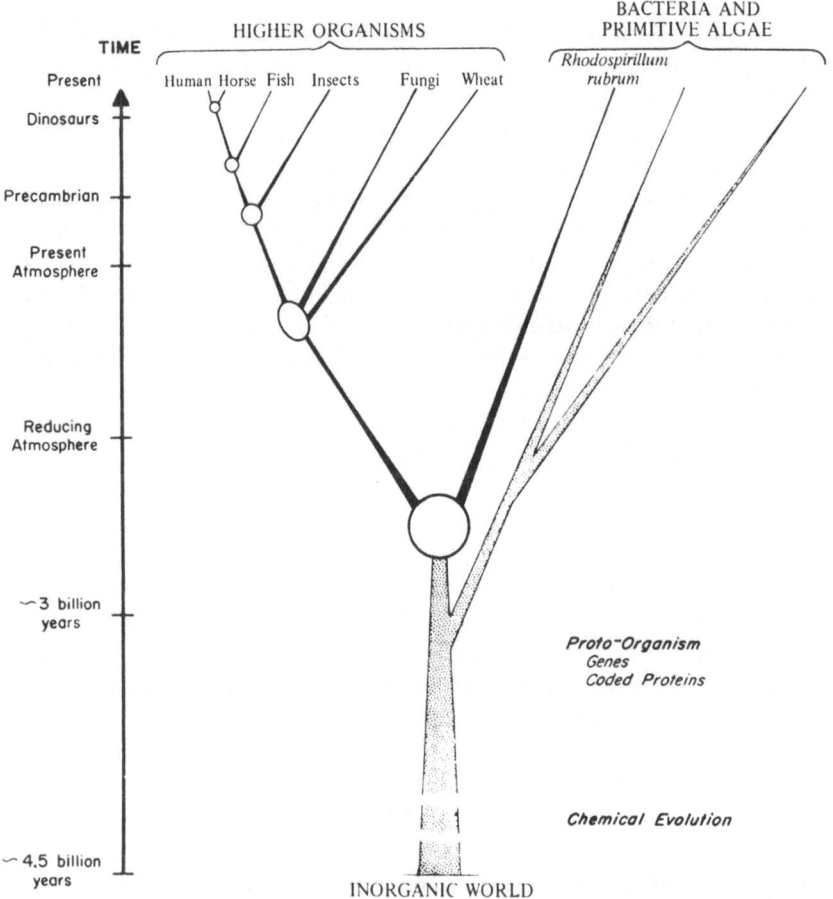

Fig. 3. Temporal aspects of the cytochrome phylogenetic tree.   Details of the upper part of the tree, shown in black, are derived from information in the cytochrome c sequences. The size of any node reflects
the uncertainty in its position.   The stippled parts of the tree are inferred from the details of biochemical and biological structures and from
geological and fossil evidence. Fossils resembling present-day bacteria
and blue-green algae have been found in chert more than three billion
years old (15, 16).   All living phyla seem to be derived from a proto-
organism which possessed many metabolic capacities, nucleic acid genes,
and protein enzymes (produced by the genetic code). Preceding the time
of the proto-organism, even more primitive organisms evolved the genetic code and the metabolism of nucleic acids and small molecules (17 -
20).   The level of the atmosphere is inferred from the oxidation states
of ancient sediments (21   - 23). (From the "Atlas of Protein Sequence
and Structure 1969").

refined understanding of the evolutionary process at the molecular level will eventually emerge.

We are interested in knowing the relative geological times associated with the events of the phylogenetic tree of cytochrome c. We estimate these times from the number of mutations on the branches. For each family of proteins, the average rate of accumulation of mutations is different. However, the risk of mutation within a family such as cytochrome c is evidently rather constant, at least for nucleated organisms, over the interval during which the species under consideration diverged (12). The variation of branch lengths which correspond to the same time interval is no more than one would expect for a random process of mutation. Therefore, in order to estimate the times of past events, we have assumed that the evolutionary rate for a given protein is reasonably constant. The proportionality constant of time per mutation must be fixed from considerations such as the divergence point of the fish and mammalian lines. Paleontological evidence (13, 14) places this particular divergence at about 400 million years ago. The cytochrome c tree places it at 11.5 accepted point mutations per 100 amino acid links (PAMs) on the average. Therefore, 400 million years correspond to 11.5 PAMs. Using the constant rate assumption, let us consider the implications for the times of divergence of the major groups.

The information in Fig. 2 is redrawn in Fig. 3 with the meaning of the branch lengths altered. Now the vertical axis is proportional to time, with the present day at the top. The positions of the nodes are derived by averaging the number of mutations on branches which represent the same time interval. The connection of the branches to the trunk was presumed to be equidistant from the bacterial and all the other sequences. From this one family of clearly related proteins, estimates of the time relationships on an extensive phylogenetic tree of life can be derived.

Because the basic metabolic processes in eukaryotes and prokaryotes are similar, we would expect that a great many families of proteins involved in these processes would possess similarities comparable to those seen in cytochrome c. Indeed, work on small catalytically active portions of many proteins confirms this. However, only two other families have been studied extensively as yet: ferredoxin, an electron transport protein, and trypsin, an enzyme which breaks down proteins.

FERREDOXIN

The evolutionary tree derived from ferredoxin is particularly interesting in connection with the study of the early evolution of life (1). Ferredoxin is found in anaerobic and photosynthetic bacteria, in blue-green and green algae, and in the plastids of higher plants. The protein is bound to iron and inorganic sulfur through some of its cysteine residues. Ferredoxin is the most electronegative metabolic protein known, with a potential close to that of molecular hydrogen. It participates in a wide variety of biochemical processes fundamental to life, including photosynthesis, nitrogen fixation, sulfate reduction, and other oxidation-reduction reactions. It may have achieved these functions very early in the differentiation of the biological kingdoms, and its structure may have been conserved strongly ever since. The evidence from ferredoxins of higher plants indicates a rate of change of the protein comparable to that of cytochrome c.

The overall topology of the evolutionary tree of ferredoxin, shown in Fig. 4, is clearly established by the sequences. The ferredoxin sequences from plant plastids are almost twice as long as those from anaerobic bacteria, whereas the one from Chromatium, a photosynthetic bacterium, is intermediate. A quantitative estimate of the evolutionary distance between the plant plastids and the bacteria in terms of point mutations is impossible from these sequences because of the many changes in length. It must be a very long distance. There has clearly been a duplication of genetic material in the bacterial line, and the two halves of the molecule are still quite similar (24 - 26). The plant line also appears to have incorporated duplications of genetic material, possibly independently of the event occurring in bacteria.

Further, the biological meaning of the first divergence on the tree is not clear. It is likely that plastids are symbionts acquired by a primitive higher organism at about the time of the divergence of plant and animal lines. The gene for plant ferredoxin may have derived either from the nuclear line or the plastid line. If it is a plastid gene, then the earliest branching point on the tree refers to the plastid--bacterium divergence rather than to the plant-nucleus--bacterium divergence.

Many questions remain to be answered regarding the history of ferredoxin. In drawing the tree, we have assumed a comparable rate of change of the sequences in bacteria and higher plants, an assumption which cannot be checked with the information presently available.

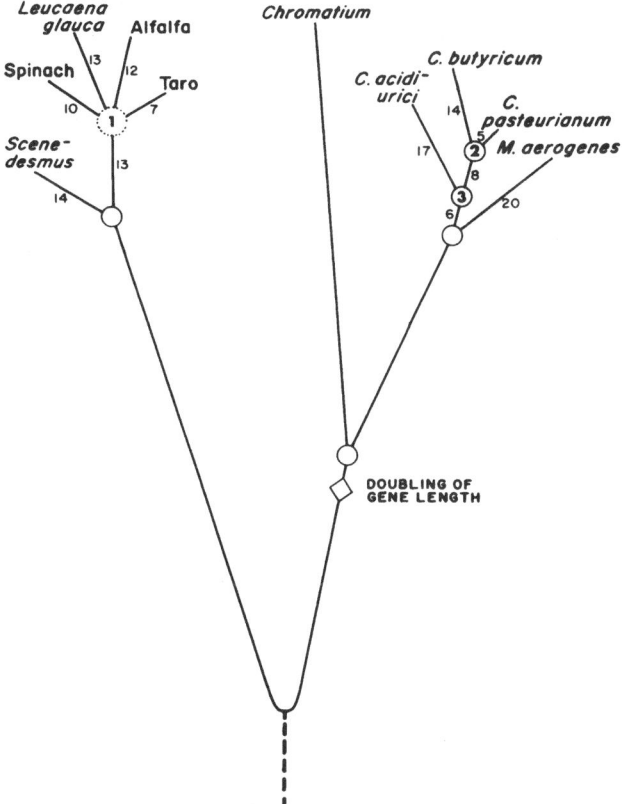

Fig. 4. Evolutionary tree of ferredoxin. The sequences from bacteria and from plants are clearly related, but very distantly so. The doubling of gene length in the bacterial line probably occurred as shown here. The details of duplications in the plant line are obscured by the many subsequent changes. Within the groups of plants and the group of nonphotosynthetic bacteria, the relative branch lengths have been estimated from the point mutations. (From the "Atlas of Protein Sequence and Structure 1972").

THE TRYPSIN FAMILY

The third family in which proteins from both the eukaryotes and prokaryotes have been sequenced is a group related to the digestive enzyme trypsin. Among the higher organisms there has been a proliferation of different enzymes, related in structure, but each having a slightly dif-

ferent catalytic specificity and physiological purpose. Trypsin, chymotrypsin, and elastase are enzymes formed in the pancreas and secreted
into the small intestine where they break down ingested proteins. Each
of these enzymes hydrolyzes specific peptide bonds in protein chains.
Thrombin is synthesized in the liver and is secreted into the blood where
it participates in the complex blood-clotting mechanism by cleaving the
protein fibrinogen. These multiple structures arose when chromosomal
aberrations occurred during replication of the germ line cells and resulted in the addition of copies of the complete gene. These copies may
occur by the duplication of a small or an extensive stretch of DNA within
a chromosome or as the result of the acceptance of a whole extra chromosome or of an entire extra set of chromosomes. All of these kinds of
aberrations are seen to occur today (27, 28). The separate gene copies
could accept different point mutations and gradually they, and the resulting protein sequences, would become quite different. The three-dimensional structures of three of these proteins, α -chymotrypsin (29), elastase (30), and trypsin (31), have been determined. It is evident that,
despite fairly extensive differences in the protein sequences, the major
aspects of the three-dimensional conformations of these molecules have
been conserved.

The proteins from the eukaryote and prokaryote superkingdoms are
more dissimilar than in the case of cytochrome c. However, statistical
methods show that it is very unlikely (P $\leq$ 0.0001) that the overall sequence similarity could have occurred by chance (1). A reasonable alignment of the sequences can be made, from which the phylogenetic tree of
Fig. 5 is derived.

If the time of the divergence of the lines leading to the mammals and
to the cartilaginous fishes is taken as about 400 million years ago, and if
the rate of change of these proteins has been constant and similar in different lines, then we can estimate that the duplications which gave rise
to the separate genes for trypsin, chymotrypsin, thrombin, and elastase
took place some 1,500 million years ago, about the time of the divergences of lines leading to the animal, green plant, and fungus kingdoms (1).
Thus we would expect the spectrum of proteases in invertebrates to be
similar to that in vertebrates. Indeed, the presence of both trypsin and
chymotrypsin has been reported in several invertebrates, including a
starfish, an insect, and a sea anemone. From this tree we can also estimate that the divergence of lines leading to bacteria and to eukaryotes
occurred about 2,600 million years ago, which agrees very well with the
estimate derived from cytochrome c and c_2 sequences. In all probability
the trypsin which occurred in this ancient common ancestor already had
the function of digesting other proteins.

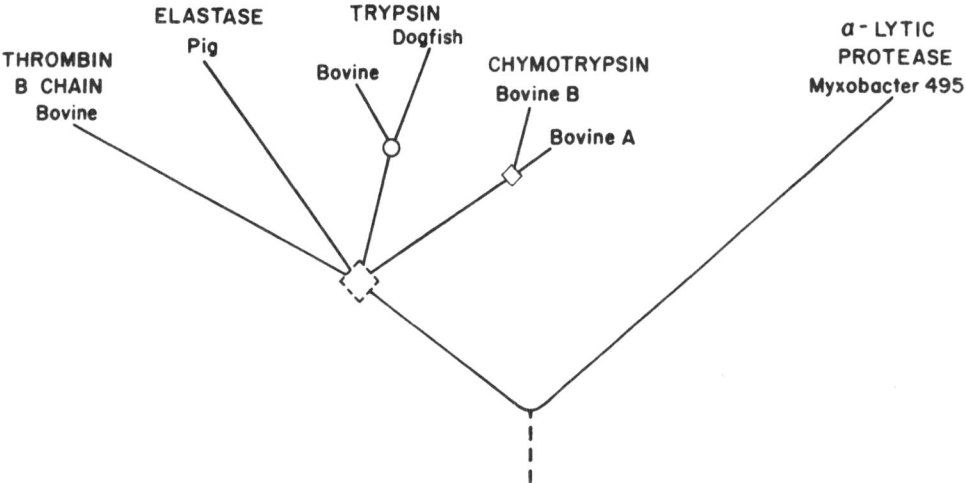

THROMBIN
B CHAIN
Bovine

ELASTASE
Pig

TRYPSIN
Dogfish

Bovine

CHYMOTRYPSIN
Bovine B

Bovine A

a - LYTIC
PROTEASE
Myxobacter 495

Fig. 5. Evolutionary tree of the trypsin family of degradative enzymes. Divergences produced by gene duplications are shown by diamonds while those produced by species divergences are shown by circles. If the time of the divergence of the lines leading to the mammals and to the cartilaginous fishes is taken as about 400 million years ago, and if the rate of change of these proteins has been constant and similar in different lines, then we can estimate that the gene duplications which gave rise to trypsin, the chymotrypsins, thrombin, and elastase took place some 1.5 billion years ago and that the divergence of lines leading to the bacteria and to the eukaryotes occurred about 2.6 billion years ago. (From the "Atlas of Protein Sequence and Structure 1972").

TRANSFER RNA

Two other biopolymer families, both ribonucleic acids, have been studied in representatives of both the eukaryotes and prokaryotes; their origin is at least as ancient as the first coded proteins of the cell. One family, the transfer ribonucleic acids (tRNAs), has been studied extensively. The tRNA molecules are the translation catalysts in the synthesis of proteins; each one is responsible for physically positioning its attached amino acid into the chain according to the sequence of nucleotides in the genetic message. These structures are in part responsible for the specificity of the genetic code. The aminoacyl-tRNA synthetases, enzymes which catalyze the preliminary attachment of one particular amino acid to a particular tRNA, are also involved in the definition of the specificity of the code. The directions for synthesis of the tRNA molecules and

of the activating enzymes are contained in chromosomal DNA. The sequences of tRNA molecules have been shaped by natural selection acting
on mutations in the genetic nucleic acids, just as in the case of proteins.
The sequences of more than 20 tRNAs from bacteria and higher organisms
with specificity for 10 of the amino acids are now known (1). These sequences are so closely similar that they can be aligned easily.

The nucleotide residues can be considered as slowly changing "characteristics" of the molecules and of the organism; from the pattern of
changes an evolutionary tree can be constructed (Fig. 6) (1). This tree
extends over a vast period of evolution, during which occurred the differentiation of types of tRNA to carry different amino acids and to recognize different codons, as well as the differentiation of species into all of
the major groups of living organisms. The present information does not
touch on many large regions of the potential knowledge inherent in the
tree. We cannot yet make a convincing case for the exact order of divergence of the different tRNA types. For both the phenylalanine and the
tyrosine sequences, it appears likely that the prokaryote-eukaryote divergence has been more recent than the differentiation of transfer RNA
types. In contrast, the sequence of the tRNA carrying valine in E. coli
is separated by considerable evolutionary distance from the sequenced
valine tRNAs in Baker's and Torula yeasts. That these tRNAs are of diverse origin is confirmed by the codons which they recognize (32 - 34).
The sequence which is known for E. coli is the major type found there and
recognizes the codons GUA and GUG best. The minor type, for which the
sequence is unknown, recognizes GUU and GUC. In yeast a minor, un-
sequenced tRNA recognizes the codons GUA and GUG best. The major
yeast type, for which the sequence is known, resembles the minor E. coli
component in recognizing GUU and to a lesser extent GUC and GUA.

All of the tRNA sequences studied are about the same length. Some
positions always contain the same nucleotide and certain regions contain
complementary bases which evidently can form hydrogen bonds; these
common features are shown in Fig. 7 (1). In all probability this structure approximates the common ancestral molecule, proto-tRNA, present
in a primeval organism before the evolution of the specificity of the code
(3). It seems unlikely that a cell could independently evolve even two
structures which would function smoothly in the coding process. However,
it is relatively easy for nature to create such similar structures by the
production of redundant genetic material (by doubling or other chromosomal aberrations) followed by the accumulation of independent mutational
changes in the separate genes (24 - 28). The development of the specificity of the genetic code seems to represent such a succession of doub-

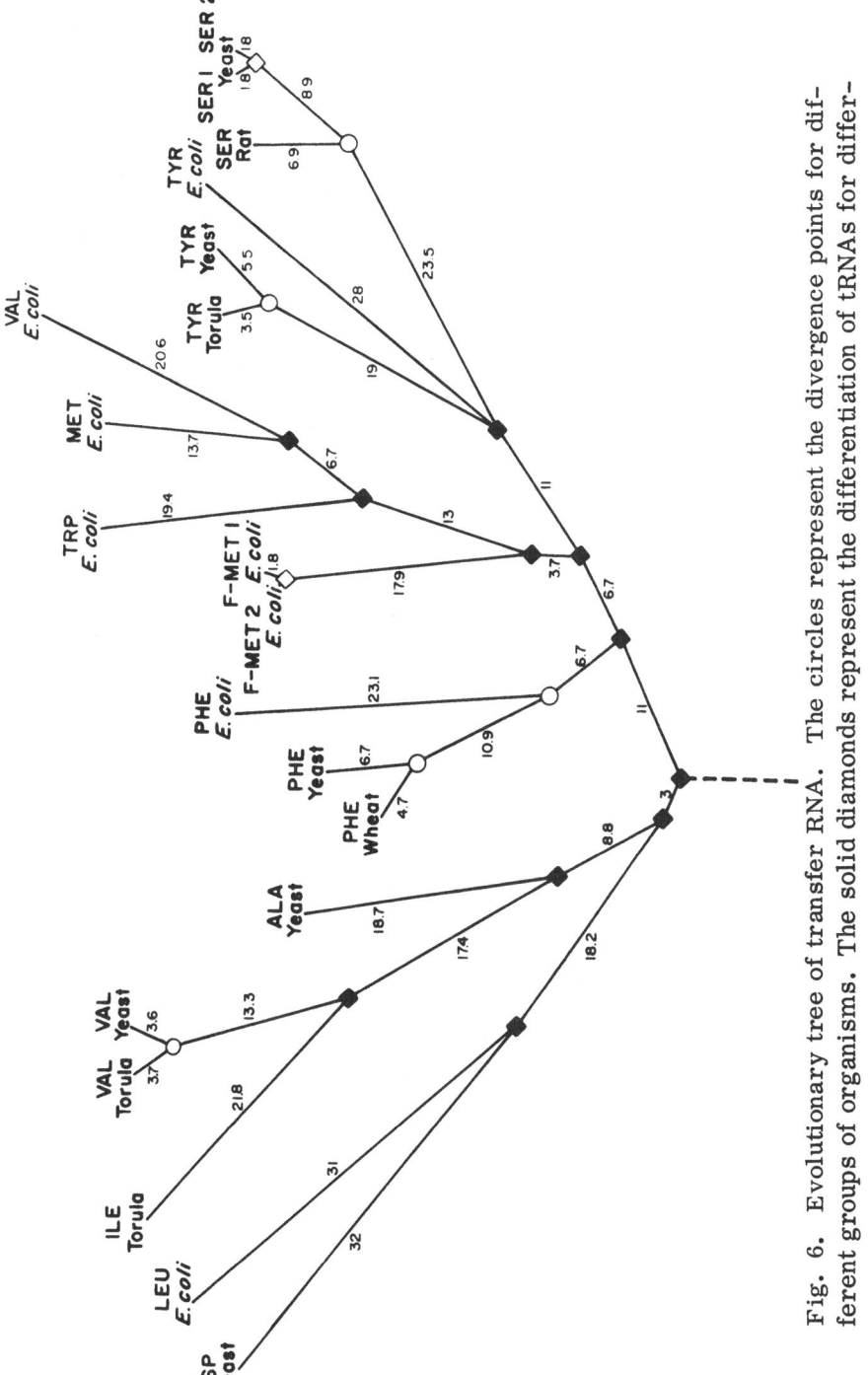

Fig. 6. Evolutionary tree of transfer RNA. The circles represent the divergence points for different groups of organisms. The solid diamonds represent the differentiation of tRNAs for different amino acids, whereas the open diamonds stand for the divergence of two forms of tRNA carrying the same amino acid. (From the "Atlas of Protein Sequence and Structure 1972").

lings and minor changes both in the genes for the tRNA molecules and in those for the activating enzymes (17 - 20).

Because the very complicated genetic code is shared by all examined living species, it almost certainly arose in the common ancestral line before the divergence of these biological species. The tRNA gene duplications and the evolution of amino acid specificity therefore took place in this primordial ancestral line. An extremely simple organism, without the ability to manufacture coded proteins, routinely synthesized a molecule with the basic tRNA structure similar to that shown in Fig. 7 (1). This simple form had evolved the ability to conserve and to propagate the structure of nucleic acids. The primitive tRNA is the oldest genetically produced molecule whose structure we can presently approximate; it most probably acted as a nonspecific catalyst, polymerizing amino acids by a mechanism similar to the one still used today. This primitive tRNA must have imparted some selective advantage to the primeval organism possessing it, so that natural selection preserved both the tRNA and the organism type. The polymerizing mechanism could hardly have arisen at any later time. Once the divergence of the specific tRNAs had occurred, it would have been almost impossible for nature to introduce a fundamental new feature; the simultaneous adaptation to this new feature by all the divergent tRNAs would be extremely unlikely.

5S RIBOSOMAL RNA

There is a fifth family of macromolecules having representatives with known sequences in bacteria and higher organisms--5S ribosomal RNA, which is a small molecule used in the non-specific part of the machinery of coding proteins. Two bacterial sequences, from Escherichia coli and Pseudomonas fluorescens, were investigated. These species are representatives of different major subgroups of aerobic bacteria; both have complete Krebs cycles and cytochrome systems including cytochrome oxidase. The human sequence is derived from the coding system of the cell nucleus.

There is an average of 48. 5% difference between the human and bacterial sequences, while the two bacterial sequences are 30% different from each other. If an estimate of the superimposed mutations is made (35), the ratio of the evolutionary distances of these two pairs is 83 to 41, or 2.0. If the overall rate of accepting mutations in these two lines was approximately the same, then the human nuclear line diverged from the bacterial lines long before the bacterial lines diverged from each

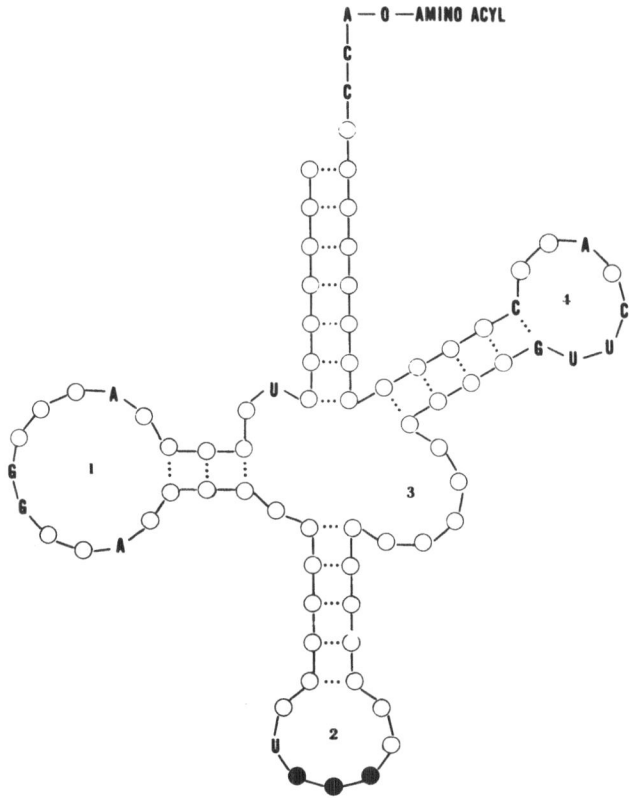

Fig. 7. Primitive transfer RNA. This molecule must have been synthesized prior to the elaboration of the genetic code and the synthesis of modern proteins. The structure shown incorporates the common features from all known tRNAs, which code for 10 of the amino acids. The letters indicate positions always filled by the same base. The circles indicate positions which are filled by more than one base. In sequences from living organisms a number of rare bases are found. These are all thought to be formed by chemical or enzymatic reactions from the unmodified bases after synthesis of the RNA chain. The dotted lines show hydrogen bonding; complementary bases are usually found in the positions connected. The solid circles represent the position of the anticodon triplet which recognizes the genetic message. The number of positions in loops 1 and 3 is variable; other regions are of constant size. There must be further interactions of the arms but the details have not yet been determined. The abbreviations for the bases are as follows: A, adenine; C, cytosine; G, guanine; and U, uracil. (From the "Atlas of Protein Sequence and Structure 1972").

other. If this rate was approximately constant in time, the interval from the present to the divergence of the eukaryotic nuclear line from the aerobic bacterial line was twice the interval to the divergence of the two major groups of aerobes.

TIME SCALE FOR BIOLOGICAL EVOLUTION

Utilizing information from two families of macromolecules for which many sequences are known, it is possible to estimate the relative times of three main events in the evolution of living things--the development of the genetic code, the divergence of bacteria from nucleated organisms, and the divergence of nucleated plants and animals.   By a careful estimate of the number of genetic changes in cytochrome c and tRNA, the divergence between nucleated organisms and bacteria was shown to be 2.6 times more remote in evolutionary distance than the divergences of nucleated organisms into green plants, animals, and fungi (35).   The divergences of these major groups of living organisms were related to an even earlier group of events in a very primitive organism--the development of the genetic code through the proliferation and distinction of genes for different types of tRNA molecules capable of transporting different amino acids.   Using only the tRNA data, comparisons were made between the tRNAs which carry different amino acids.   The ratio of the evolutionary distance between tRNA types to the evolutionary distance between prokaryotes and eukaryotes is 1.2.   In Fig. 8 this differentiation is placed at a distance proportional to the scale of the other divergences.   By a comparison of the tRNA sequences of different amino acid specificity, it is possible to show that the total amount of evolutionary change of the tRNA sequences is not perceptibly different in prokaryotes and in eukaryotes.   If one makes the gross approximation that the risk of change was constant through geological time, then time becomes proportional to the evolutionary distance.   Extrapolating from the divergence of the ancestors of plants and animals at 1.1 billion years ago, we calculate that the lines to bacteria and nucleated organisms diverged some 2.9 billion years ago and that the genetic code evolved approximately 3.4 billion years ago.

POTENTIAL INFORMATION

Just as tRNA molecules constitute a family which must have arisen through gene duplication and subsequent adaptation very early in the history of life, so must there be other such families. The protein enzymes

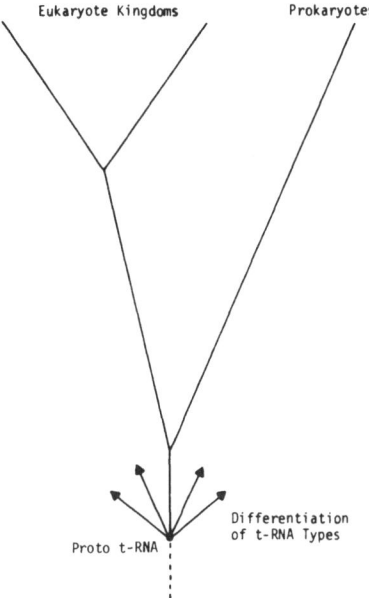

Fig. 8. The evolutionary distances between eukaryote kingdoms, between eukaryotes and prokaryotes, and between tRNA types (35). Not shown are the further differentiations of each type of tRNA with the divergences of groups of organisms.

which catalyze the attachment of the amino acids to the tRNAs may also constitute such a family. The enzymes which control the replication of DNA and of RNA and the transcription of RNA from DNA may be related. No sequences are presently known for these ancient proteins. Further, there is much sequence information still preserved in the metabolically functional nucleic acid sequences of the cell, such as the ribosomes and the control sections of DNA. The ribosomes are intimately bound to structural proteins which may be of very ancient origin. These may very well reveal more about the biochemistry of life contemporaneous with the development of the genetic code and even preceding it.

Eventually it should be possible to extrapolate far back into these primeval eras using our knowledge of macromolecular structures and the principles of the evolutionary process. It is our hope that, when the sequences of ribosomal RNAs, more tRNAs, DNA, and their associated proteins have been worked out, it will be possible to connect smoothly the details of the evolution and structure of the earliest living organisms with the processes and products of chemical evolution.

ACKNOWLEDGEMENTS

This paper was supported by Contract NASW-2288 from the National Aeronautics and Space Administration and NIH Grants GM-08710 from the National Institutes of General Medical Science and RR-05681 from the General Research Support Branch, Division of Research Facilities and Resources.

REFERENCES

1. Dayhoff, M. O., ed. "Atlas of Protein Sequence and Structure 1972," Natl. Biomedical Res. Fnd., Washington, D. C., 1972.
2. Dayhoff, M. O., and Eck, R. V., in "Atlas of Protein Sequence and Structure 1969," M. O. Dayhoff, ed., Vol. 4, pp. 1-5, Natl. Biomedical Res. Fnd., Silver Spring, Md., 1969.
3. Dayhoff, M. O., and McLaughlin, P. J., in "Atlas of Protein Sequence and Structure 1969," M. O. Dayhoff, ed., Vol. 4, pp. 89-94 (Chap. 11: Transfer RNA), Natl. Biomedical Res. Fnd., Silver Spring, Md., 1969.
4. Sagan, L., J. Theor. Biol. 14, p. 225 (1967).
5. Margulis, L., Science 161, 1020 (1968).
6. Margulis, L., "Origin of Eukaryotic Cells," Yale Univ. Press, New Haven, 1970.
7. Dayhoff, M. O., Park, C. M., and McLaughlin, P. J., in "Atlas of Protein Sequence and Structure 1972," M.O. Dayhoff, ed. Vol. 5, pp. 7-16 (Chap. 2: Cytochrome c) Natl. Biomedical Res. Fnd., Washington, D. C., 1972.
8. Young, J. Z., "The Life of Vertebrates," 2nd ed., Oxford Univ. Press, New York, 1962.
9. Dayhoff, M. O., Eck, R. V., and Park, C. M., in "Atlas of Protein Sequence and Structure 1969," M. O. Dayhoff, ed., Vol. 4, pp. 75-83, (Ch. 9: A Model of Evolutionary Change in Proteins), Natl. Biomedical Res. Fnd., Silver Spring, Md., 1969.
10. Hirschmann, R., Nutt, R. F., Verber, D. F., Vitali, R. A., Varga, S. L., Jacob, T. A., Holly, F. W., and Denkewalter, R. G., J. Am. Chem. Soc. 91, 507 (1969).
11. Gutte, B., and Merrifield, R. B., J. Am. Chem. Soc. 91, 501 (1969).
12. Kimura, M., Proc. Natl. Acad. Sci. U.S. 63, 1181 (1969).
13. Berry, W. B. N., "Growth of a Prehistoric Time Scale," p. 9, W. H. Freeman, San Francisco, 1968.

14. Romer, A. S., "Vertebrate Paleontology," 3rd ed., Univ. Chicago Press, Chicago, 1966.
15. Barghoorn, E. S., and Schopf, J. W., Science 152, 758 (1966).
16. Barghoorn, E. S., and Tyler, S. A., Science 147, 563 (1965).
17. Jukes, T. H., Biochem. Biophys. Res. Commun. 24, 744 (1966).
18. Eck, R. V., and Dayhoff, M. O., "Atlas of Protein Sequence and Structure 1966," pp. 141-147, Natl. Biomedical Res. Fnd., Silver Spring, Md., 1966.
19. Crick, F. H. C., J. Mol. Biol. 38, 367 (1968).
20. Orgel, L. E., J. Mol. Biol. 38, 381 (1968).
21. Rutten, M. G., Space Life Sci. 1, 1 (1970).
22. Rutten, M. G., The Geological Aspects of the Origin of Life on Earth," Elsevier, New York, 1962.
23. Calvin, M., "Chemical Evolution," Oxford Univ. Press, New York, 1969.
24. Eck, R. V., and Dayhoff, M. O., Science 152, 363 (1966).
25. Jukes, T. H., "Molecules and Evolution," Columbia Univ. Press, New York, 1966.
26. Matsubara, H., Jukes, T. H., and Cantor, C. R., Brookhaven Symp. Biol. 21 ("Structure, Function and Evolution in Proteins") 201 (1969).
27. Atkin, N. B., and Ohno, S., Chromosoma 23, 10 (1967).
28. Ohno, S., Wolf, U., and Atkin, N. B., Hereditas 59, 169 (1968).
29. Birktoft, J. J., Matthews, B. W., and Blow, D. M., Biochem. Biophys. Res. Commun. 36, 131 (1969).
30. Shotton, D. M., and Watson, H. C., Nature 225, 811 (1970).
31. Stroud, R. M., Kay, L. M., and Dickerson, R. E., Cold Spring Harbor Symp. Quant. Biol. 36, 125 (1972).
32. Kellogg, D. A., Doctor, B. P., Loebel, J. E., and Nirenberg, M. W., Proc. Natl. Acad. Sci. U.S. 55, 912 (1966).
33. Mirzabekov, A. D., Grunberger, D., Krutilina, A. I., Holy, A., Bayev, A. A., and Sorm, F., Biochim. Biophys. Acta. 166, 75 (1968).
34. Soll, D., Cherayil, J., Jones, D. S., Faulkner, R. D., Hampel, A., Bock, R. M., and Khorana, H. G., Cold Spring Harbor Symp. Quant. Biol. 31, 51 (1966).
35. McLaughlin, P. J., and Dayhoff, M. O., Science 168, 1469 (1970).

Received 17 November, 1971

TRYPSINOGEN ACTIVATION PEPTIDES: AN EXAMPLE

OF MOLECULAR EPIGENESIS*

Marcel Florkin and Suzanne Bricteux-Grégoire

Laboratoire de Biochimie, Institut Léon Fredericq
Université de Liège, Belgique

For many years phylogeny has been based on morphological data; thousands and thousands of observations and fossil records have permitted the formulation of what we now call "classical phylogeny."

The discovery of the genetic role of DNA, the deciphering of the genetic code and the now widely-accepted idea that mutations affecting the nucleotides of DNA are the primary cause of evolution, have led the way to a new concept of phylogeny based on purely biochemical data. Phylogenetic trees have been built from amino acid sequences (1) which are, except in a few isolated cases, in remarkable agreement with classical phylogenetic trees. Already in 1944 the concept of descent with modification at the molecular level had been formulated by one of us (2): "The study of biochemical characteristics depends upon techniques which are frequently complicated, and such a study is more difficult to accomplish than direct observation of morphological characters. Nevertheless, had naturalists started from these, rather than from morphological observations they would have been bound to conceive the idea of evolution of animals" (3).

In the present state of our knowledge, the biochemical conception of an organism distinguishes three levels of biochemical realities: the in-

---

*This work was supported by a research grant from the Fonds de la Recherche fondamentale collective, contract n°786.

formation molecules of DNA; their effectors, the proteins, synthetized under their indirect influence; and the metabolic pathways and structural aspects based on the nature of the proteins. We may accept as a theory that evolution at the phenotype level, due to modifications of DNA resulting from point mutations, gene duplication, recombinations, crossing-over, etc., is the result of such changes in the proteinic effectors as have been detected in the case of cytochromes, hemoglobins, fibrinopeptides, trypsinopeptides, etc. We may also accept as a theory that biochemical evolution, or descent with biochemical modifications, results from molecular epigenesis at the level of proteins, which results from molecular epigenesis at the level of nucleic acids (4).

When we consider the nature and the frequency in time of point mutations at the level of the primary structure of proteins, we are certainly not studying evolution, any more than when we study, for instance, morphogenesis. What we study is molecular epigenesis which is only one of the many components of the origin of the production of modification, which appear at the level of the populations of organisms. The concept of molecular epigenesis must be distinguished from molecular evolution which comprises modifications conce rning the whole metabolic fabric of an organism. Evolution itself has a sense only at the level of populations.

If there is now a nearly unanimous agreement that point mutations affecting DNA and translated as amino-acid substitutions are one of the causes of the distribution of present-day sequences of homologous proteins in different species, the role of natural selection or selection pressure has been the matter of much discussion and much discrepancy. Opinions on this subject have been recently confronted in an international conference on biochemical evolution held in Liège (5), and many fruitful discussions have followed each conference. It seems to us, however, that in many cases the apparent discrepancy is only a matter of words. Many tenants of different theories could be reconciled if they would admit that selection pressures do not act in isolation on any individual compound. They act rather as a whole on biochemical sequences of related compounds within the cell which affect the fitness of the individual in its environment (6). R. L. Watts (7) states that the selective importance of a particular character, considered separately, can only be assessed within a framework of understanding of the entire constitution of the organism and that a new gene which is effectively neutral in its particular genetic background might become fixed either through random fluctuation in gene frequencies or by chance association with other genes having high selective advantage. We think that a proper search for so-called "molecular adaptation" must start at the level of the relation organism-environment, at the level not only of the organism but also at the level of the "biochemical continuum" (4). It is our opinion that apart from mutations that are definitely favourable, for instance when a new enzyme emerges, increasing the fitness of the organism in its environment, or in other words giving this organism more opportunity to reproduce and have an important offspring, a significant number of mutations can be fixed in a population by chance. Certain empirical features have led independently several authors (8, 9, 10) to believe that the random contribution is considerable. This of course supposes polymorphism. Numerous cases of polymorphism have been detected in hemoglobins. Dayhoff (1) estimates that several times as many undetected interchanges exist in the population studied. Because most studies are based on electrophoretic methods, only amino acid interchanges which produce charge differences are detected. We do not see any reason to consider hemoglobin as a particular case. Polymorphism must be as frequent in other proteins as it is in hemoglobins. It has simply not been detected yet because the number of individuals which have been studied is far more important for hemoglobin than for any other proteins. The fact that polymorphism has not been detected by Margoliash et al. (11) in eighteen cases of human cytochrome C or in twelve cases of horse and twelve cases of donkey cytochrome C is quite unconvincing, since far too few individuals have been critically examined. The above authors calculate that the probability of a randomly sampled individual of the mammalian species being homozygous in the cytochrome C gene is 0.99, which means that there is only a 1 % chance that any given individual would carry an unusual cytochrome C. They conclude that even if most mutations in the cytochrome C gene are selectively neutral, this does not require a substantial amount of polymorphism. Clearly the test of polymorphism so far carried out would not have detected it at such a low level. Let us add that a clear-cut case of polymorphism has been detected by Petra and Neurath (12) in bovine carboxypeptidase.

Let us return now to molecular epigenesis, which at the present time is the easiest way to get a glance at the molecular evolution of species. To compare homologous molecules along the branches of phylogenetic trees raises certain methodological objections. When we compare a macromolecule in an organism living today with homologous macromolecules in another organism also living today but belonging to a more primitive category, do we not unduly neglect the fact that the more primitive species themselves have changed since the branch on which we find the more specialized species has separated? For instance, has not actual pig trypsinogen had the same time to evolve as bovine trypsinogen since the departure of the latter from the common line? To this we may answer that it appears from comparison of primary structures of hemoglobins, cytochromes, or insulin on reasonably short segments of phylogeny, that the more the organisms are changed in the course of time, the more different the structures of the macromolecules concerned. All organisms have changed in the course of time, but in a limited linear sequence of phylogeny, those which have changed less from the organismic viewpoint are also those which have changed less from the molecular point of view.

The comparison of the sequences of trypsinogen activation peptides offers a fair example of molecular epigenesis. Most of these sequences have been determined in the present authors' laboratory.

MECHANISMS OF THE ACTIVATION OF TRYPSINOGEN

Trypsinogen is a pancreatic zymogen. Its activation into trypsin is accompanied by the splitting of an N-terminal peptide in which a particular sequence of four successive aspartic acid residues has been found, just preceding the strategic Lys-Ile bond broken during activation. This particular sequence has been found in all species studied so far, except in the case of the primitive lungfish Protopterus aethiopicus (13). This accumulation of negative charges had led Neurath and Dixon (14) to imagine the well-known model of the activation of trypsinogen in which the four Asp, by interaction with positively charged residues elsewhere along the chain, or by hydrogen bonding, maintain the particular configuration of the zymogen. The hydrolysis of the Lys-Ile bond and the concomitant liberation of the N-terminal peptide would allow the interaction of His 46 and Ser 183, creating the active center of trypsin. Delaage et al. (15) and Abita et al. (16) have studied the kinetic constants of trypsinogen and model peptides and have concluded differently that the activation peptide must not be in interaction with the rest of the macromolecule but must be like a tail, floating freely at the surface of the macromolecule. The four Asp, instead of promoting the rupture of the Lys-Ile bond, slow down its hydrolysis and thus play a protective role, preventing the accidental production of trypsin. The latter being a universal activator of all pancreatic zymogens, it is clear that its concentration has to be kept to a minimum during intracellular transport, during storage in the zymogen granule and also in the pancreatic juice. This may not be easy to achieve, since trypsinogen is very abundant in pancreatic secretions and its activation is autocatalytic. Thus the accidental production of minute amounts of trypsin could have deleterious effects on the pancreatic zymogen system.

The fact that the strategic Lys-Ile bond is nevertheless the only extensively broken bond during activation of the native enzyme, in the presence of calcium ions, implies that the other trypsin-susceptible bonds are deeply buried inside the native macromolecule.

CLASSICAL PHYLOGENY OF THE ARTIODACTYLS

To the order Artiodactyla, or ungulates with an even number of toes, belong different species of domesticated animals such as the pig, the ox, the goat and the sheep, and a few wild species like the wild boar and the deer. All these species are evolutionary close since their common ancestor lived some seventy million years ago (17). The family Camelidae also belong to the Artiodactyla. Thanks to the kind cooperation of Dr. Vet. Mortelmans and Dr. Van Den Bergh, Director of the Antwerp Zoo, who gave us a dromedary pancreas, we were able to isolate dromedary trypsinogen and study the activation peptide. The departure of the Camelidae branch from the common line is subsequent to the Suidae but prior to the Cervidae and Bovidae (17).

ISOLATION OF TRYPSINOGEN AND CHARACTERIZATION OF ACTIVATION PEPTIDES

Trypsinogen has been extracted by acid from pancreas, precipitated by ammonium sulfate and purified by chromatography on CM-cellulose. The activation peptide has been isolated by column chromatography on Dowex 50-X2 with volatile buffer, whenever at least 100 mg of pure trypsinogen was available, or by a fingerprint procedure adapted to the microscale (18) which necessitates about 15 mg of trypsinogen. All the particular procedures have been described in more detailed papers and the reader is referred to the references listed in Table I for additional information.

The sequences of all known trypsinogen activation peptides are listed in Table I. The peptides from two species of primitive fishes, determined in Neurath's laboratory, have been included for comparison. Although these are phylogenetically remote from the artiodactyls, the four successive Asp are found in the dogfish, but not in the lungfish where an accumulation of three acidic residues is nevertheless found. This particular sequence undoubtedly plays a role in the maintenance of the particular conformation of the inactive precursor, since it has been found in all species examined so far.

TABLE I. Sequence of N-Terminal Activation Peptides from two Primitive Fishes, Eight Artiodactyls, and the Horse

Species	Sequence	Reference
Lungfish (Protopterus aethiopicus)	Phe-Pro-Ile-Glu-Glu-Asp-Lys	13
Dogfish (Squalus acanthias)	Ala-Pro-Asp-Asp-Asp-Asp-Lys	19
Horse (Equus caballus)	Ser-Ser-Thr-Asp-Asp-Asp-Lys	20
Wild boar (Sus scrofa)	Phe-Pro-Thr-Asp-Asp-Asp-Lys	21
Pig (Sus scrofa)	Phe-Pro-Thr-Asp-Asp-Asp-Lys	22
Dromedary (Camelus dromedarius)	Val-Pro-Ile-Asp-Asp-Asp-Lys	23
Red deer (Cervus elaphus)	Phe-Pro-Val-Asp-Asp-Asp-Lys Val-Asp-Asp-Asp-Lys	24
Roe deer (Capreolus capreolus)	Phe-Pro-Val-Asp-Asp-Asp-Lys Val-Asp-Asp-Asp-Lys	24
Sheep (Ovis aries)	Phe-Pro-Val-Asp-Asp-Asp-Lys Val-Asp-Asp-Asp-Lys	25
Goat (Capra hircus)	Phe-Pro-Val-Asp-Asp-Asp-Lys Val-Asp-Asp-Asp-Lys	26
Ox (Bos taurus)	Val-Asp-Asp-Asp-Lys	27

The peptides listed in Table I, at least those belonging to artiodactyls, are produced by the activation of cationic proteins. In fact, an anionic trypsinogen has also been isolated from the bovine species (28) and from the pig (29, 30). The isolation of an anionic form of ovine trypsinogen is in progress in our laboratory. An anionic trypsinogen is probably present in all listed species but is generally destroyed by the acid extraction procedure.

The comparison of the peptides listed in Table I shows that the more closely related are the species morphologically, the more similar are the peptides. The wild boar and the pig have exactly the same activation peptide, which is not surprising since the pig is derived from the wild boar by domestication and this domestication, according to Aeuner (31), is not older than 2,500 years. Four species, the red deer, the roe deer, the sheep, and the goat, also have exactly the same peptides. Two activation peptides have been encountered in each of the four above species, one of them two residues shorter than the other. The possible origin of these two peptides has been discussed in previous papers (24, 25). The hypothesis of a duplication of the trypsinogen gene, followed by a deletion of two codons in one of the genes, has been invoked to explain the presence of two trypsinogens differing in their N-terminal sequence, but otherwise very similar, since we have never succeeded in separating them by any of the numerous methods essayed. If these genes were allelic, the possibility exists that some individual would be homozygous for one of the genes. Having in mind the success of Petra and Neurath (12) with carboxypeptidase, a screening experiment was started with trypsinogen isolated from individual sheep pancreas. As a first approach five individual sheep were studied, but all five of them exhibited the two N-terminal peptides, however in significant different ratio (32). This led us to suppose that the alternate hypothesis, of one unique gene coding for the longest trypsinogen and the shorter variety appearing by artificial reduction of the protein during the extraction procedure, was in fact more correct. In favour of the latter hypothesis, preliminary results seem to indicate that a sample of trypsinogen isolated from an acetone powder of mixed sheep pancreas exhibits the N-terminal corresponding to the longest peptide only. This question must remain open until further results are obtained. [Subsequent experiments have not confirmed that trypsinogen isolated from an acetone powder of sheep pancreas had an N-terminal corresponding to the longest peptide only. In fact both N-terminal sequences were present in the same ratio in acetone-extracted or acid-treated trypsinogen. (Added in proof.)]

The variable part of the peptides can be compared from a phylogenetic viewpoint. Species such as dogfish (Squalus acanthias) and lungfish

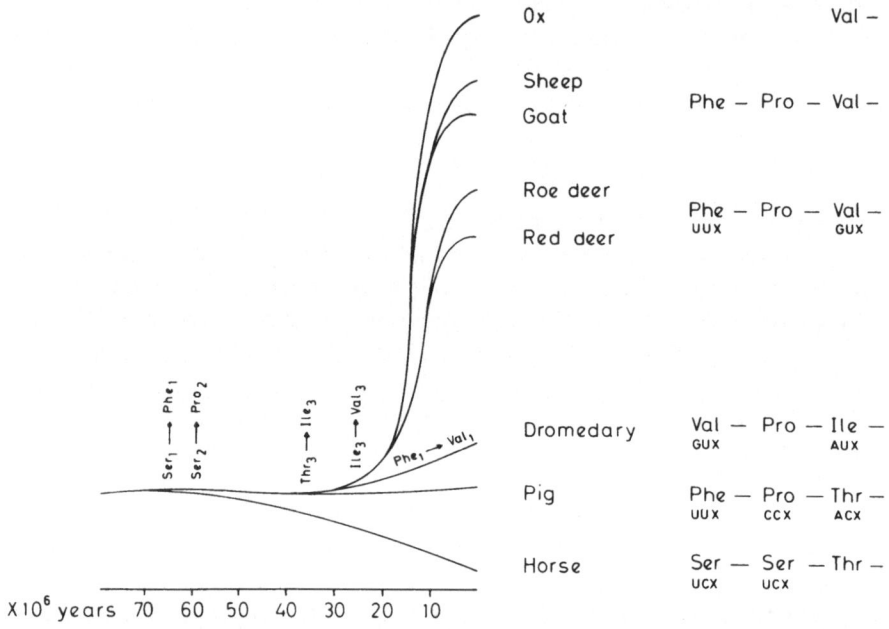

Fig. 1. Classical phylogenic tree of the artiodactyls and the horse and sequence of the variable part of trypsinogen activation peptide.

(Protopterus aethiopicus) are too distant from the artiodactyls to be directly compared, except of course to notice the persistence of the sequence of successive acidic residues which one may expect to find in most of all the intermediary species, and which are certainly in relation with the particular properties of the zymogen (14, 15, 16). The results are too scarce to permit the construction of phylogenetic trees based on purely biochemical data. But a classical phylogenetic tree can be drawn (Fig. 1) and the species compared on the basis of the sequence of their trypsinogen activation peptide. A look at Fig. 1, where the variable part of the sequence of the peptide has been written beside the species, shows that animals considered to be closely related show greater resemblances in their peptides, and that for instance the two deer, the sheep, and the goat have exactly the same peptides. Most of the differences between closely related species can be accounted for by a one-base substitution in the corresponding RNA code word. Since the species compared are closely related and since the peptides are not very different, one can follow the reflect of the successive point mutations that have affected the DNA through the amino acid substitutions. A tentative explanation of how things happened is proposed in Fig. 1.

The possible point mutations of the first codon, from the horse to the deer, sheep and goat are: UCX (horse) → UUX (pig) → GUX (dromedary) → UUX (deer, sheep, goat). This implies a back mutation phenylalanine → valine → phenylalanine. It seems more plausible , however, that phenylalanine has been conserved from the pig to the more recent artiodactyls and that the mutation phenylalanine → valine has occurred only in the Camelidae branch, subsequently to the departure of this branch from the common line. This mutation would thus be of more recent origin.

The second residue, proline, is common to all artiodactyls. A single mutation from UCX coding for serine in the horse accounts for the presence of proline (CCX) in all subsequent species.

The possible point mutations having affected the third codon are ACX (horse, pig) → AUX (dromedary) → GUX (deer, sheep, goat).

A deletion of the first two codons has occurred in the bovine species. This is only a tentative satisfactory explanation of the possible mutations that have affected the DNA coding for trypsinogen. One must always be careful in interpreting amino acid substitutions in terms of point mutations, since it is always possible to explain the substitution of a residue by another by successive point mutations, if intermediaries are postulated. We feel however that molecular epigenesis at the level of proteins can only be translated into molecular epigenesis at the level of nucleic acid if a minimum number of single base substitutions has to be invoked between closely-related species. The dromedary sequence was the last to be determined. Before that, we had to postulate the existence of an intermediary between the pig and the deer, since a single point mutation cannot account for the change of the third residue threonine coded by ACX into valine coded by GUX. This intermediary had to be either isoleucine (coded by AUX) or alanine (coded by GCX). The dromedary with isoleucine in the third position was thus the "missing link" that confirmed our hypothesis.

REFERENCES

1. Dayhoff, M. O. , "Atlas of Protein Sequence and Structure," Vol. 4. National Biomedical Research Foundation, Silver Spring, Maryland, 1969.
2. Florkin, M. ,"L'Evolution Biochimique," Masson, Paris, 1944.

3. Florkin, M., "Biochemical Evolution" (translated by S. Morgulis), Academic Press, New York, 1949.
4. Florkin, M., in "Biochemical Evolution and the Origin of Life," (E. Schoffeniels, ed.) p. 368. North-Holland Publishing Company, Amsterdam, 1971.
5. "Biochemical Evolution and the Origin of Life," Proc. of the International Conference on Biochemical Evolution, (E. Schoffeniels, ed.). North-Holland Publishing Company, Amsterdam, 1971).
6. Swain, T., in "Biochemical Evolution and the Origin cf Life," (E. Schoffeniels, ed.) p. 280. North-Holland Publishing Company, Amsterdam, 1971.
7. Watts, R. L., in "Biochemical Evolution and the Origin of Life," (E. Schoffeniels, ed.) p. 18. North-Holland Publishing Company, Amsterdam, 1971.
8. King, J. L., and Jukes, T. H., Science 164, 788 (1969).
9. Kimura, M., Nature 217, 624 (1968).
10. Vogel, H., and Zuckerkandl, E., in "Biochemical Evolution and the Origin of Life" (E. Schoffeniels, ed.) p. 352. North-Holland Publishing Company, Amsterdam, 1971.
11. Margoliash, E., Fitch, W. M., and Dickerson, R. E., in "Biochemical Evolution and the Origin of Life" (E. Schoffeniels, ed.) p. 89. North-Holland Publishing Company, Amsterdam, 1971.
12. Petra, P. H., and Neurath, H., Biochemistry 8, 5029 (1969).
13. Hermodson, M. A., Tye, R. W., Reeck, G. R., Neurath, H., and Walsh, K. A., FEBS Letters 14, 222 (1971).
14. Neurath, H., and Dixon, G. H., Federation Proc. 16, 791 (1957).
15. Delaage, M., Desnuelle, P., Lazdunski, M., Bricas, E., and Savrda, J., Biochem. Biophys. Res. Commun. 29, 235 (1967).
16. Abita, J. P., Delaage, M., Lazdunski, M., and Savrda, J., Europ. J. Biochem. 8, 314 (1969).
17. Thenius, E., "Phylogenie der Mammalia," Walter de Gruyter and Co., Berlin, 1969.
18. Schyns, R., Arch. Internat. Physiol. Biochim. 78, 415 (1970).
19. Bradshaw, R. A., Neurath, H., Tye, R. W., Walsh, K. A., and Winter, W. P., Nature 226, 237 (1970).
20. Harris, C.I., and Hoffmann, T., Biochem. J. 114, 82P (1969).
21. Bricteux-Grégoire, S., Schyns, R., and Florkin, M., Arch. Internat. Physiol. Biochim. 77, 544 (1969).
22. Charles, M., Rovery, M., Guidoni, A., and Desnuelle, P., Biochim. Biophys. Acta 69, 115 (1963).
23. Bricteux-Grégoire, S., Schyns, R., and Florkin, M., Biochim. Biophys. Acta 251, 79 (1971).

24. Bricteux-Grégoire, S., Schyns, R., and Florkin, M., in "Bio-
    chemical Evolution and the Origin of Life" (E. Schoffeniels, ed.)
    p. 130. North-Holland Publishing Company, Amsterdam, 1971.
25. Schyns, R., Bricteux-Grégoire, S., and Florkin, M., Biochim.
    Biophys. Acta 175, 97 (1969).
26. Bricteux-Grégoire, S., Schyns, R., and Florkin, M., Biochim.
    Biophys. Acta 229, 123 (1971).
27. Davie, E. W., and Neurath, H., J. Biol. Chem. 212, 515 (1955).
28. Puigserver, A., and Desnuelle, P., Biochim. Biophys. Acta 236,
    499 (1971).
29. Desnuelle, P., Gratecos, D., Charles, M., Peanasky, R., Baratti,
    J., and Rovery, M., in "Structure-Function Relationships of Pro-
    teolytic Enzymes" (P. Desnuelle, H. Neurath, and M. Ottesen, eds.)
    p. 21. Munksgaard, Copenhagen, 1970.
30. Voytek, P., and Gjessing, E. C., J. Biol. Chem. 246, 508 (1971).
31. Zeuner, F. E., "A History of Domesticated Animals," Hutchinson,
    London, 1963.
32. Bricteux-Grégoire, S., Schyns, R., and Florkin, M., Comp. Bio-
    chem. Physiol. 42B, 23 (1972).

Received 18 November, 1971

SOLVENT EFFECTS ON ENZYMES: IMPLICATIONS FOR

EXTRATERRESTRIAL LIFE

M. R. Heinrich

Planetary Biology Division, NASA
Ames Research Center, Moffett Field, California 94035

Water plays many roles during the course of an enzymatic reaction. Its importance in maintaining the structural integrity of proteins has been emphasized repeatedly; water is also directly involved in many reactions as either a reactant or product. Additionally, there are more subtle effects of water: influencing the ionization of charged groups, directing tertiary structure through interaction with hydrophobic and hydrophilic side-chains, binding at definite sites, and displacement during the course of reaction (1, 2).

The special place of water in enzymatic reactions is to be expected, in view of the long history of biological evolution in the presence of a great excess of water. The earlier, pre-biotic phases of evolution, during which the simple organic molecules were formed and then polymerized into polypeptides, polynucleotides, polysaccharides, etc., are more of a puzzle. The condensation and polymerization of small molecules in almost every case involves a removal of water, a reaction which would not be expected to go well in aqueous solution. Although dehydrating agents have been used to condense monomers to peptides and polynucleotides in aqueous solution (3), both the yield and the product molecular weights are low. The only effective procedure for making high molecular weight products in good yield is that described by Sidney W. Fox and associates (4). This is an anhydrous system, using heat to condense dry amino acids to form polypeptides. The products have been studied extensively; they have molecular weights of 10,000 and higher, contain the amino acids which were present in the starting mixture, are not completely random in composition or sequence, and demonstrate catalytic

activity (5). Fox has suggested that at least this phase of evolution occurred in the absence of water, possibly with volcanic heat (6).

Present-day enzymes have obviously evolved to function with water
as the principal component of the environment. The properties of water
are not unique, however, as Pimentel, et al. have pointed out (7). There
are many inorganic and organic solvents which resemble water in some
of the properties which are thought to be important in living systems.
It thus becomes of interest to determine how the structure, catalytic
activity, specificity, and stability of an enzyme are altered when some
or all of the water in the medium is replaced by another solvent.

One of the questions which may be asked in such studies concerns
the minimum amount of water which is necessary to maintain an enzyme
in an active state. An approach to this question by Skujins and McLaren
(8) involved the exposure of samples of a lyophilized mixture of urease
and urea-^{14}C to atmospheres containing known amounts of water vapor.
Hydrolysis was followed by the appearance of $^{14}CO_2$. It was found that
there was no hydrolysis of substrate if the amount of water present was
less than that equivalent to 1.3 moles per mole of side-chain polar
groups on the enzyme. If it is assumed that each polar group tightly
binds one molecule of water, then an excess of water above that amount
is required for enzymatic activity.

The water requirement may also be studied by placing the enzyme
in a non-aqueous solvent, to which known amounts of water are added.
Dastoli, et al. (9) suspended crystalline α -chymotrypsin in methylene
chloride containing 0.25% (0.14M) water. This preparation was found to
hydrolyze acetyltyrosine ethyl ester in a 2-hour incubation, and to react
slowly with two active site inhibitors. In other studies, Dastoli and
Price (10, 11) suspended lyophilized xanthine oxidase in several anhydrous nonpolar solvents, and studied the kinetics of the anaerobic oxidation of crotonaldehyde or hydroquinone. Enzyme activity (V_{max}) in the
organic solvents ranged from 14 to 0 per cent of that in aqueous buffer.
Highest activity was found in acetone, benzene, thiophene, and toluene;
solvents which supported no activity included ethanol, cyclohexane, and
p-xylene. Substrate affinity was decreased in most of the solvents, with
K_m ranging from 2 to 70 times higher than that for aqueous solution. By
analysis, the solvents contained millimolar levels of water, but these
amounts did not correspond to the enzyme activity in each solvent. Because the enzyme responded differently in each organic solvent, it is
obvious that the solvents did not act merely as inert supports. The
authors found no correlation between enzyme activity and parameters

such as viscosity, dipole moment, or dielectric constant. They concluded that the important factor for this enzyme was the general suitability
of the solvent for supporting redox reactions of free radicals. Two other
enzymes, peroxidase and catalase, have also been reported to function
in organic solvents (12). It is apparent from these and other examples
that some terrestrial enzymes maintain their active conformation when
water is replaced almost completely by another solvent.

In this laboratory, the role of water has been studied by exposing
enzymes to various concentrations of water-miscible organic solvents.
Changes in kinetic constants, specificity, and stability indicated the influence of the solvent on the enzyme. In the study of over 40 enzymes,
and many solvents, no case was found which was insensitive to replacement of a portion of the water by another solvent. In other words, no
solvent was found which was "inert"--as we consider water to be--and
which did not alter some property of the enzyme system. It was found
that not all of the effects decreased enzyme activity. It is interesting
that some enzymes which have evolved to very high efficiencies in an
aqueous environment can be stimulated to higher activity by other solvents.

The enzymes chosen for detailed study were trypsin, chymotrypsin,
and carboxypeptidases A and B; one of the most interesting solvents was
dimethylsulfoxide, which will dissolve most substrates and many proteins (13). When the esterase activity of trypsin was assayed in a series
of concentrations of dimethylsulfoxide (DMSO), it was found that enzyme
activity <u>increased</u> with increasing DMSO concentration (14). A maximum
of about 400% increase occurred at 30% DMSO when tosylarginine methyl
ester (TAME) was the substrate; higher concentrations of DMSO gave
less stimulation, and the solvent became inhibitory above 50% (Fig. 1).
Similar findings were reported independently by Rammler (15). The
hydrolysis of other ester substrates, benzoylarginine methyl and ethyl
esters (BAME, BAEE), was also stimulated although not to the same extent. The peptidase activity, on the other hand, was not increased in the
presence of DMSO. Thus the simple substitution of 30% of the water in
the environment by DMSO greatly increased the esterase activity, without appreciably affecting the peptidase activity of trypsin. These effects
were found over the pH range 6.5 to 8.5, with a maximum stimulation at
about pH 7. Kinetic studies suggested that, although the rate of the reaction (V_{max}) increased, the affinity for the substrate decreased.

The effects of DMSO on trypsin activity cannot be explained merely
as the consequence of removing water from the medium, because there

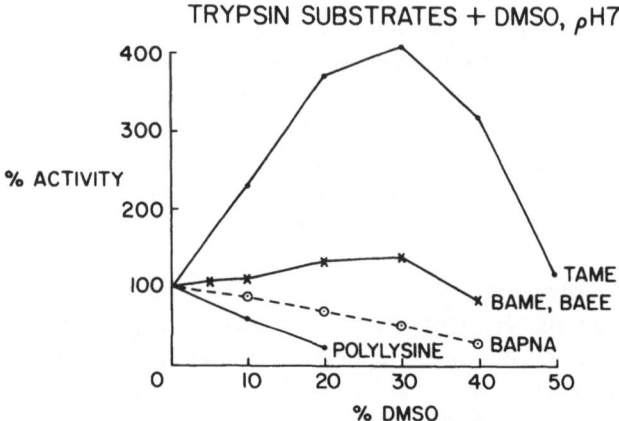

Fig. 1. Influence of DMSO concentration (v/v) on trypsin activity. Titrimetric assays were performed in 20 ml of 0.1 M NaCl
at pH 7.0, using 0.01 or 0.02 mg trypsin for esters and 0.3 mg
trypsin for polylysine. Substrate concentrations were 10^{-3} M
for esters and 0.25 mg/ml for polylysine (Sigma, mol. wt.
3000). Activity toward BAPNA was assayed spectrophotometrically at pH 7 in 0.05M Tris buffer, with 8×10^{-4}M BAPNA and
0.0125 mg/ml of trypsin [Erlanger, B., et al., Arch. Biochem.
Biophys. 95, 271 (1961)]. TAME: tosylarginine methyl ester;
BAME: benzoylarginine methyl ester; BAEE: benzoylarginine
ethyl ester; BAPNA: benzoylarginine-p-nitroanilide.

is considerable specificity in the solvent effect (Fig. 2). Some of the
other solvents which produce similar effects with trypsin include dimethylpropionamide, dimethylformamide, and acetone; solvents which
do not increase esterase activity include dimethylsulfide, dimethyla-
mine, and formamide.

Further evidence of the specificity of this effect is found in the surprising fact that chymotrypsin, known to be closely related to trypsin in
sequence and conformation, is inhibited rather than stimulated by DMSO
and related solvents. This is true of both esterase and peptidase activity, at all concentrations of DMSO; the inhibition appears to be competitive. Two other well-characterized proteolytic enzymes, carboxypeptidase A and B, show DMSO effects opposite to those of trypsin: peptidase
activity is increased, and esterase decreased.

The increase in activity noted for certain substrates of trypsin and
carboxypeptidase in the presence of DMSO is by no means a general

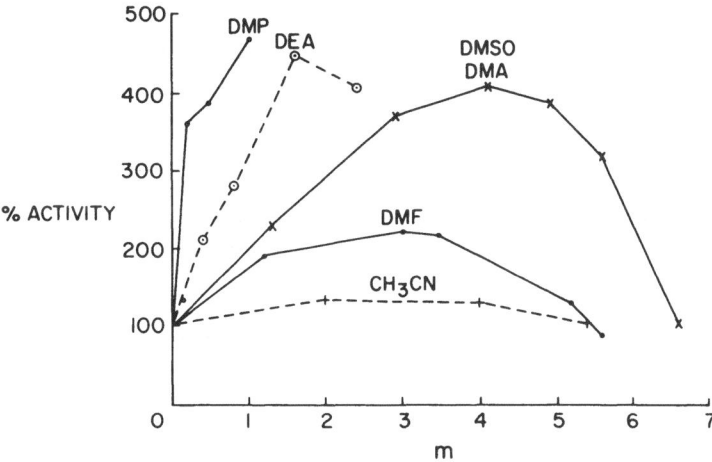

Fig. 2. Esterase Activity of Trypsin in Organic Solvents. Substrate: TAME. Assays as in Fig. 1. Molar concentration of
solvents. DMP: dimethylpropionamide, DEA: diethylacetamide,
DMSO: dimethylsulfoxide, DMA: dimethylacetamide, DMF: dimethylformamide, CH_3CN: acetonitrile.

phenomenon.   The majority of 40 other enzymes tested showed inhibition by DMSO at the 20% level, 8 showed no change in activity, and only
4 were activated (16).   There appeared to be no correlation between the
type of enzyme and its response to the solvent.

     Another characteristic of the interaction between trypsin and DMSO
is the marked heat stability of the enzyme in this solvent. Stability was
measured by heating a solution of trypsin (2 mg/ml) in the solvent under
study, and removing aliquots at intervals for assay with TAME or
benzoylarginine-p-nitroanilide (BAPNA).   The stability was expressed
as "half-life", i.e., time required to reduce activity to 50% of the starting activity.   At 60°, using TAME as substrate, the half-life of trypsin
in water was 1/2 hr.; in 50% DMSO, 9 hrs.; in 100% DMSO, over 24 hrs.
When the enzyme was tested at 100°, the protection by DMSO was not as
dramatic, but the half-life was approximately doubled in both 50% and
100% DMSO. Trypsin is more stable in 0.001N HC1 than in water; when
stability tests were carried out in 0.001N HC1 plus DMSO, the lifetimes
were longer, and the protection by DMSO was still present.   The above
results were obtained when the esterase activity of trypsin was assayed
with TAME.   When BAPNA was used as substrate, to follow peptidase

activity, there was much less protection against heat denaturation--the
half-life was slightly more than doubled in 50% DMSO at 60°.

It is interesting to examine present views of trypsin structure and
function for possible explanations of the diverse nature of the effects of
these solvents. (a) The "hydrophobic slit" proposed by Mares-Guia and
Shaw (17) would be a logical site for organic solvents to act. Seydoux,
et al. (18) suggest competition between DMF and alcohols for this non-
polar site. Although this area may be involved in the results described
in the present study, the specificity of the solvent effects indicates that
other regions are also involved. (b) Several reports have suggested the
involvement of tyrosine in the active center of trypsin (17, 19-21), and
work from this laboratory also indicates that interpretation. In the nitration of trypsin with tetranitromethane, e.g., two tyrosines are rapidly nitrated in aqueous solution. When the reaction is carried out in 50%
DMSO, only one tyrosine reacts rapidly (22), demonstrating a difference
in the environment of the tyrosines in this solvent. The mono- and di-
nitrotrypsins also have different enzymatic properties. On the other
hand, nitration of three tyrosines in trypsin does not eliminate the stimulation of esterase activity by DMSO, suggesting that other parts of the
molecule are involved. (c) Conformational changes in organic solvents,
known to occur in many proteins (23, 24), could account for altered
activity as well as the difference in nitration discussed above. DMSO,
however, is known as a poor denaturant; the conformation of lysozyme
is changed only above 70% DMSO (25). In addition, Herskovits (26) has
found in spectral perturbation studies that not all residues in a protein
are exposed to DMSO. (d) The several levels of specificity observed--
for enzyme, substrate, and solvent--make it obvious that some specificity center on the enzyme is involved. The stimulation of trypsin and
competitive inhibition of chymotrypsin by the same solvents suggest the
active site; this could also explain the protection against heat denaturation. On the other hand, if the effect were mediated through the active
site only, the difference between ester and peptide substrates would not
be expected.

It is apparent from this brief summary of the results that the way
in which certain organic solvents alter the activity of trypsin is not yet
explained. At the very least, these studies point up the utility of the
solvents as a tool for probing enzyme function, and the fact that solvents
other than water should be investigated as media for controlling and directing enzyme reactions.

It might have been expected that because all the enzymes which are available for study have evolved in an aqueous environment, all would be adversely affected by any solvent except water. The studies above, and many others (27, 28), have shown that the results are actually unpredictable; some enzymes are inhibited, some unaffected, and some stimulated by organic solvents. This range of response may be related to the original cellular environment of the enzymes--some are found in dilute solution, and some occur in lipid-protein membranes with low water content. At this time there is not enough information to determine whether or not that kind of correlation exists.

With the beginning of the planetary exploration phase of the space program, it has become necessary to give consideration to both the survival of terrestrial organisms on other planets and to the properties of any organisms native to those planets. It is essential to take whatever precautions are necessary to avoid spreading earth-type organisms to another planet until it has been explored for the presence of life, and until the characteristics of any indigenous life forms that are found have been determined. The precautions that are required will, of course, depend upon the chances of terrestrial microorganisms surviving under the conditions existing on the other planet. Those conditions will also determine the characteristics of the native forms, so that it is important to have as much information as possible about the planets. The only major environmental parameters that are known, even for the closer planets, are temperature, atmospheric composition, and pressure. Venus is hot (ca. 400°C), with a dense atmosphere; Mars has temperatures somewhat lower than those on Earth (-70° to +20°C daily at the equator), and a very thin atmosphere; in both cases the atmosphere is predominantly carbon dioxide, with little or no oxygen. What is most pertinent to this discussion is the scarcity of water on Venus and Mars-- a concentration of water vapor probably 0.001 of that on Earth. On Venus, with its high temperature, a pressure of 170 atmospheres would be required for the existence of liquid water; it is not yet known if surface pressures are that high. On Mars, it has been suggested that liquid water might be found in special circumstances: in a permafrost layer, with small areas melted by geothermal activity; oases of water in favorable locations such as deep valleys; or ponds saturated with salts, thereby lowering the vapor pressure, and holding some water as hydrates. If these conditions exist, then the survival of earth organisms is very likely, because we know of many bacteria which grow in the cold (29), or in high salt concentrations (30, 31). Furthermore, if those conditions exist, it would prove that Mars had a history of water. In that case the probability would be much higher that life did develop, and that it would generally resemble terrestrial life, including the use of water as a solvent.

There is another, more intriguing possibility: on a planet with little or no water, the reactions which went on in the primitive atmosphere would produce compounds quite different from those made on Earth and then incorporated into living systems. If some of those compounds were liquids, it is possible that they could have served as solvents, even for the development of life. In that case, the life systems on that planet could vary from those on Earth anywhere from slightly (e.g., different amino acids in the proteins) to drastically (e.g., biocatalysts are non-protein). The possibility that variations like that could exist is given support by studies which show that even terrestrial enzymes are more versatile than generally realized. Examples include the finding of Fahnestock and Rich (32) that the protein-synthesizing system of E. coli ribosomes can also synthesize polyesters from α -hydroxy acids related to normal amino acids, as well as the previously mentioned instances of enzyme activity at very low water concentrations, and the tolerance of enzymes for solvents other than water. If planetary exploration does discover examples of life forms which use chemically different systems for catalysis, heredity, or energy utilization, it would provide a whole new biochemistry for our study.

REFERENCES

1. Steitz, T. A., Henderson, R., and Blow, D. M., J. Mol. Biol. 46, 337 (1969).
2. Blake, C. C. F., Mair, G. A., North, A. C. T., Phillips, D. C., and Sarma, V. R., Proc. Roy. Soc. Ser. B 167, 365 (1967).
3. Ibanez, J. D., Kimball, A. P., and Oro, J., Science 173, 444 (1971).
4. Fox, S. W. and Harada, K., Science 128, 1214 (1958).
5. Rohlfing, D. L. and Fox, S. W., Arch. Biochem. Biophys. 118, 122 (1967).
6. Fox, S. W., Nature 201, 336 (1964).
7. Pimentel, G. C., Atwood, K. C., Gaffron, H., Hartline, H. K., Jukes, T. H., Pollard, E. C., and Sagan, C., in "Biology and the Exploration of Mars," p. 243. National Academy of Sciences, Washington, 1966.
8. Skujins, J. J. and McLaren, A. D., Science 158, 1569 (1967).
9. Dastoli, F. R., Musto, N. A., and Price, S., Arch. Biochem. Biophys. 115, 44 (1966).
10. Dastoli, F. R. and Price, S., Arch. Biochem. Biophys. 118, 163 (1967).
11. Dastoli, F. R. and Price, S., Arch. Biochem. Biophys. 122, 289 (1967).

12. Siegel, S. M. and Roberts, K., Space Life Sciences 1, 131 (1968).
13. Rees, E. D. and Singer, S. J., Arch. Biochem. Biophys. 63, 144 (1956).
14. Heinrich, M. R. and Mack, R., Federation Proc. 27, 786 (1968).
15. Rammler, D. H., Ann. N. Y. Acad. Sci. 141, 291 (1967).
16. Bugna, E. and Heinrich, M. R., to be published.
17. Mares-Guia, M. and Shaw, E., J. Biol. Chem. 240, 1579 (1965).
18. Seydoux, F., Yon, J., and Nemethy, G., Biochim. Biophys. Acta 171, 145 (1969).
19. Inada, Y., Kamata, M., Matsushima, A., and Shibata, K., Biochim. Biophys. Acta 81, 323 (1964).
20. Herskovits, T. T. and Villanueva, G. B., Arch. Biochem. Biophys. 131, 321 (1969).
21. Mares-Guia, M. and Figueiredo, A. F. S., Biochemistry 9, 3223 (1970).
22. Mack, R. and Heinrich, M. R., Federation Proc. 30, 1077 (1971).
23. Tanford, C., Adv. Prot. Chem. 23, 121 (1968).
24. Timasheff, S. N. and Inoue, H., Biochemistry 7, 2501 (1968).
25. Hamaguchi, K., J. Biochem. (Tokyo) 56, 441 (1964).
26. Herskovits, T. T., J. Biol. Chem. 240, 628 (1965).
27. Timasheff, S. N., Accounts Chem. Res. 3, 62 (1970).
28. Inagami, T. and Sturtevant, J. M., Biochim. Biophys. Acta 38, 64 (1960).
29. Deal, P. H., Cryobiology 7, 107 (1970).
30. Hochstein, L. I. and Dalton, B. P., J. Bacteriol. 95, 37 (1968).
31. Lanyi, J. K. and Stevenson, M. J., J. Biol. Chem. 245, 4074 (1970).
32. Fahnestock, S. and Rich, A., Science 173, 340 (1971).

Received 8 November, 1971

PROTOCELLS AND CELLS

PROTEIN-LIPID FILMS AS PROTOTYPES

OF BIOLOGICAL MEMBRANES

A. I. Oparin, G. A. Deborin, and N. D. Yanopolskaya

Bakh Institute of Biochemistry, USSR Academy of Sciences,
Moscow, USSR

Due to extensive theoretical and experimental researches into the origin of life performed during the last decades (1-3), it has become an undoubted fact that life has evolved in the course of a long-term chemical evolution of matter (4). Spectacular achievements in the field of experimental abiogenic synthesis of different classes of organic compounds, vitally important for life, have convincingly shown that the materialistic approach to the elucidation of origin of life holds great promise (5).

The peculiar property of life is the individuality of single living organisms. This, in turn, assumes the existence of a certain boundary, a delimiting surface, a membrane and, finally, an envelope separating the living organism from the surrounding medium and helping it to maintain its individuality. It is now suggested that most membranes of contemporary organisms are structurally uniform and operate on the basis of general principles. This is an indication of a very ancient origin of membranes. Primary membranes may have originated at a remote period of time when chemical evolution resulted in the development of multimolecular prebiological systems.

In the course of evolution, metabolic processes occurring in the prebiological systems became more and more complicated and perfect and eventually gave rise to biological metabolism. This process was evidently accompanied by a kind of selection which involved adaptation of the composition, structure and function of surface layers, delimiting the systems from the surrounding medium, to the environmental parameters and their

343

final transformation into the present-day membranes. The development and evolution of the cell membrane should be examined together with the development and evolution of biological metabolism; this approach is in conformity with the general principle of unity of structure and function.

As is known, surface active substances at phase boundaries form surface films of a unique molecular structure. In addition to proteins and polysaccharides, substances of lipid origin play an important role in the spatial organization of the cell. The lipids can spontaneously orient themselves in water solutions in the form of mono- and bimolecular layers as well as micelles, requiring no specific mechanisms. Of great significance are certain phospholipids, particularly lecithin, aggregating into micelles that are similar to the layers which are capable of involving other lipids and sterols, thus producing phase changes. This fact allowed the conclusion that the formation of lipid molecules together with proteins and nucleic acids may have played a decisive role in the origin of life. Low molecular weight fatty acids can be abiogenically synthesized together with amino acids with the aid of both electric discharges and other energy-providing factors. At present, extensive investigations concerning pathways of the primary abiogenic synthesis of high molecular weight fatty acids and other lipids are in progress.

Wilson (6) has demonstrated experimentally that by abiogenic synthesis in electric discharge surface-active hydrocarbons appear. It can be therefore supposed that solubilization in the primary hydrosphere of the Earth of organic compounds containing in their molecules both hydrophobic and hydrophilic groups should have inevitably resulted in a spontaneous self-assembly of a surface film at the boundary, delimiting the hydrosphere and atmosphere. The same films may separate initial multimolecular systems (coacervates, microspheres, etc.) from the surrounding solution.

Thus, self-assembly of surface films of organic substances is inherent in their chemical nature, being non-specific and unrelated in any way to their biochemical properties. It depends exclusively on their chemical structure, presence of certain groups in the molecule, energy-providing factors at the phase boundary, and environmental parameters. Since the physico-chemical mechanisms causing the formation of surface films are inherent in the very chemical nature of these compounds, they seem to have undergone slight changes during evolution. All this justifies the use of data derived from studying surface films of proteins and lipids isolated from present-day organisms to develop models of primary surface structures regarding them as molecular prototypes of a more complex structural organization peculiar to biological membranes.

One of the crucial consequences of the development of structural organization is spatial separation of chemical substances involved in metabolism. It is well known how significant is the pH difference on both sides of the biological membrane, salt concentration, composition and concentration of enzymically active proteins, substrates, inhibitors, sugars, etc. The selective transfer of these components is of key importance for the regulation of cell metabolism. It has been shown, for instance, that Ehrlich ascite carcinoma cells absorb hemoglobin molecules from the surrounding medium (7). The first step in the absorption is protein binding to the cell membrane based presumably on an electrostatic interaction of a positively charged protein with a negatively charged membrane. Neifakh and Vasilets (8) have demonstrated that enzymes "squeeze" into the tumour cell through its external membrane. The authors believe that this is a case of passive diffusion. The experiments by Brachet et al. (9) have shown that ribonuclease molecules penetrate into the nuclei of living cells from the cytoplasm.

Many biochemists make great efforts in order to simulate these processes. This simulation may eventually help to reveal the simplest molecular mechanisms underlying specific functions of the biological membrane at the lowest level of its organization.

It was recently demonstrated in our laboratory that the separation with an artificial lipid membrane of ribonuclease and ribonucleic acid solutions kept under optimal conditions made it possible to observe the development of the enzymic process of RNA decomposition in the substrate-containing compartment of the apparatus (10). Similar results were obtained in experiments with trypsin-serum albumin and trypsin-casein systems (11). The presence of small amounts of the enzyme capable of maintaining its action was enzymatically found on the other side of the membrane. It is interesting to note that in the absence of substrate no enzyme could be detected in this compartment of the apparatus, if a pure solvent but not a substrate occurred on the other side of the membrane.

The apparatus in which the experiments were carried out is presented in Fig. 1 and consists of two chambers--E and S--separated with a membrane. The description of the apparatus, techniques of membrane preparation, and experimental procedures can be found elsewhere (10).

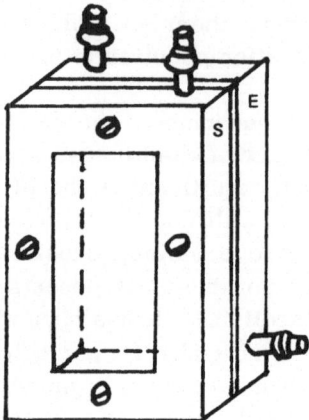

Fig. 1. The diagrammatic scheme of experimental chambers.

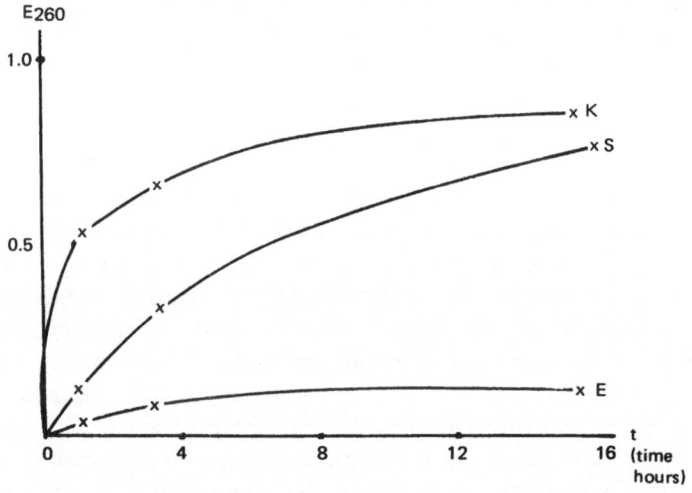

Fig. 2. Kinetic curves of RNA enzymatic breakdown.
A_{260}, absorbance at 260 nm; t, time in hours. Course
of the process in the chamber: S, with substrate; E, with
enzyme. K, control.

Fig. 2 gives kinetic curves of enzymic RNA hydrolysis achieved in our experiments. The ordinate shows absorption at 260 nm and the abscissa time in hours. Absorbance at 260 nm corresponds to the content of nucleotides in the supernatant following precipitation of protein and insoluble polynucleotides, according to the method of Fiers and Stocks (12). As can be seen from Fig. 2, in chamber S, originally filled with the substrate, the amount of nucleotides increased steadily with time reaching within 16 hours the maximum value which was close to that in the control test tube experiment (curve K). In chamber E, originally filled with the enzyme solution, the absorbance increased insignificantly, probably due to mononucleotide diffusion from chamber S formed during RNA hydrolysis. This gives evidence that enzymic decomposition of RNA gave rise mainly to oligonucleotides which, unlike mononucleotides, were incapable of diffusion through the lipid membrane. It can be seen in Fig. 2 that there is a definite difference between the experimental and control curves; the rate of the enzymic process in the test tube was higher than in the apparatus, as a result of which the plateau on the kinetic curve was reached sooner. This difference can be explained by the fact that in the presence of a lipid membrane the process should be limited by the rate of the enzyme transfer across the membrane.

Table I gives results of experiments in which after a 3 hour incubation of the RNA-RNase system in the apparatus the solutions were transferred to test tubes, incubated again for 18 hours at 37°C, and then ana-

TABLE I. Absorbance Increase in Solutions from Chambers E and S after Additional 18 hour Incubation in Test Tubes.

A_{260} after 3 hour in the apparatus		A_{260} after additional incubation in test tubes		Increase in A_{260}	
S	E	S	E	S	E
0.530	0.100	0.90	0.12	0.37	0.02
0.550	0.100	0.82	0.13	0.29	0.03
0.535	0.135	0.78	0.14	0.23	0.01
0.500	0.155	0.80	0.16	0.30	0.05
0.530	0.180	0.79	0.19	0.26	0.07
				m = 0.30	m = 0.04

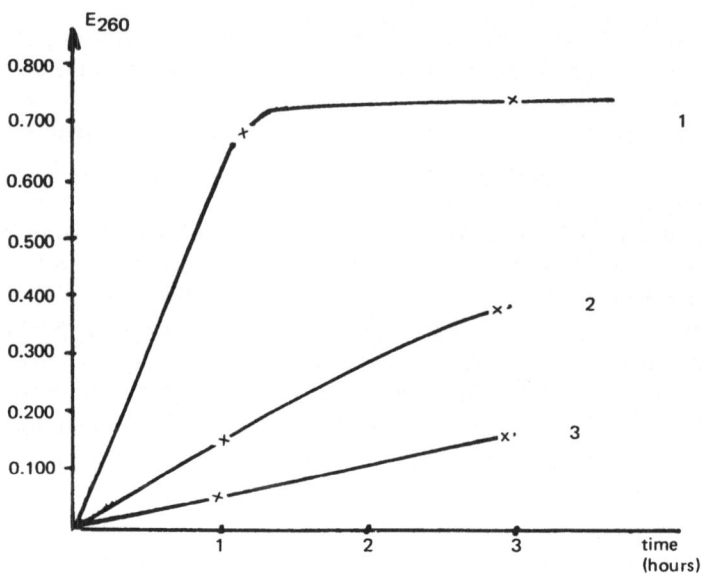

Fig. 3.   Kinetic curves of RNA enzymic breakdown on cephalin-cholesterol and olive oil membranes. 1, test tube; 2, cephalin-cholesterol membrane; 3, olive oil membrane.

lyzed. The tabulated data suggest that under our experimental conditions ribonuclease probably passed through the lipid membrane into the substrate solution and continued there its enzymic action.

It should be noted that the enzyme inactivated by a pH shift towards the acidic region lost its capacity to pass through the membrane into the substrate containing chamber.   This capacity was restored after the pH was returned to the initial level. According to Scheraga et al. (13), a pH shift towards the acidic region disrupts the helical structure of certain sites of the ribonuclease molecule, thus disturbing its active center.   It can, therefore, be assumed that the destruction of the native conformation of the enzyme in our experiments led to the loss of capacity of the enzyme either to form a Michaelis complex or to form complexes with membrane lipids.   In both cases, the enzyme lost its capacity to penetrate through the membrane to reach substrate.

One of the most significant problems in this type of experiment is the role of the lipid moiety in the phenomena observed. Fig. 3 shows kinetic curves of RNA hydrolysis by ribonuclease for two cases: the membrane consisting of (1) a mixture of fatty acid glycerides (olive oil) and (2) an equimolar cephalin-cholesterol mixture. It can be seen that the cephalin-cholesterol membrane yielded higher hydrolysis rates, which may be attributed to the greater affinity of the protein for phospholipids than glicerides. This is confirmed by the data of Rzhekhin and Krasilnikov (14).

It is of importance that the major amounts of the membrane-bound enzyme can be removed from the membrane by washing it once or twice with the solvent. In addition, the membrane also has firmly-bound enzyme which cannot be removed by washing and starts functioning when the membranes are placed into Petri dishes containing a substrate solution. It should be mentioned that the amount of the enzyme absorbed on the membrane was several times greater if the membrane separated enzyme and substrate than when it separated enzyme and solvent.

The relationship between the trypsin transport through the membrane towards its substrate (albumin, casein) and membrane thickness was also studied. It was found that transport decreases with an increase of the membrane thickness, showing linear proportionality between enzyme diffusion and thickness of the lipid layer.

Specific experiments were performed with the purpose of determining the site of localization of the absorbed enzyme on the membrane. After careful washing with water the enzyme-absorbing membrane was placed into another apparatus. The membrane was facing the substrate in one experiment with the side that previously contacted the enzyme, and in the other, with the side that contacted the solvent. The apparatus was incubated at $37^{\circ}C$ and then the hydrolysis was assayed. The data derived for the RNA-RNase system are given in Table II. A conclusion can be

TABLE II. $A_{260} - A_k$. In all cases the membranes being taken out from the chambers were washed with water three times.

No. of experiment	The side of the membrane which previously faced the enzyme is facing the substrate	The side of the membrane which previously faced water is facing the substrate	The membrane is in the Petri dish containing the substrate
1	0.550	0.280	0.575
2	0.600	0.220	0.600
3	0.620	0.260	0.620
4	0.580	0.265	0.575

drawn that in the enzyme-water chamber the enzyme is mainly absorbed on the side of the membrane facing the enzyme. This is indicated by the fact that absorbance values in the dish, where the substrate was in contact with both sides of the membrane, and in the chamber, where it was in contact with only one side (which previously faced the enzyme), were very close.

On the basis of the above experimental findings suggestions can be made regarding a possible mechanism of ribonuclease transport through the lipid membrane under the described conditions (Fig. 4). The enzyme forms with membrane lipids an inactive protein-lipid complex which under the influence of diffusion is distributed all through the membrane. The substrate solution, in contact with the other side of the membrane, forms a triple complex of substrate--enzyme--lipid which then dissociates to release the Michaelis complex into the substrate-containing chamber. The subsequent breakdown of this complex follows ordinary laws of enzyme kinetics. According to the advanced hypothesis, the protein-lipid complex may be considered as a transport form which assures the

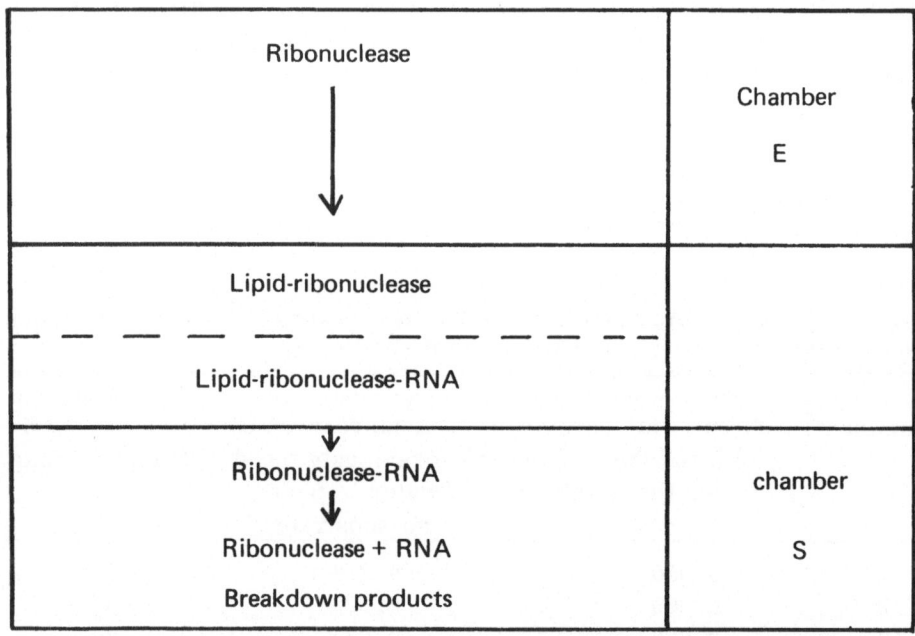

Fig. 4. Scheme of RNase transport through the lipid membrane.

transfer of a water-soluble enzyme across the lipid membrane from one aqueous solution to another. It can be supposed that a similar mechanism works in the case of the trypsin-serum albumin or trypsin-casein system.

Our hypothesis finds support in the data accumulated by Monnier (15) who showed that, if a drop of linseed oil which, similarly to olive oil, is a mixture of fatty acid triglycerides, separates two liquid phases, such a membrane, as well as natural biological membranes, displays greater permeability for K ions than for Na ions.   Membranes of a cephalin-cholesterol mixture have electroconductivity, water content, and water permeability that reversibly decrease under the influence of $CaCl_2$ and increase under the influence of KC1 (16).

On the basis of the literature data and our own findings it can be suggested that the appearance of a lipid membrane of the simplest composition separating solutions of protein or protein-like polymer and corresponding substrate may have resulted, due to certain chemical and physicochemical requirements, in the fact that the membrane has developed a capacity of controlling the penetration of primary catalytically active polypeptides and proteins to their substrates.   Of great importance may have been the presence of other compounds in the environment, e. g. , ions, inhibitors, activators, denaturants, chelates, etc. During subsequent evolution these processes acquired the pattern that was dictated by the arising metabolism and which eventually led to the development of biological membranes and their inherent functions (17).

It is interesting to note that although our model is far from being perfect, it helps to reveal certain features typical of biological membranes at the molecular level.   These data are in favour of the above mentioned concept that protein and lipid films are likely to have been the prototype from which biological membranes of the modern type have developed in the course of evolution.

REFERENCES

1. Oparin, A. , "Genesis and Evolutionary Development of Life," Academic Press, 1968.
2. Bernal, J. D. , "The Origin of Life," Weldenfeld and Nicolson, London, 1967.
3. Fox, S. , Naturwissenschaften 56, 1 (1969).
4. Calvin, M. , "Chemical Evolution--Molecular Evolution Towards the Origin of Living Systems on the Earth and Elsewhere, " Oxford Univ. Press, New York, 1969.

5. Fox, S. W., "The Origin of Prebiological Systems," Academic Press, New York, 1965.
6. Wilson, T., Nature 188, 4755, 51, (1960).
7. Choi, J., and Kay, E., Canad. J. Biochem. 46, 12 (1968).
8. Neifakh, S., Vasilets, V., Cytolgia 7, 3, 347, (1965).
9. Brachet, J., Protoplasma 63, 1, 86, (1967).
10. Deborin, G., Torkhovskaya, T., Tyurina, I., and Oparin, A., Federation Proc. 25, 5, 863-868, 1966.
11. Deborin, G., Yanopolskaya, N., and Oparin, A., J. Evol. Biokh. Fisiol. 2, (1971).
12. Fiers, W., and Stocks, J., Naturwissenschaften 44, 115 (1957).
13. Hermans, J., and Scheraga, H., J. Am. Chem. Soc. 83, 3283 (1961).
14. Rzhekhin, V., and Krasilnikov, V., Prikl. Biokh. Mikrob. 1, 6 (1965).
15. Monnier, A., J. Gen. Physiol. 51, 5 (1968).
16. Litch, G., and Tobias, J., J. Cell. Comp. Physiol. 63, 225 (1964).
17. Deborin, G., Proc. 2nd All-Union Congress of Biochem., Tashkent, 1969.

Received 8 February 1971

MODELLING OF STRUCTURE AND FUNCTIONAL

UNITY ON COACERVATE SYSTEMS

K. B. Serebrovskaya and G. I. Lozovaya

Bakh Institute of Biochemistry, USSR Academy of Sciences, Moscow
and
Institute of Botany, Academy of Sciences of the Ukrainian SSR, Kiev

The present paper is based on two ideas advanced by Professor Sydney Fox a few years ago.

The first concept assumes the possibility of applying the biogenetic law to evolution at the prebiological level (1). If this assumption is true, the biogenetic law which unites chemical and biological evolution may help elucidate the mechanisms that underlay the development of primary metabolism, by means of extrapolating the data on metabolism in contemporary living organisms.

Another concept postulates that the material of which the supramolecular organizations (e. g., microspheres) are made influences their functions and properties (2). With respect to the lipid-containing structures we are interested that this principle should have the following significance: If pigment evolution led to its hydrophobization (3), the lipid component responsible for the hydrophobic nature of the pigment carrier should have played a definite part in the photochemical process.

Finally, the occurrence of ATPase activity in histidine-containing proteinoids detected by Fox (4) suggests that the utilization of light energy on the primeval Earth may have been achieved through reversing the ATPase reaction in lipid-protoprotein complexes which, according to Gaffron (5), had been formed prior to an organized cell and existed per se for a certain period of time. This type of the phosphorylation mechanism is vividly discussed at present in scientific periodicals (6).

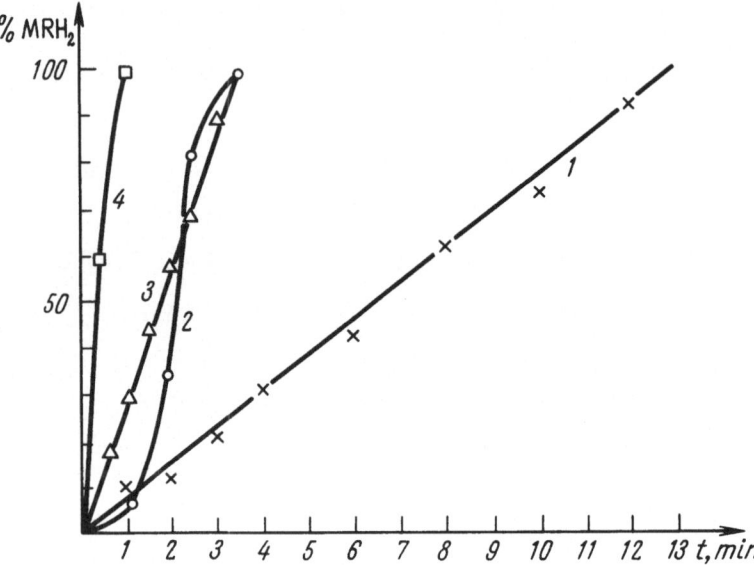

Fig. 1. Photosensitization of the oxido-reductive reaction by protoporphyrin IX (oxidation of methyl red
with ascorbic acid). 1, pigment in water; 2, pigment
+ protein; 3, pigment + gum water; 4, pigment + coacervate.

The development of the photosynthetic capability in probionts was the
turning point in chemical evolution since it formed the basis for a direct
utilization of long wavelength solar radiation and a subsequent oxygen accumulation in the Earth's atmosphere.   The substances which may have
played the role of photosensitizers could have been porphyrin pigments.
The possibility of their abiogenic synthesis on primary Earth has been
demonstrated by several authors (8-10).

Abiogenically synthesized protoporphyrin, e.g. protoporphyrin IX,
had an insignificant solubility in neutral and, according to Haldane (11),
even in alkaline waters of the primary hydrosphere, because it needed a
sufficiently acidic medium to be fully dissolved (9).   Nevertheless, the
presence of abiogenically synthesized protoproteins should have accelerated its disaggregation due to formation of pigment-protein complexes
(12). Their combination with other polymers gave rise to the emergence
of coacervate systems where the pigment acquired a substantial photochemical activity (Fig. 1).

Another possibility for a primary disaggregation and orientation of this hydrophilic pigment may have been the utilization of micellae of various abiogenically synthesized surface-active substances, the former carrying charges on their surface. The hydrophilic pigment can be incorporated into such a micella through one pathway only, i.e., absorption on its hydrophilic surface. The interaction of a pigment-impregnated micella with a protoprotein resulted in the formation of a pigment-containing lipid-protein coacervate (13).

The pigment showing a sufficient hydrophobic capacity has essentially different potentialities. An interaction of such a pigment with surface-active substances brings about a mixed pigment-lipid micella in which the pigment is noticeably oriented (14).

The interaction of the micella containing a built-in pigment with a protoprotein results in the formation of coacervate systems arranged as a chloroplast membrane (15). This coacervate is based on a surface-active substance micella whose hydrophilic groups interact with the porphyrin nucleus of the hydrophobic pigment. The phytol residue improves the correct orientation of the pigment when interacting with the hydrophobic moiety of the micella (16). As micellae themselves are randomly oriented within the coacervate, the orientation of the pigment appears insufficient to assure high efficiency of sensitization. An important role in increasing the sensitizing capacity of the pigment is also played by the concentrating effect of the coacervate structure as a whole (17).

In Table I are data comparing activities of the pigments protoporphyrin IX (hydrophilic) and protochlorophyll (diphilic) incorporated into lipoprotein coacervates. The activity of the hydrophilic pigment is only 1.5 times higher than its concentrating effect in a coacervate drop, whereas activity of the diphilic pigment exceeds the effect by almost one order of magnitude. This is probably associated with a better orientation of the diphilic pigment of surface-active substance micellae.

According to Pulman and Pulman (18), selection of substances with respect to their stability preceded the prebiological evolution of the systems. However, the selection regarding an increase of the hydrophobic capacity should have developed while the systems existed. Only the systems which assured the maximum orientation of the pigment that resulted in the maximum efficiency of sensitization should have occupied key positions. At earlier stages of prebiological evolution the chief part could have been played by protein systems incorporating hydrophobic sensitizers. Later the higher was the portion of the long wave-length radiation,

TABLE I.   Comparison of Hydrophilic and Diphilic Pigments in Coacervate Drops.

Parameter	Drop/equilibrated liquid ratio	
	Protoporphyrin IX (hydrophilic)	Protochlorophyll (diphilic)
Pigment concentration per volume unit	29.5	20.5
Sensitization rate per volume unit	45.9	177.0
Acceleration effect	1.5	8.9

the greater importance was acquired by diphilic pigments capable of being built into a protein-lipid structure according to a specific orientation.

The scheme of the processes in systems with both hydrophilic and diphilic pigments is as follows:

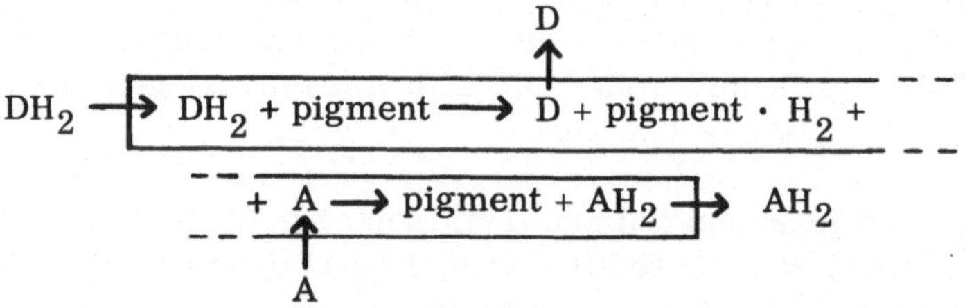

In compliance with the hypothesis advanced by Oparin (19), the typical specificity for a living state is participation of substances constituting the structure in the assimilation-dissimilation process.

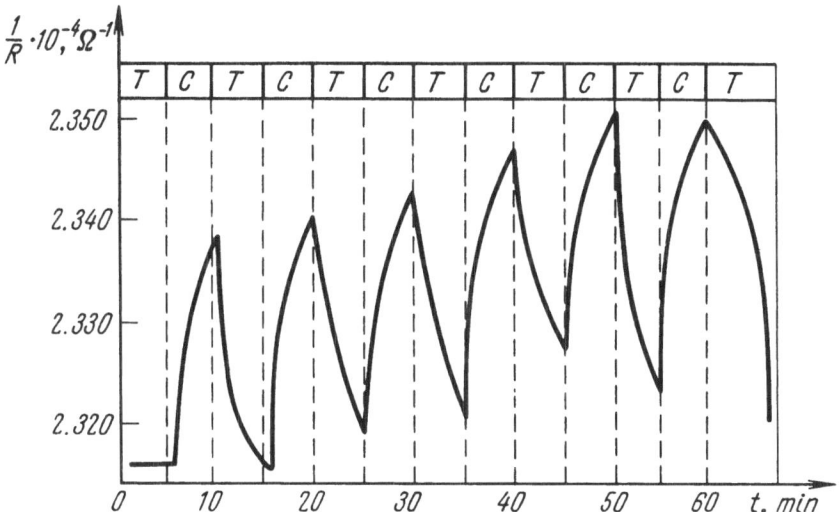

Fig. 2. Photochemical activity of the sodium oleate: chlorophyll: water system during light darkness alternations.

A few years ago we revealed involvement of structural lipids forming the supramolecular organization in the oxido-reductive reaction (20). Studies in vacuo of the surface active substance: water system containing chlorophyll during light-dark alternations demonstrated a distinct change of conductance (Fig. 2). It is assumed that in the light the pigment gives a proton over to the lipid which returns it back to the pigment in the dark.

An examination of the effect of proton donors and acceptors on the photoconductance of the pigment: surface-active substance: water system (21) showed a fall and rise of conductance, respectively. This phenomenon can be explained by the fact that unsaturated fatty acid residues of the tested surface-active substances may act as a donor or an acceptor in the oxido-reductive reaction. In direct experiments we showed palmitate dehydration under the influence of UV-light and proton transfer from its molecule to the acceptor (for instance, riboflavin) as well as lecithin dehydration in response to the white light effect (22).

On the basis of these data the scheme of photosensitization in a coacervate system involving a structural lipid can be built as follows:

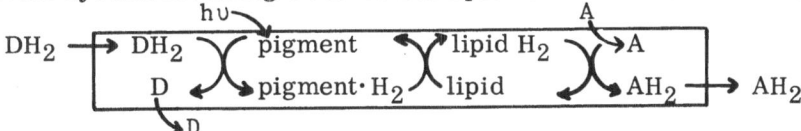

It cannot be ruled out that, prior to the emergence of pigments, abiogenically synthesized surface-active substances could have participated in oxido-reductive reactions. Due to the lack of an ozone screen, an intensive photolysis of water took place on the primary Earth (23); therefore, water could act as proton donor with respect to the UV-dehydrated surface-active substance:

$$\begin{array}{ccc}
\text{UV-}h\nu & & \\
\downarrow & \text{UV-}h\nu & \\
\downarrow \; -H^+ & \downarrow \; -H^+ & \\
H_2O \rightarrow SAS & \longrightarrow & A \text{ (flavin, chinone, etc.)}
\end{array}$$

The protein component of the pigment-lipid-protein complex apparently plays an important role in the utilization of light energy (24). At the end of 1930's, Oparin and his co-workers (25) put forth a hypothesis concerning the reversible action of enzymes within a living cell. At present this hypothesis has been used and expanded by Racker and his colleagues (26) to interpret the mechanism of oxidative phosphorylation in mitochondria in regard to the reversal of the ATPase reaction.

This idea has been confirmed by our data on the reversal of activity of enzymes incorporated into lyzosomes (27). The histidine-containing proteinoids having an ATPase activity, which have been obtained by Fox (4), are of particular interest from the above point of view. It can be supposed that histidine-proteinoids absorbed on primary surface-active substances could also perform etherification of various derivatives to form both polyphosphates and ATP.

With an increasing oxygen accumulation in the Earth's atmosphere, UV-induced direct photolysis of water has become impossible. The secondary development of photolysis in green photosynthesizing organisms can be explained by the function of the pigment-lipoprotein complex as a photoelectric cell (28). The potential difference between chlorophyll and carotene molecules is 0.3 v; therefore, a combination of several complexes might yield 1.2 v, the value required for water dissociation. The estimates indicate that it is solely chlorophyll which could be a component of this cell because the photopotential of other pigments (for example, bacteriochlorophyll) is substantially lower.

Thus, evolution tended: 1) to find a way of utilizing a cheap proton donor, i.e., water, 2) to attain hydrophobization of pigments in order to assure their optimal orientation on the structure which facilitates complete utilization of the incident light and 3) to provide elongation of the chain of donors and acceptors required for a maximally complete reduction in the oxygen atmosphere, of structural lipids (to eliminate, if possible, the formation of their peroxides).

An enhancement of the hydrophobic property of ATPase-active proteins in the course of evolution may have contributed to their better adjustment as components of the pigment: protein: lipid complex. This may have resulted in an improvement of the photosensitization function of the oxido-reductive reaction and the capacity for the most effective storage of light energy in macroergic phosphate bonds.

ACKNOWLEDGMENTS

We are greatly indebted to Academician A. I. Oparin for his exceptional interest to our work and to the organizers of the volume for the honourable invitation to contribute to it.

REFERENCES

1. Fox, S. W., in "The Origin of Life on the Earth" (Intern. Symp., 1957). Academic Press, New York, 1958.
2. Fox, S. W., "The Origins of Prebiological Systems," Academic Press, New York, 1965.
3. Granic, S., Proc. 5th Intern. Congr. Biochem., Moscow, vol. 9 (N. Sisakian, ed.). Pergamon Press, London, 1961.
4. Fox, S. W., Biochem. Biophys. Acta 160, 246, (1968).
5. Gaffron, H., in "Horizons in Biochemistry" (M. Kasha and B. Pullman, eds.), p. 590. Academic Press, New York, 1962.
6. Boyer, P. D., in "Molecular Biology," p. 227. Nauka, Moscow, 1964 (in Russian).
7. Sols, A., and Grisolia, S. (eds.), "Metabolic Regulation and Enzyme Activity," FEBS 19, 1970.
8. Szutka, A., Hazel, J. E., and McNabb, W. M., Radiation Res. 10, 597 (1959).
9. Krasnovsky, A. A., and Umrikhina, A. V., DAN SSSR 155, 691 (1964).

10. Umrikhina, A. V., in "Problems of Origin and Essence of Life." Nauka, Moscow (in Russian) (in press).
11. Haldane, J. B. S., New Biology 16, 12 (1954).
12. Oparin, A. I., Serebrovskaya, K. B., and Lozovaya, G. I., DAN SSSR 179, 1240 (1968).
13. Oparin, A. I., Lozovaya, G. I., and Serebrovskaya, K. B., Jurnal Evol. Bioch. Fiziol. 5, 2 (1969).
14. Serebrovskaya, K. B., Lozovaya, G. I., Balaevskaya, T. O., Jurnal Evol. Bioch. Fisiol. 2, 4 (1966).
15. Bakker, H. A., Proc. Acad. Sci. Amsterdam 37, 688 (1934).
16. Benson, A. A., and Weiner, T. E., Amer. J. Bot. 54, 389 (1967).
17. Evreinova, T. N., "Substance Concentration and Enzyme Action in Coacervates," Nauka, Moscow, 1966 (in Russian).
18. Pullman, B., in "Biogenese," p. 162, Massons, Paris, 1967.
19. Oparin, A. I., Izv. AN SSSR, OMEN, ser. biol., NG, 1733 (1937).
20. Serebrovskaya, K. B., Korneeva, G. A., Lutsik, T. K., and Samsonova, V. M., Biofizika 15, 973 (1970).
21. Serebrovskaya, K. B., and Korneeva, G. A., Biofizika (in press).
22. Serebrovskaya, K. B., "Coacervates and the Protoplasm," Nauka, Moscow, 1971 (in Russian).
23. Krasnovsky, A. A., in "Origin of Life on Earth," Izd. AN SSSR, Moscow, 1959 (in Russian).
24. Baltscheffsky, H., "Third Intern. Conf. Origin of Life," Pont-a-Musson, France, 1970.
25. Oparin, A. I., Erg. Enzymf. 3, 57 (1934); Kursanov, A. L., in "Enzymes," Izd. AN SSSR, Moscow-Leningrad, 1940 (in Russian).
26. Racker, E., "Mechanisms in Bioenergetics," Academic Press, New York, 1965.
27. Serebrovskaya, K. B., in "Reaction Mechanisms and Control Properties of Phosphotransferases." Leipzig, GDR, 1971.
28. Komissarov, G. G., in "The Mechanism of Respiration, Photosynthesis and Nitrogen Fixation." Nauka, Moscow, 1967 (in Russian).

Received 1 July, 1971

COACERVATE SYSTEMS AND EVOLUTION OF MATTER

ON THE EARTH

T. N. Evreinova, T. W. Mamontova, and V. N. Karnaukhov

Moscow University, Lomonosov Biological Faculty,
Department of Plant Biochemistry and Biophysical Institute,
Academy of Sciences of the USSR, Moscow, USSR

INTRODUCTION

The origin of life and the artificial synthesis of life are among the great problems of science. The famous investigations of Professor Sidney Fox and his colleagues on proteinoids and microspheres are a tribute to these problems (1).

According to A. I. Oparin's theory of the origin of life (2), there were some initial stages of chemical evolution during the prebiological period. Many different molecules formed by abiogenic syntheses (3, 4), and the ancient ocean was the "primary soup." Some molecules were also concentrated between layers of aluminosilicate clays in the lagoons of the ancient ocean [by John Bernal (5)].

The association of molecules, however, could have taken place in microspheres (S. Fox), lipoprotein bubbles (R. Goldeicr) and coacervate drops (A. I. Oparin). Other precell models were possible. The place of precell models in the evolution of matter is shown in Diagram 1 (2, 6).

Coacervate systems were studied by H. G. Bungenberg de Jong at the beginning of this century. Now more than 200 hydrophilic systems are known, and they may consist of two or more different molecules. They may be divided by their chemical composition into ten large groups. Most coacervates have been obtained from biological macromolecules

Homo sapiens
(1 M.Y.)
↑
MAMMALIANS, BIRDS   ANGIOSPERMAE
            (140 M.Y.; Cainozoic)

REPTILIANS   ↑   GYMNOSPERMAE
            (225 M.Y.; Mesozoic)

AMPHIBIANS   ↑   EQUISETINAE

FISHES   LYCOPODINAE
            (570 M.Y.; Plaeozoic)

INVERTEBRATES   ↑

PROTOZOA   ALGAE
            (1900 M.Y.; Proterozoic)
                    ↑
            PRIMITIVE CELLS
        (2700 M.Y.; Archeozoic)
                    ↑
                   LIFE
        (3000 - 3500 M.Y.; Catarcheozoic)
                    ↑
               PROTOCELLS
(microspheres, bubbles, coacervate drops, others)
                    ↑
                POLYMERS
(polypeptides, polynucleotides, polysaccharides)
                    ↑
           ORGANIC COMPOUNDS
(hydrocarbons, amino acids, organic acids, lipids, others)
                    ↑
              PLANET EARTH
              (5000 M.Y.)

            COSMIC MATTER

Diagram 1. Evolution of matter. Numbers denote millions of years ago.

that exist in protoplasm. There are coacervate drops in protoplasm of living cells. The hydrophilic coacervate systems consist of drops and equilibrium liquid. The common property of any coacervate system is the cooperation, or association, of molecules in the coacervate drops, which are 0.5 - 640 µ in diameter. Only a relatively few polymer molecules exist in the equilibrium liquid, coacervate drops may absorb other molecules from the equilibrium liquid, and this property of drops is used in practice (5).

The purpose of this paper is to show how the cooperation of molecules in the coacervate affects the size and chemical composition of the individual coacervate drops.

EXPERIMENTAL

Coacervate systems were obtained from aqueous solutions (0.1% to 0.5%) of acidic and basic proteins, RNA, DNA, carbohydrates, and enzymes (phosphorylase, polyphenoloxidase, etc.) and their substrates. Some low molecular weight substances took part in these systems. Twenty different systems were studied; most existed at the pH range (5.0 - 7.5) characteristic of protoplasm (5).

The size of individual coacervate drops was measured by an AB-analysator of particles. More than 1000 drops were measured in every system. The dry mass of individual coacervate drops was determined by interference microscopy. The coacervate drops in the microscope field with interference fringes in visible light is shown in Fig. 1. The measurements were made with monochromatic light at 545 nm. The limit of measuring was 1×10^{-14}g. The quinone content of individual drops was found by means of a recording Cytospectrophotometer MUV-4-5. The limit of the analysis was 10^{-13} to 10^{-15}g. The absorbtion spectra of drop, equilibrium liquid, and glass are shown in Fig. 2. The errors of all methods were 2 - 5 per cent of the value found. Details of composition of coacervate systems and optical methods and calculations for homogenous and heterogenous drops have been reported (5,7,8). Some of the results are illustrated in Figs. 1-4 and Tables 1-3.

RESULTS AND DISCUSSION

The size distribution of coacervate drops is a function of the charges and isoelectric pH of the molecules involved (Fig. 3). For example, isoelectric pH values of the molecules involved are:

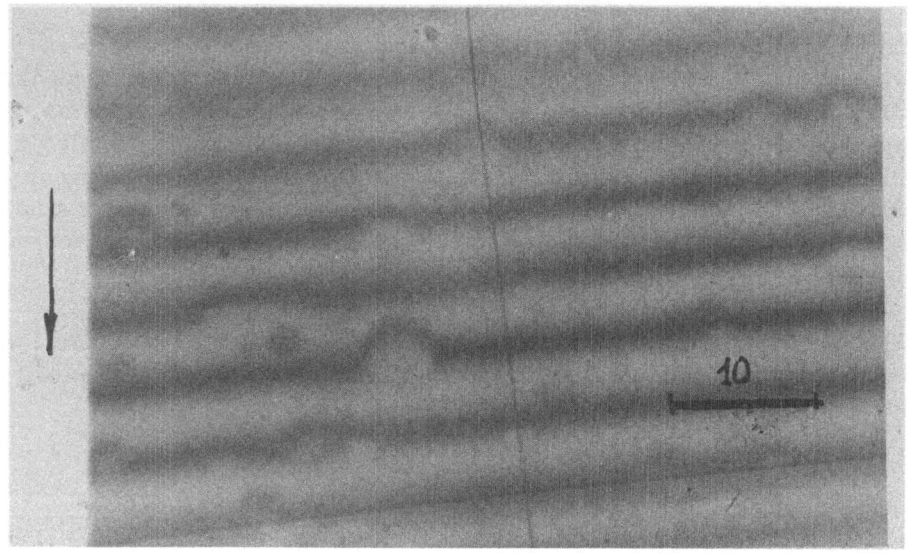

Fig. 1.   Coacervate drops under the interference microscope
with interference fringes (visible light).

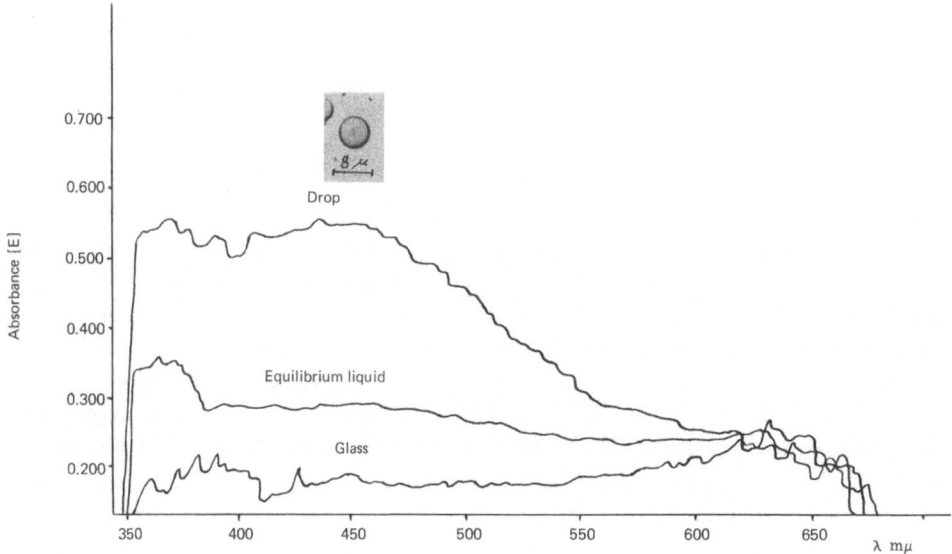

Fig. 2.   The distribution of oxidized o-dianisidine in the coacervate system: perioxidase-
histone-DNA-o-dianisidine (red./oxd).

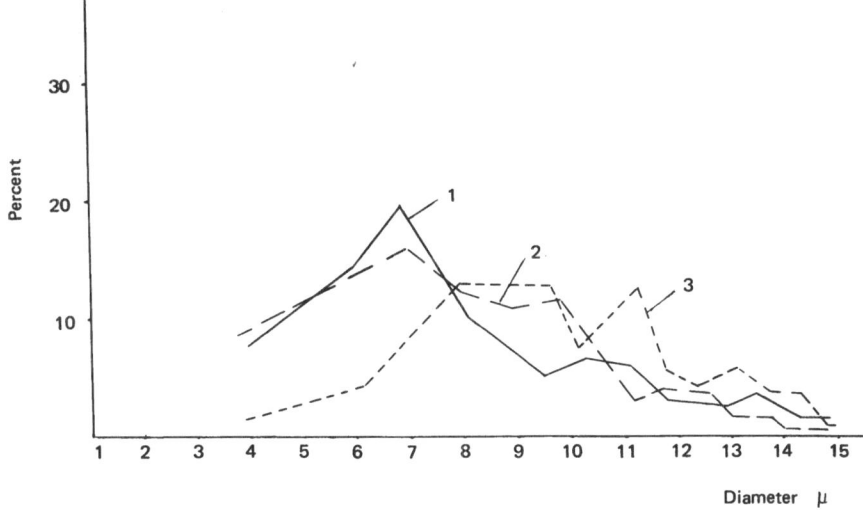

Fig. 3. Size distribution of drops in coacervate systems. The ordinate represents the percentage of the total number of drops. Curve 1, clupeine-DNA; curve 2, histone-DNA; curve 3, histone-serum albumin.

Molecules	Isoelectric pH
Serum albumin	5.0-5.5
DNA (RNA)	1.1-1.2
Histone	9.5
Clupein	12

The histone-DNA preparation contains more small drops than the histone-serum albumin system, because the difference in isoelectric pH values between histone and DNA is higher.

The total volume of drops in both types of systems is equal, but the small-droplet systems have the greater surface area.

Tiny drops contain a higher dry mass per unit volume than large ones. The concentration of molecules in coacervate drops is 10 to 400 times that in the initial solution used to form the coacervates (Table I).

The distribution of molecules inside the drops is heterogenous. For example, the concentration of DNA and histone in the vacuoles of the drops is only 0.1% to 0.25%. The concentration of clumps inside the drops may be 87%.

TABLE I.   Dry Weight and Concentration of Dry Mass in Individual Coacervate Drops.

No.	Diameter, 10^{-4}cm	Volume, 10^{-12}cm^3	Dry Mass, 10^{-12}g	Conc., %	K_c^*
		Protein–Carbohydrate			
		(histone–gum arabic, pH 5.5 – 6.0)			
1	3.30	18.8	9.7	52	94
2	6.16	122.1	53.0	43	79
3	8.36	305.3	102.3	38	61
		Protein–Protein			
		(histone–serum albumin, pH 5.5 – 6.0)			
4	2.20	5.6	2.4	44	44
5	3.96	32.4	11.3	35	35
6	5.50	86.9	15.6	18	18
		Protein–Nucleic Acid			
		(histone–DNA, pH 6.8 – 7.6)			
7	2.2	5.6	2.3	42	84
8	4.4	44.6	11.4	25	51
9	16.72	2446.5	169.0	7	14
		(clupeine–DNA, pH 7.6 – 8.2)			
10	1.76	2.8	2.3	56	431
11	4.04	35.3	9.1	26	185
12	9.24	412.9	56.2	14	85
	Polyphenol Oxidase–Histone–Gum Arabic–Quinones, pH 6.0				
13	5.33	73.67	21.04	26.45	53
14	7.77	244.15	42.65	17.47	35
15	9.55	453.65	69.22	15.28	31

$^*K_c$ = concentration in drop/concentration in initial solution.

TABLE II. Quinones in Individual Coacervate Drops.

No.	Diameter, 10^{-4}cm	Volume, 10^{-12}cm^3	Conc. of quinones $\mu M/ml$
	Peroxidase-Histone-DNA-Pyrogallol-Quinones, pH 6.0		
1.	2.00	4.2	6000
2.	4.4	44.6	860
3.	6.3	130.8	430
	Peroxidase-Histone-Gum Arabic-Pyrogallol-Quinones, pH 6.0		
4.	5.6	98.85	301
5.	24	7229.95	12.9
	Peroxidase-Histone-DNA-o-Dianisidine-Quinones, pH 6.0		
6.	4.55	49.11	76
7.	4.62	51.80	72
8.	5.95	108.2	54
9.	6.37	128.3	37

Reaction of coacervate droplets with enzymes changed their size, structure, and stability (5, 7). Drops in the more frequently described coacervate systems settle on the bottom of the vessel and disappear; they are transformed into a thin layer. For example, the drops which consist of serum albumin-histone disappear in 30 minutes. Other systems are more stable, the first being prepared in 1967. This system consisted of polyphenoloxidase (E.C. 1.10.3.1)-histone-gum arabic-pyrocatechol (as substrate) at pH 6.0 and 16 - 25°C. Polyphenoloxidase oxidized pyrocatechol to quinones.

The drops containing the quinones also settle, but they do not disappear and are stable for three years or more. Stable coacervate systems can also be prepared with peroxidase, pyrogallol, and H_2O_2 (or o-dianisidine), instead of polyphenoloxidase, its substrates.

T. N. EVREINOVA, T. W. MAMONTOVA, V. N. KARNAUKHOV

TABLE III. Weight and Concentration of Dry Mass of Various Materials.

Material	Diameter, cm	Volume, cm^3	Dry mass, g	Conc. %
Bacteria	$2.5x10^{-4}$	$8.5x10^{-12}$	$2.5x10^{-12}$	30–25
Mammalian cells	$2.5x10^{-3}$	$8.5x10^{-9}$	$2.1x10^{-9}$	25–10
Amoeba	$1.0x10^{-2}$	$5.2x10^{-7}$	$7.8x10^{-8}$	15–10
Coacervate drops	$2.4x10^{-4}$	$7.4x10^{-12}$	$2.5x10^{-12}$	34–7
	$1.0x10^{-2}$	$5.3x10^{-17}$	$3.5x10^{-8}$	

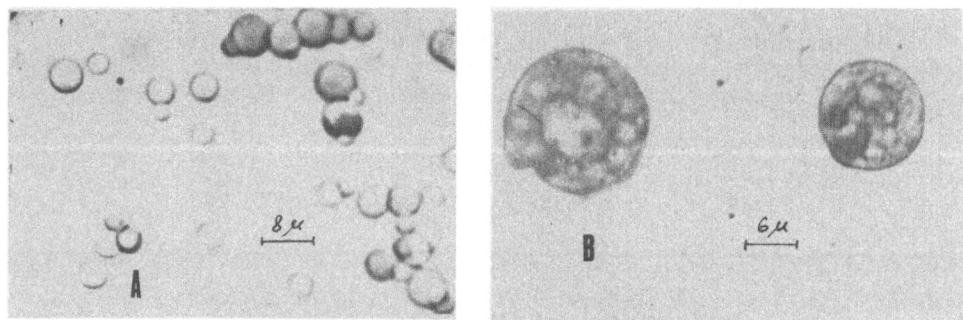

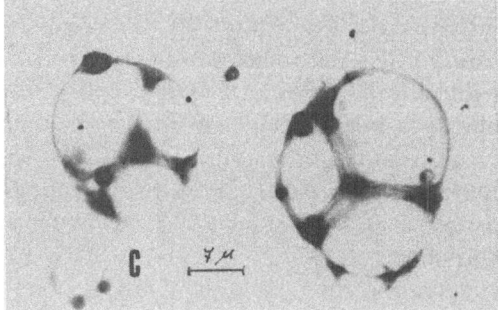

Fig. 4. The structure of coacervate drops at different concentrations of acetate buffer (M).   A, 0.068 M; B, 0.23 M; C, 0.27 M.

The details of compositions of systems and the action of enzymes were described in our papers (8 - 10). The content of quinones in individual drops were 0.003 - 1/4% (Table II). Only a relatively few molecules of quinones are in the equilibrium liquid (Fig. 2).

For example, 1 molecule of quinone was associated with 2 molecules of histone in the drop, which was 6.3 in diameter.

Different concentrations of acetate buffer (per ml) changed the distribution of molecules inside of drops (Fig. 4). The drops without quinones disappeared under such conditions.

There are many different coacervate drops. Their diameter, content of solids, and molecular content may be the same as in bacteria, mammalian cells, or amoeba (Table III). More than 50% of the space in living cells is taken up by protein and nucleic acid molecules, and the same result may be obtained with coacervate drops. This does not mean that coacervate drops and cells are the same thing. However, we would like to emphasize only that both cells and coacervate drops may have similar properties. It seems this phenomenon is the result of the same compound macromolecules which exist in both cytoplasm and coacervate drops.

ACKNOWLEDGEMENT

We are very grateful to Professor Fox and we wish him much success in his investigations. The authors would like to thank Academicians A. I. Oparin and G. M. Frank.

REFERENCES

1. Fox, S. W., Quarterly J. Florida Academy of Sciences 31, 1 (1968).
2. Oparin, A. I., "The Beginning and the Origin of Life on the Earth," Medizine, Moscow, 1966.
3. Pavlovskaja, T. E., Pasinskii, A. C., Sedorov, W. S., and Ladiginskaja, A. J., in "Abiogenes and the Initial Stages Evolution of Life," p. 41, Nauka, Moscow, 1968.
4. Ponnamperuma, C., "Chemical Evolution and the Origin of Life," New York State Journal of Medicine 70, 1169 (1970).
5. Evreinova, T. N., "Concentration of Substances and the Action of Enzymes in Coacervate Systems," Nauka, Moscow, 1966.

6. Wologdin, A. T., "The Origin and Evolution of Life on the Earth,"
    The Knowledge, Moscow, 1970.
7. Evreinova, T. N., and Bailey, A., <u>Dokl</u>. <u>Acad</u>. <u>Sci</u>. <u>USSR</u> 723,
    179 (1968).
8. Evreinova, T. N., Karnaukhov, V. N., Mamontova, T. W., and
    Ivanizki, G. R., <u>J</u>. <u>Colloid</u> <u>and</u> <u>Interface</u> <u>Science</u> 30, 18, (1971).
9. Evreinova, T. N., Bailey, A., Mamontova, T. W., and Garber,
    M. M., <u>J</u>. <u>Evolutionary</u> <u>Biochem</u>. <u>USSR</u> 1, 31 (1971).
10. Evreinova, T. N., Mamontova, T. W., Karnaukhov, V. N., and
    Dudaev, A. N., <u>in</u> "Chemical Evolution and the Origin of Life"
    (R. Buvet and C. Ponnamperuma, eds.), p. 337. North-Holland,
    Amsterdam-London, 1971.

Received 25 January, 1971

CONJUGATION OF PROTEINOID MICROSPHERES: A MODEL

OF PRIMORDIAL RECOMBINATION

Laura L. Hsu

Institute of Molecular and Cellular Evolution
University of Miami, Coral Gables, Florida

INTRODUCTION

A recent publication (1) showed the formation of specialized junctional structures following contact between proteinoid microspheres (Fig. 1). Small packets of proteinoid materials, called endoparticles, are formed within the microspheres either spontaneously or by appropriate treatment of the solutions. The endoparticles have been shown to pass through junctions connecting microspheres (Fig. 2). Experiments described below show the ability of endoparticles to meet across junctions and fuse through the communicative portals. The "recombinant" endoparticles are able to behave as condensation nuclei around which new proteinoid materials are deposited to form normal sized microspheres.

EXPERIMENTAL

Further studies of the communication system in proteinoid microspheres show that endoparticles are able to combine with one another when they have the opportunity to collide or meet within the boundaries of the conjugating microspheres. Fig. 3 A-C show the coupling of endoparticles from two different microspheres across the conjunctive space. The "recombinant" couple are shown to leave ensemble from the confines of the "parental" microspheres into the medium.

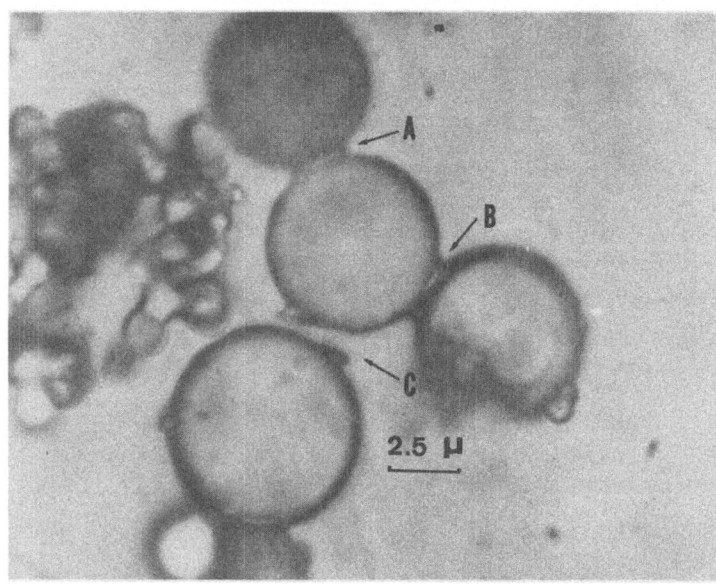

Fig. 1. The connecting structures between proteinoid microspheres are
shown in various stages of breakage. The microspheres were prepared
by heating 25 mg of 2:2:1-proteinoid per ml of distilled water and decanting. The solution was allowed to cool slowly and was then aged 3 months
at room temperature. A is intact, B shows fracture, C has come asun-
der. From reference 1 with permission from North-Holland Publishing
Co.

When endoparticles, whether single, coupled, or in small clusters,
are suspended in a cooling proteinoid solution, they serve as accretion
centers around which new proteinoid materials in the medium are deposited until normal sized microspheres are formed (Fig. 4). The materials and methods for accretion experiments are described in detail
in a previous paper (2) and those for carrying out communication studies
may be found in reference (1). Endoparticles are extruded from their
parental microspheres, stained with 1/10 strength Gram's crystal violet, and allowed to incubate in fresh, particle-free accreting medium.
The accreting medium is prepared by boiling a suspension of appropriate proteinoid in distilled water (25 mg proteinoid/ml dist. water),
allowing the solution to cool, and filtering at 45°C (2). The seeded accreting medium is allowed to cool slowly from 45°C to room temperature.

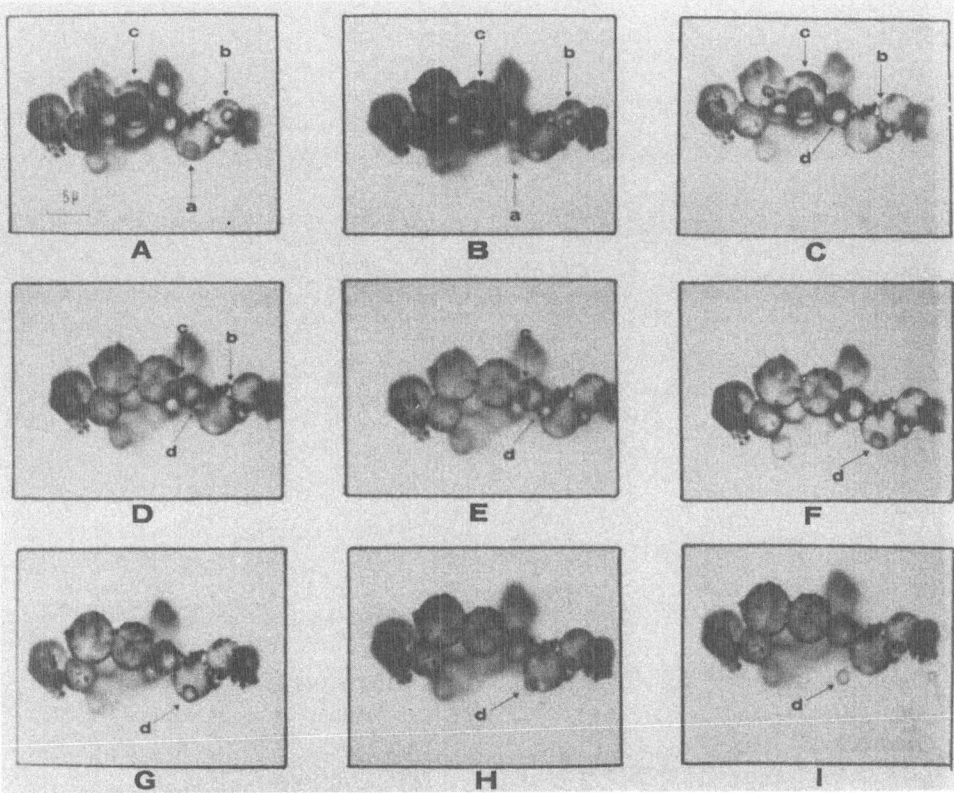

Fig. 2. Microspheres in this series of photomicrographs were prepared by boiling a mixture of 25 mg of 2:2:1-proteinoid per ml of distilled water, decanting, and allowing the solution to cool slowly to room temperature. Incubation at room temperature for 2 weeks followed. Microspheres were then stained with crystal violet and stabilized with iodine solution. Controlled solution of the interior was accomplished by allowing distilled water to stream through a sample on a microscopic slide. This series of photomicrographs was taken at 2 min intervals approximately 20 min after the onset of solution. Endoparticle (a) escapes into the environment (Figs. A-B), while endoparticle (d), (Figs. C-I) originated from a microsphere immediately below, is seen to exit through the same route and portal as used by endoparticle (a). The process of formation of endoparticle (c) (Figs. A-C) shows the existence of a communicative tunnel prior to solution and the subsequent formation of discrete endoparticles. Endoparticle (c) is shown entering a neighboring microsphere (Figs. D-E). The movement of endoparticle (b) can be traced in Figs. A-D. From reference 1 with permission of North-Holland Publishing Co.

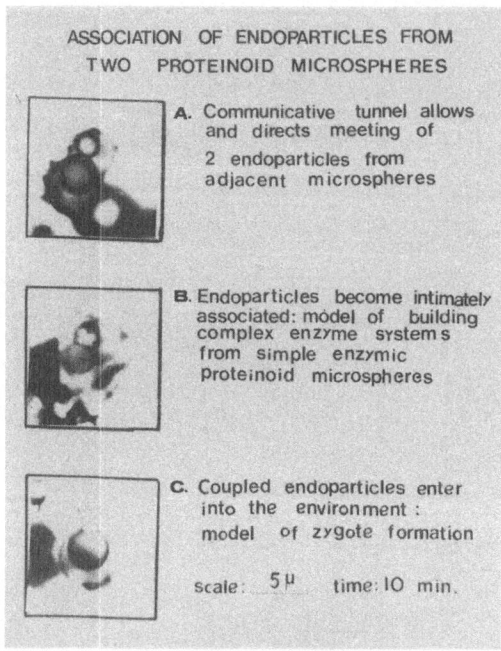

Fig. 3. Endoparticles from neighboring microspheres reach across the connection junction (A), associate (B), and enter into the medium as a couple (C).

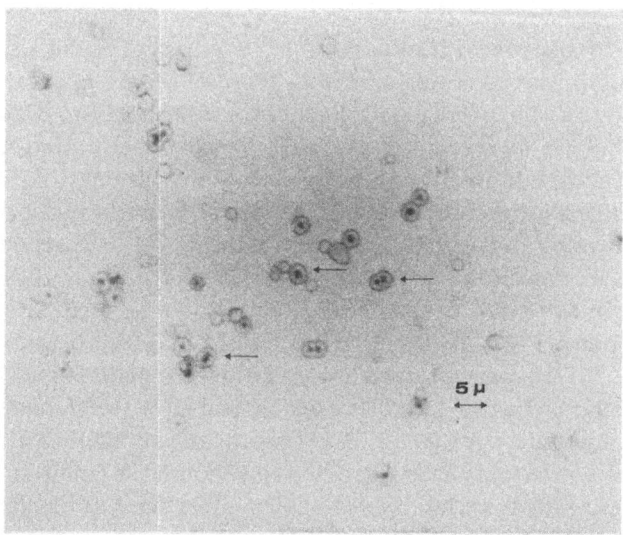

Fig. 4. New proteinoids in solution accumulate around stained endoparticles to form normal sized microspheres. Microspheres with 1, 2, or 3 endoparticles as accretion nuclei may be seen. Some microspheres are formed without endoparticles as accreting seeds.

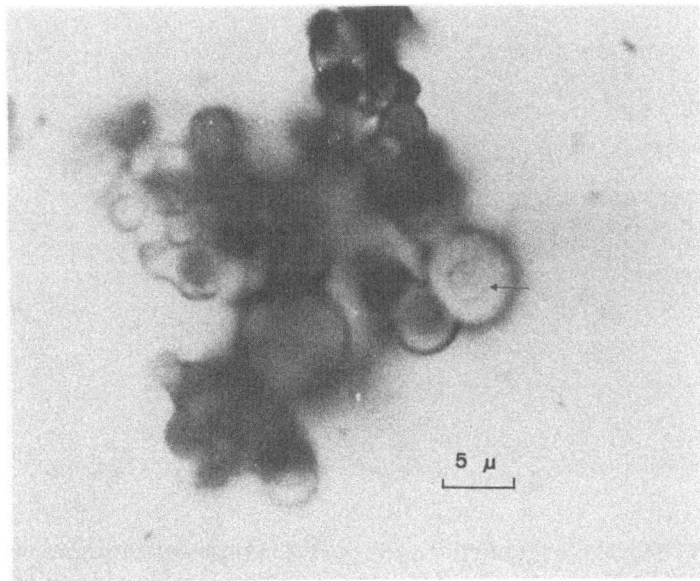

Fig. 5. A newly formed microsphere shows clear separation of the endoparticle (an accretion nucleus) from the proteinoid outer layer.

Upon accumulating new proteinoids from the medium, the endoparticles, behaving as condensation nuclei, are shown to progressively in-
filtrate into the newly accreted proteinoid layers. Fig. 5 show clear
demarcations between endoparticle and the accreted proteinoid. With
time, there is a gradual dissipation of such boundaries, and the endo-
particle-nucleus becomes a diffused, stained area within the new microsphere (Fig. 6).

DISCUSSION

The demonstration of the physical existence of communicative junctions between proteinoid microspheres has opened up new perspectives
expanding the repertoire of activities exhibited by these microsystems.
The significance of this communicating system can be best appreciated
only in conjunction with a consideration of the nature of the material being transferred. Proteinoids exhibit an impressive array of qualities,
many of which are in common with contemporary proteins (3) (Table 1),
and the microspheres formed from proteinoids have many cell-like properties (3) (Table 2). The conditions and reactants used in the preparation of proteinoids and microspheres are imputed to have been relevant
to the primitive Earth (3, 4).

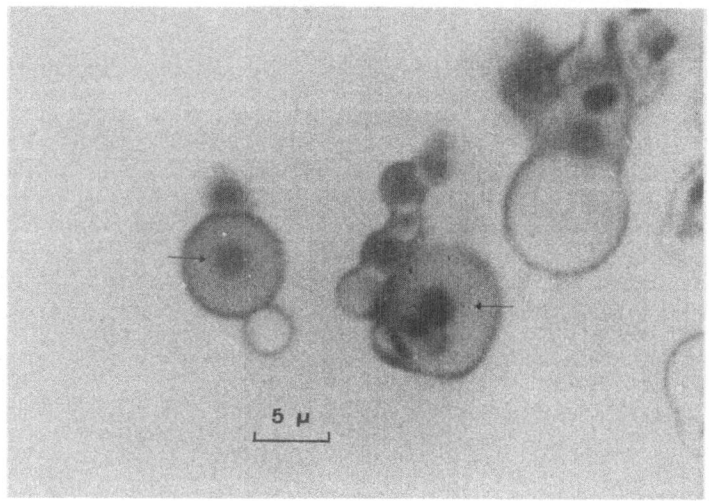

Fig. 6. Stained endoparticles behaving as condensation nuclei show indistinct outlines indicating diffusion into the peripheral proteinoid layers.

TABLE I.   Properties Common to Thermal Poly- α-amino Acids and
Proteins.   From reference 3 with permission from Springer-Verlag.

Qualitative composition
Quantitative composition (except serine, threonine, aspartic acid or lysine)
Range of molecular weights (4,000 - 10,000)
Color tests
Inclusion of nonamino acid groups
Solubilities
Salting-in and salting-out properties
Precipitability by protein reagents
Hypochromicity
Infrared absorption maxima
Recoverability of amino acids on mineral acid hydrolysis
Susceptibility to proteolytic enzymes
Some catalytic activity
Nutritive quality
Tendency to assemble into microparticulate systems.
Limited heterogeneity
Quantitative composition (except serine, threonine)
Many catalytic activities
Inactivatability by heating in aqueous solution
Hormonal activity (melanocyte stimulation)
Binding of polynucleotides (by basic proteinoids)

TABLE II. Properties of Proteinoid Microparticles. From reference 3 with permission from Springer-Verlag.

Stability (to standing, centrifugation, sectioning)
Microscopic size
Variability in shape
Uniformity of size
Numerousness
Stainability
Producibility as gram-positive or gram-negative
Osmotic type of property in atonic solutions
Structured boundary
Ultrastructure (electron microscope)
Selective passage of molecules through boundary
Catalytic activity
Patterns of association
Budding and fission
Motility
A number of catalytic activities
Growth by accretion
Ability to propagate through budding and growth by accretion
Binding of polynucleotides in various ways

What is taking place is the transfer among cell-like microsystems packets of informational macromolecules which are capable of self-ordered assembly (5 - 7) and certain enzyme-like activities (8). Conjugation between and among compatible microspheres followed by the production of recombinant-type microspheres is reminiscent of recombination in contemporary biological systems. A next question is whether this model of recombination can provide a variety of microspheres with sufficiently great differences to allow natural selection to operate.   One approach toward an answer would be a study of patterns of conjugal compatibility among microspheres composed of different proteinoids.   This kind of information would be of special interest in view of the fact that different proteinoids have been shown to possess different arrays of catalytic capacities (8).   Successful "marriages" between and among potentially supplementary or synergistic proteinoids seem a conceivable means by which more complex enzyme systems might have evolved from simpler ones.

REFERENCES

1. Hsu, L. L., Brooke, S., and Fox, S. W., A Model of Primordial Communication, Currents in Modern Biology 4, 12 (1971).
2. Fox, S. W., McCauley, R. J., and Wood, A., Comp. Biochem. Physiol. 20, 773 (1967).
3. Fox, S. W., Naturwissenschaften 56, 1 (1969).
4. Fox, S. W., and Windsor, C. R., Science 170, 984 (1970).
5. Fox, S. W., and Nakashima, T., Biochim. Biophys. Acta 140, 155 (1967).
6. Usdin, V. R., Mitz, M. A., and Killos, P. J., Arch. Biochem. Biophys. 122, 258 (1967).
7. Dose, K., and Zaki, L., in "Chemical Evolution and the Origin of Life" (R. Buvet and C. Ponnamperuma, eds.), p. 263. North-Holland, Amsterdam-London, 1971.
8. Rohlfing, D. L., and Fox, S. W., Advances in Catalysis 20, 373 (1969).

Ed. Note: As for all papers in this volume, this paper was reviewed by critics enlisted by the editors. Although the Institute for Molecular and Cellular Evolution has its own internal mechanism for review of papers, that was bypassed in this case for obvious reasons, but with official permission.

Received 2 August, 1971

HYDROTHERMALLY ASSOCIATED ORGANIC MICROSPHERES

FROM S. W. AFRICA AND ELSEWHERE

George Mueller

Institute for Molecular and Cellular Evolution
University of Miami
521 Anastasia Avenue, Coral Gables, Florida 33134

INTRODUCTION

The bulk of the organic matter constituting the terrestrial carbosphere is present in a more or less finely divided state within water-deposited sediments. The oldest known sedimentary rocks, of some 3.1 to 3.5 Bi. Y. Radiogenic Age, are from the Swaziland System of South Africa. Some structures in these sedimentary shales and cherts have been interpreted as possible micro-fossils (1), although the possibility that the particles in question may be inorganic microconcretions or organic artifices cannot be excluded. The small quantities of amino acids detected in the rocks (1) may be of ultimate pre-biological or biological origins, or they may be more recent contaminants.

Cherts from the sedimentary Gunflint formation of Canada, of approximately 2.1 Bi. Y. Radiogenic Age, proved to contain well preserved structures with detailed morphology which is similar to certain contemporaneous algae (2 - 3). The hitherto accumulated volume of morphological and chemical evidences seem to support the hypothesis of a biogenic origin of these Canadian sediments.

The carbonaceous meteorites contain finely divided organic matter which cannot be resolved with the optical microscope. This organic matter was interpreted as an abiogenic sediment from the transient atmosphere of the asteroidal parent bodies (4). Small quantities of finely divided

organic matter including amino acids have been reported recently from
the lunar fines by Fox et al. (5), Hare et al. (6), and others.

All the previously mentioned terrestrial and extraterrestrial finely
divided organic phases have the common features of sedimentation from
hydrospheres or atmospheres or transient atmospheres, and of being diluted by a relatively larger quantity of inorganic co-sedimented particulate matter.

Scanty data exists so far on a second type of occurrence of organic
substances, which are associated with hydrothermal deposits. Hydrothermal deposits are generally interpreted as condensates of volatiles
which escaped from the relatively hotter interior of the earth through
fissures and other zones of permeability within the rocks (7).   The surface vents of such hydrothermal deposits are the hot springs and the
fumarolas.

The organic matter within certain hydrothermal deposits is usually
clearly distinguished from the finely divided sedimentary organic matter,
in forming sporadically distributed, microscopically or macroscopically
visible particles of relatively pure organic substances with less than 10%
inorganic ashes.

This paper describes in some detail the theoretically most interesting hydrothermally associated organic substances within veins traversing
Precambrian formations from S. W. Africa. Brief descriptions of hydrothermally associated bitumens from other, mostly younger, formations follow. The significance of the observations in relation to the problem of the origin of life is discussed in the theoretical part of the paper.

ORGANIC INCLUSIONS IN HYDROTHERMAL QUARTZ FROM PRECAMBRIAN FORMATIONS OF S. W. AFRICA

According to Martin (8) the veins in the vicinity of Geiaus Farm, Klein
Karas district, ascend through a broad diabase dike of Precambrian age,
which is intruded into older, non-carbonaceous metamorphic rocks of the
Kheis system.   It was found in the course of field work by the author in
1967 that the veins both in the Geiaus Farm locality, and in the hitherto
not described Althorn Farm locality of the Klein Karas district, contain
an older generation of compact, turbid calcite, younger generation of
scalenohedrons of colorless to brown calcite, aggregates of pale blue
fluorite, and often doubly terminated quartz crystals, up to 15 cm long.

The volume-percentage of organic inclusions, which are present under elevated pressure, was determined on the integrating microscopical stage in randomly selected crystals from Geiaus Farm with the following results: quartz, 0.63%; fluorite, 0.18%; calcite, 0.07%.   Loose quartz crystals with organic inclusions have been purchased from mineral dealers at Windhoek, S. W. Africa, from the following unspecified Precambrian localities: A) Klein, Karas district, two further occurrences, the specimens from which differ from the foregoing localities in the yellow and grey-brown coloration of the quartz, respectively; B) extremely clear quartz crystals from the Berseba Native Reserve; C) a single 25 cm long, freely grown quartz crystal from the Gruenau district.

It was found that the crystals of quartz from all the aforementioned localities contain water, gas, and four main types of organic substances. Elemental analyses of the total organic substance within carefully cleaned and crushed quartz crystals from Geiaus Farm have been made by a commercial firm, and Dr. C. B. Moore of Arizona State University has made elemental analyses of crystals from the Althorn Farm locality.   None of the analyzed elements could be detected by the commercial firm from two blanks of quartz from Brazil.   The results are summarized in Table I below.

TABLE I.   Elemental Composition of Organic Inclusions in Quartz from S. W. Africa.

Locality	C	H	N	S
Althorn Farm	0.041%	N.D.	0.008%	0.003%
Geiaus Farm	0.26 %	0.07%	N.D.	N.D.

According to a personal communication of Dr. G. W. Hodgson of the University of Chalgary, Canada, the fluorescence and other indications for porphyrins are more intensive from the extracts of the central portions of a quartz crystal from Althorn Farm than from the surface of the same crystal. According to Kvenvolden and Roedder (9) of the mass spectrometer-gas chromatograph (MSGC) diagrams of the organic substances within quartz crystals from Geiaus Farm and Aukam contain normal alkenes, isoprenoids, and other organic molecules.

For the purposes of study of the distinct types of organic substances, which show sharp phase boundaries against each other in the inclusions, the quartz crystals were cut into plan-parallel slabs, and the intact inclusions investigated with microscopical methods. Free organic matter from opened inclusions was also investigated. The properties of the four main types of organic substances are described below, in order of their decreasing mean abundances in the inclusions.

1) Vaselinous oil. The substance is a yellow viscous liquid with bluish-white fluorescence, and with an M. P. range of -35° to $+2^OC$. The oil floats on both the watery phase of the inclusions and on distilled water. It is soluble in chloroform. The I. R. and U. V. Spectra indicate a predominantly paraffinic structure.

2) Black bitumen. The hard solid is completely opaque, black and non-fluorescent, with a partial M. P. between 350° and 450^OC. It is partially soluble in chloroform, and its density ranges between 1.1 and 1.2. The I. R. and U. V. spectra of chloroform extracts indicate a prevalently aromatic structure.

3) Elaterite. The elaterite type substances are from elastic to resilent resinous consistency, of reddish brown color and brown fluorescence, with M. P. ranging between 100° and 150^OC. The density is close to unity, and the particles show a greater or lesser degree of solubility in chloroform. The spot-spectrograms in visible and near U. V. light of the minute particles, made with the instrument of Stroher and Wolken (10), show a well defined maximum absorption at 4500 ^OA which is indicative of the presence of appreciable quantities of molecules with conjugated double bonds.

4) Brown bitumens. The group of substances has a narrow range of all the hitherto determined physical and chemical properties. Their color ranges from buff to brown and their fluorescence from orange to brick red. On heating on the heatable microscopical stage the particles partially melt between 80° and 150^OC, and the material chars around 350^OC. The particles swell in alcohol or chloroform and they are weakly birefringent. The spot-spectrograms show a pronounced absorption between 3500 and 4000 ^OA, which is indicative of a marked concentration of lipids.

Rare, sporadically distributed microspheres of oil, black bitumen, and elaterite have been observed. The relatively more abundant microspheres of the brown bitumen within oil or water-filled inclusions show structures which seem to resemble closely those of coacervate droplets

(11, 12), proteinoid microspheres (13, 14), and microorganisms. These structures are:

A) Microspheres with vacuoles filled with water or fine watery emulsions (Fig. 1).

B) Microspheres covered with sheaths of darker material which are occasionally ruptured by buds (Fig. 1).

C) Microspheres with darker cores, often covered by bud-like spheroids (Fig. 2).

D) Grape-bunch-like colonies and chains of microspheres, and more rarely, prolate spheroids (Fig. 3).

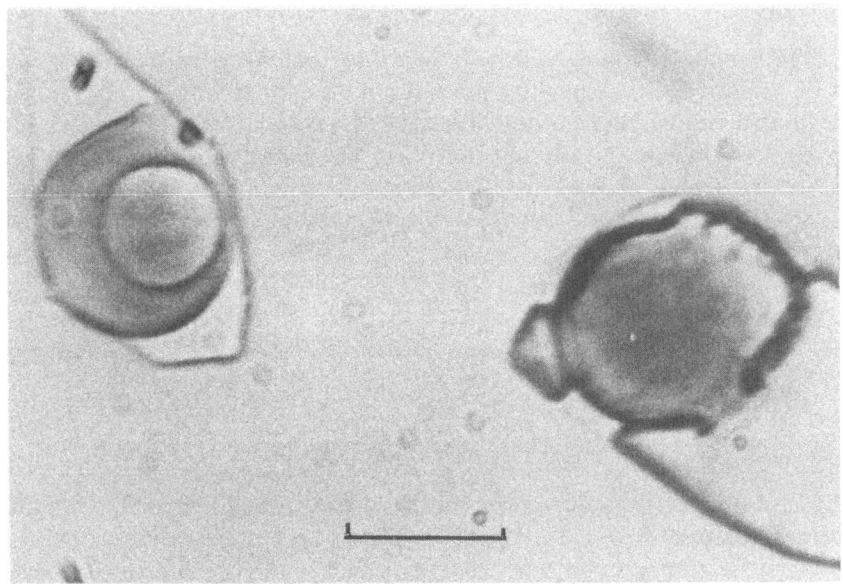

Fig. 1. Photomicrograph of two inclusions in a quartz crystal from Althorn Farm, S. W. Africa. Left: water-filled inclusion with microsphere of brown bitumen with a large vacuole of water, and a small twinned microsphere. Right: microsphere with bud (left) and incipient bud (right); both of these structures seem to be protruding from ruptures of a dark brown sheath which covers the light brown microsphere. (The scale shown on each of the photomicrograph represents 10 microns).

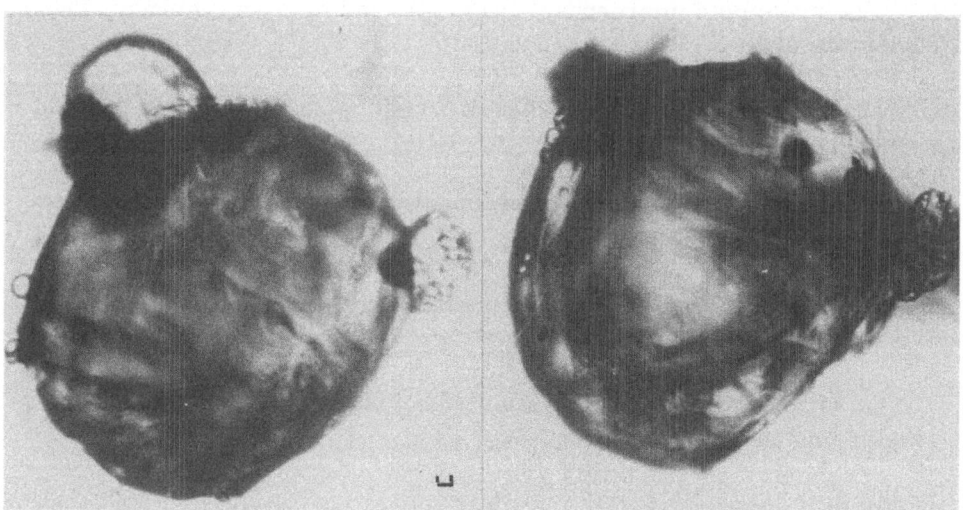

Fig. 2. Two photomicrographs of the same complex microsphere with focus on top surface (left) and on central portions (right). Note: A) Spheres and hemispheres of black bitumen within the microsphere of brown bitumen. B) Two buds of pale brown color, of the first generation. C) Minute colorless buds of a second generation which are mainly concentrated on the first generation buds.   D) Central core of darker material.   Althorn Farm.

E) Gigantic two or three-dimensional lattices of quasi-equidimensional microspheres in close-to-hexagonal or cubic packing (Fig. 4).

F) More or less regularly spaced spirals and spiral flaws (Fig. 5).

G) Some rare structures include combinations of threads, tubules, and microspheres.

Some of the oils from the Geiaus Farm locality precipitate microspheres with the optical properties of brown bitumens on cooling to -10°C, which redissolve at room temperatures.

A proportion of the microspheres is anchored to the walls of the inclusions.   Others are in free Brownian movement in the water, or more rarely, in the oil. Some of the particles are in rapid oscillatory or circular movement on intensive illumination, a phenomenon which does not occur on heating to 200°C.   Close-packed, soft microspheres tend to

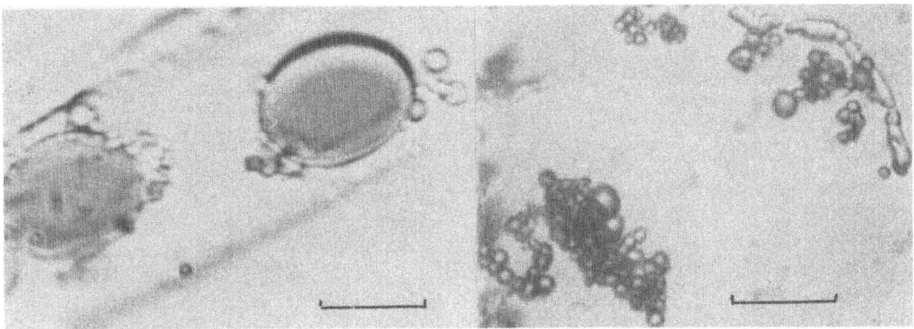

Fig. 3. Chains of microspheres and prolate spheroids. Left photomicrograph: Part of a water-filled inclusion with two prolate spheroids and groups of smaller microspheres of brown bitumen. Note striking similarity of dimensions and eccentricities of the two microspheres. Right photomicrograph: Part of large water-filled inclusion which has about 200 floating and anchored groups of microspheres, comprising mostly irregular colonies and chains. The chain at top right consists of a more or less regular succession of globular and spheroidal units, which tend to resemble certain contemporaneous algae and also the structures in cherts from the Precambrian gunflint formation of Canada. Althorn Farm.

squeeze each other into polygonal patterns, and they often show bud-like protrusions and distortion. Actual division into two units or coalescence, however, has not yet been proved with time lapse photomicrographs.

HYDROTHERMALLY ASSOCIATED ORGANIC SUBSTANCES FROM OTHER LOCALITIES

From the year 1948 onwards the author investigated some 20 localities in Europe, the Americas, and Africa for which a possible hydrothermal association of the bitumens was indicated in the almost exclusively 19th century literature, summarized by Dana and Ford (16) and other textbooks on mineralogy.   About 60 other hydrothermal deposits throughout the seven continents were searched for organic substances of which there was no record in the literature, and bitumens have been detected in 8 of the localities in Chile and Peru.   The features of the hydrothermally associated organic substances which have been studied in some detail so far,

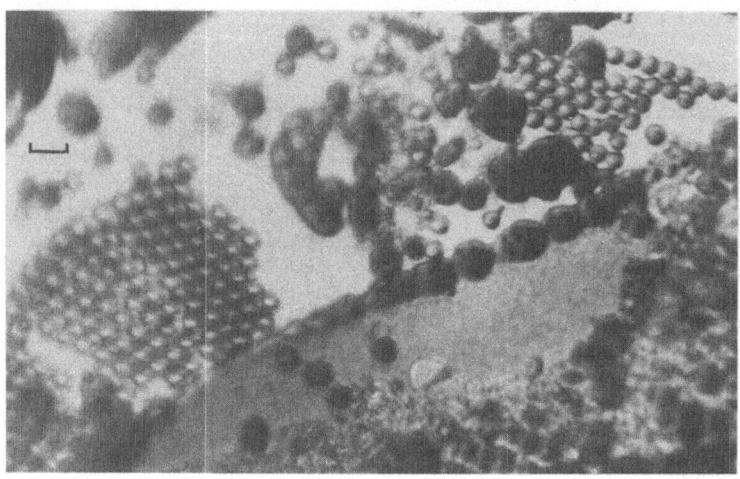

Fig. 4.   Part of large water filled inclusion with films (mainly bottom right) and microspheres of diverse shades, sizes, and types of aggregation of the brown bitumen.   Note generation of approximately 4.5 micron diameter light buff microspheres, which form gigantic quasi-hexagonal and cubic two dimensional lattices. Berseba Native Reserve.

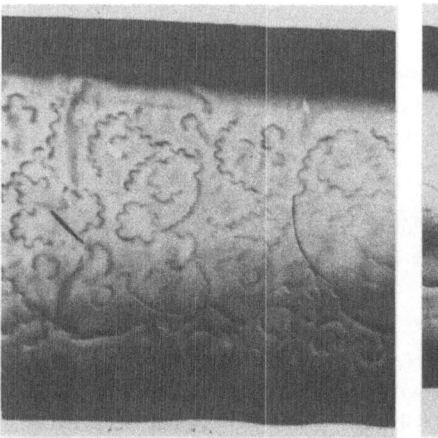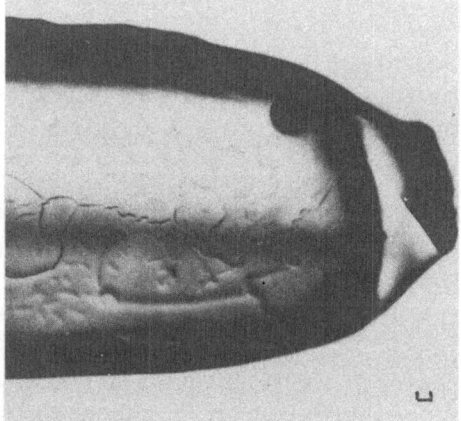

Fig. 5.   Central portion (left photomicrograph) and one of the extremities (right photomicrograph) of a 1.5 mm long rod of brown bitumen, anchored in a water-filled inclusion. Note more or less regularly spaced spirals and a single globule of black bitumen. Althorn Farm.

are briefly described below in order of decreasing geological age of the formations traversed by the veins.

1) Gold deposits from the Barberton area of S. Africa. The gold bearing veins traverse the Onverwacht and Figtree shales and cherts of 3.1 to 3.4 Bi. Y. of Radiogenic Age. A small quantity of massive black quartz could be collected on some old tips of the mines, in 1965. Thin sections of the non-fluorescent quartz reveal under the microscope opaque particles within the 1 to 5 micron diameter range, which are often of globular habits. Some of the minute opaque particles have been also detected in water and gas filled inclusions with diameters of up to 20 microns. The small size of the individual inclusions and their sporadical distribution within the hitherto collected material prevented more detailed investigation of these particles, which may be organic or inorganic.

2) Hydrothermal veins traversing pegmatites of Besmer Mine, Brit, Ontario, Canada. Shows of oil and hardened resinous materials were previously described (17, 18) from calcite-pyrite bearing veins, which traverse a pegmatite situated in later Precambrian metarocks of the locality in question. It was observed by the author, in the course of field investigations in 1969, that the organic substances seem to appear exclusively in those portions of the hydrothermal veins that traverse the thucholite-bearing zones of the pegmatite. This indicates the possibility that the organic substances in question generated through the hydrothermal hydrogenation, fractionation, etc., of the thucholite of magmatic origin. The veins contain a mainly paraffinic oil and microspheres, with occasional buds and short chain aggregations, of a substance with optical properties close to that of the brown bitumen from the S. W. African localities. As in the case of the S. W. African brown bitumen, the microspectrograms of these microspheres revealed the presence of appreciable quantities of lipids. Most of the organic material within the veins is interstitial, although some of the microspheres of the brown bitumen are overgrown by calcite.

3) The "Herkimer Diamonds" from Middleville, New York. The Silurian dolomite of the locality is honeycombed with cavities mineralized with dolomite, calcite, and quartz. Microscopically and macroscopically detectable organic inclusions are restricted to the quartz crystals. The bulk of the interstitial and included fragments of organic substances have properties within the range of those of black bitumens. About one percent of the up to 10 cm. long quartz crystals proved to show fluorescence in U. V. light. The inclusions in such fluorescent crystals contain, be-

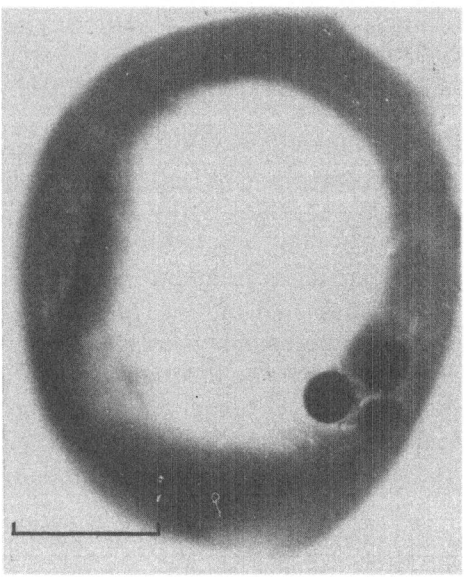

Fig. 6.   Minute balloon of blue-
white fluorescent vaseline which
freely floats in the watery phase of
a large inclusion of a quartz crystal from Middleville, New York.
Note three size-controlled microspheres of dark brown bitumen
within the lining of the balloon.

sides the fragments of the black bitumen, small quantities of a yellow oil
and simple microspheres (within a rather well-defined diameter range
between 5 and 10 microns) of a dark brown bitumen (Fig. 6).

4) The North Derbyshire Orefield.   Detailed field investigations in
N. Derbyshire and Staffordshire, England, revealed well over 50 hydrothermal veins with associated bitumens. The veins traverse Lower Carboniferous shales and limestones (19, 20). The fact that in several veins
organic substances in 100 - 1000 g quantities can be readily collected
facilitates the detailed chemical work.   An other favorable feature of the
bitumens associated with the veins of the orefield is their considerable
range of variability, which seems to be interrelated to the changes in the
hydrothermal mineral composition of the vein and the type of biogenic
sediments traversed by the vein.   Thus it was observed that mainly cal-

citic veins, traversing kerogen-containing shales within the Mill Close Mine and other areas of the orefield, contain a single oil of asphaltic base, which tends to turn to solid asphalt and finally to asphaltite with the increase of the fluorite and sulphide ore percentage of the veins.

In the Castleton and the Matlock areas, where the sediments proved to be finely divided coal type substances, the calcitic veins contain a variety of black bitumen, the Derbyshire variety of which was termed "carbonite" (19, 20). The high fluorite and sulphide ore percentage veins of the same areas contain a suit of five distinct organic substances which show invariably sharp phase boundaries against each other. These veins contain: a resinous variety of carbonite; two types of elaterite, namely, the elastic, low-boiling substance which was originally termed "elaterite" (16) and extruded rods of the resilient, resinous bernalite (20) vaselinous oil; and a brown bitumen-type resin, which was termed foxite (20). The results of elemental analyses, functional analyses of fractions and extracts, X-ray crystallography, I. R. and U. V. spectroscopy, and some MSGC work seem to support the original hypothesis (19) that each one of the organic substances in the vein contains the bulk of one main type of organic molecule. Thus carbonite is of mainly aromatic structure; elaterite and bernalite are low and high molecular olefinic substances respectively, the vaselinous oil is an almost pure crystallizable mixture of paraffins, and foxite contains the bulk of the organic acids and the nitrogen-containing organic molecules. Hitherto unpublished work of the author with the cooperation of Mr. C. Windsor and Dr. Jun-ichi Oh-Hashi (of the Institute of Molecular Evolution) revealed 1.55% organic nitrogen; 17.82% -COOH group, and 170.48 ppm. amino acids in carefully cleaned, sterile samples of foxite from Derbyshire, whereas the other four types of organic substances associated with the foxite proved to contain only about one tenth or less of these biologically important constituents. It was found that only the brown bitumen-type substances form microspheres within the inclusions or within the interstitial organic matter. A brown bitumen-type substance was observed to form patterns of microparticles parallel with the main crystallographical orientations of fluorite hosts (21).

5) The Illinois-Kentucky Orefield. The fluorite-lead-zinc vein and bedding replacement deposits are within lower Carboniferous, Mississippian limestones, shales, and sandstones of the orefield (22). It was observed by Roedder (23) that the oil-filled inclusions in fluorite contain brown solid organic particles which are arrayed in patterns parallel with the 100 and other crystallographical directions of the fluorite host. According to hitherto unpublished work of the author, the particles of the brown bitumen-type substance lining the walls of the inclusions are single

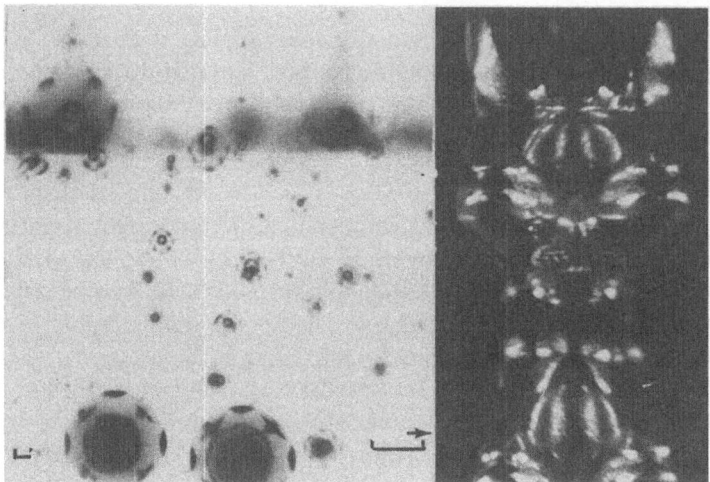

Fig. 7.   Crystallographically controlled patterns of particles of a brown bitumen–type substance lining oil-filled inclusions in fluorites from Mahoney Ozark Mine, Illinois.   Left photomicrograph: Top view and side view (top) of globular, bottle, and tubule shaped inclusions with particles forming regular patterns on the walls in the 100, 110 and 111-parallel directions of the host.   Right photomicrograph: Portion of the profile of a long tubule with elongation parallel with the 100 direction of the fluorite host, polarized light, crossed nicolls.   All the particles in the elaborate pattern lining this tubule extinguish at $45^O \pm 1^O$, and therefore in the parallel position of the illustration they show maximum luminosity.

crystals, the optic extinction positions of which are related within $\pm 2^O$ to the crystallographical orientation of the wall of the host (Fig. 7).   Occasional quartz crystals associated with the fluorite contain the particles of brown bitumen within a filling of mainly paraffinic oil, but the inclusions in this host are randomly distributed and oriented.   Some of the oil fillings of the inclusions contain also a colorless crystallized substance of low birefringence which dissolves in the oil on heating to $100^O C$ and reprecipitates on cooling to room temperature.

6) Dolomite deposits within the Maule district of Chile. The body of dolomite is emplaced in lavas of Mesozoic age, and it is traversed by small veins with calcite, pyrite, and chalcopyrite, according to results of hitherto unpublished field observations of the author. The veins contain sporadically distributed oil shows, thin films of a black bitumen, and globules of a brown bitumen; all these organic substances show sharp phase boundaries against each other.   Rare microspheres within the 20

micron diameter range of the brown bitumen have been observed lining some of the vugs of the veins.

7) Mercury veins from Californian Localities. The mercury veins of California traverse formations of Tertiary age and contain sporadically distributed, well crystallized, readily sublimable organic substances of which curtisite has been more or less precisely identified as consisting mainly of condensed aromatics (24). Similar, likely condensed aromatic substances from California include aragotite, prosepnyite, and napalite (16).

In addition, the deposits studied by the author proved to contain paraffinic oil and bituminous substances with physical and optical properties within the range of those of the brown bitumens. According to Bailey (25) froth veins are present in Abbott Quicksilver Mine, Lake County, and elsewhere in California. The filling of these veins consist of innumerable droplets of oil and microspheres of the brown bituminous substance, covered, and cemented together with opal or quartz. According to microscopical studies of the author most of the microspheres and globules within these veins are simple. The diameters range from 10 to 1000 microns without any trend of size control.

DISCUSSION OF THE EVOLUTIONAL SIGNIFICANCE OF THE HYDROTHERMALLY ASSOCIATED ORGANIC SUBSTANCES AND MICROSPHERES

The genesis of the hydrothermally associated organic substances and organic microspheres may be explained by one of the following two hypotheses:

A) The organic substances are products of hydrothermal mobilization (distillation, melting-out, etc.) of the sediments which surround the hydrothermal veins. On cooling of the hydrothermal fluids, a portion of the ultimately biogenic mobilizates would condense into the organic particles. This process seems to resemble, at least in its main physical aspects, the experimental procedure of Oparin (11, 12) in his production of coacervate droplets from pre-existent biological material, such as egg albumen.

B) The organic substances of ultimate abiogenic origin reached the vein with the juvenile volatiles, and the organic particles precipitated on cooling of the solution in the course of their ascent. This process seems to resemble (when ignoring probable differences in the chemical composition of the natural organic particles) the experimental procedure devised originally by Fox (13 - 15) for the synthesis of proteinoid microspheres. According to this procedure amino acids (which may be abiogenic) can be condensed to proteinoids between 120 and 200°C, and the solution of these proteinoids in boiling water readily precipitates the microspheres on cooling to room temperature. It is interesting to note in this connection that the observed homogenization temperatures of the inclusions in the S. W. African quartz crystals (9, and results of the author) indicate a minimum trapping temperature of about 120° to 160°C.

The hitherto obtained inconclusive data on the geology of the S. W. African localities seem to favor hypothesis B. The rocks which immediately surround the veins at Geiaus and Althorn Farms are normal igneous diabases, and the metamorphic rocks flanking the two diabase dikes in question seem to contain no appreciable amounts of carbonaceous matter. It is unlikely that the organic substances with densities mostly below unity would have migrated downwards from the presently eroded overlaying sediments. The geology of the Canadian deposits similarly indicates a juvenile origin of the associated organic substances. In the case of the rest of the hitherto studied deposits, which traverse mainly sediments, an ultimate biogenic mobilizate origin of the organic substances is a relatively more likely alternative, although the possibility of the presence of juvenile substances which entered the vein with the hydrothermal fluids cannot be excluded.

Chemical criteria for the determination of biogenic or abiogenic origin of a given organic substance were discussed (4, 26, 27). Such chemical criteria include the presence of specific biomolecules, the odd preference of the paraffins, elemental composition, C 12/13 ratios, etc. These have been considered without any firm conclusions or agreements as to chemical criteria which would decide between biogenic or abiogenic origin of a given substance. Kvenvolden and Roedder (9) came to the conclusion that the presence of isoprenoids within the organic inclusions from the S. W. African localities suggested a biogenic origin, whereas the C 12/13 ratio of the same substances tends to fall outside the range of biogenic organics. A systematic comparison of geological and chemical data of a diverse range of hydrothermally associated organic substances with synthetic substances and bio-products may yield in the future more conclusive criteria as to the determination of ultimate abiogenic or biogenic origin.

Observations made so far about certain general trends of interrelations between the assemblage of hydrothermal minerals in the veins and the associated carbonaceous complex may serve as vague guidelines for more detailed work in the future. It seems, at least as a first approximation, that the properties of the hydrothermally associated organic substances primarily depend on the temperature of formation of the hydrothermal minerals present in the vein, and they are influenced to a lesser extent by the genetical type and age of the country rock surrounding the vein.

The veins with a predominantly calcitic gangue [generally believed, on the basis of inorganic geological thermometers (27, 28) and organic geological thermometers (20), to be indicative of a relatively low temperature of formation--possibly below $100^{\circ}C$] contain, without any hitherto observed exceptions, a single organic substance which may be an oil, asphalt, or asphaltite. With the increase of the percentage of fluorite and pyrite, and other hydrothermal minerals for which a relatively higher temperature of condensation within the 100° to $400^{\circ}C$ temperature range is postulated (28), there is a trend of subdivision of organics into two or more phases. This trend may be absent, or it may be relatively little pronounced if the mean composition of the carbonaceous complex is highly saturated with H/C ratios around 2.0. In the case of a lower value of H/C ratio for the mean of the carbonaceous complex (19), the indications are that two or more substances may separate; such differentiates may consist of mainly aromatic, mainly olefinic, mainly paraffinic, and mainly polar molecules. This process of differentiation occurs through polycondensation, fractionation, and unmixing of an originally homogeneous carbonaceous complex. The number and exact chemical type of the separated organic substances would depend on the composition of the original carbonaceous complex and on the chemical and temperature-pressure conditions which prevailed in the course of mineralization.

It appears from the foregoing considerations that organic substances of both ultimate abiogenic and biogenic origin may tend to undergo chemical changes, approaching the equilibrium conditions which prevail within a given hydrothermal vein. The processes involved seem to be missing links in chemical evolution which may lead eventually to the emergence of the living organism. These possible advances in chemical evolution within hydrothermal veins may be briefly summarized below as follows:

1) Formation of masses of organic substances up to 99% or greater purity, on a microscopic and macroscopic scale.

2) The trend of concentration of organic substances on the surface and within the interior of the optically rotating quartz crystals, as observed from S. W. Africa and elsewhere, may have a significance in the production of optically active substances.

3) Separation of the brown bitumen-type substances, which, according to spectroscopical and chemical work on the S. N. African and the Derbyshire deposits, tend to concentrate the following types of constituents of possible significance in the pathways of chemical evolution: A) Lipids and other polar molecules. (It may be noted in this connection that the specific molecules which form the bulk of the living organism all belong to the broad group of polar organic substances.) B) Nitrogen-containing organic molecules. C) Amino acids, as observed in the brown bitumens of Derbyshire.

4) Formation of strikingly biomorphic particles from the brown bitumen, which seem to have the most biosimilar chemistry of all the organic substances in the high temperature veins.

5) The evolutional significance of crystallographically ordered organic particles in inclusions in fluorites from the Derbyshire and the Illinois-Kentucky orefields cannot be as yet assessed. The structures seem to be examples of self-assembly of matter, without any hitherto established inter-connections to living processes.

The interrelations between the properties of the organic substances and the associated hydrothermal minerals in the veins in S. W. Africa and elsewhere were previously discussed in greater detail by the author (29 - 30). It may be concluded that systematic studies in the future on a greater number and diversity of bitumen bearing hydrothermal veins may lead towards a more precise reconstruction of the genetical history of hydrothermally associated carbonaceous complexes.

The presently available data outlined in this chapter seems to stress the possible evolutional significance of processing of organic substances within hydrothermal vents. This supports the theory of Fox (27), according to which life may have originated through processes which are similar to the synthesis of proteinoid microspheres and which may have occurred in the vicinity of fumaroles and hot springs. The fumaroles and hot springs are generally interpreted as the surface vents of the ascending hydrothermal fluids, as indicated in the introduction of this paper. A proportion of the biomorphic organic particles which generated in the hydrothermal veins of S. W. Africa and other localities becomes preserved as inclusions in the crystals of the gangue minerals. The bulk of the organic particles, however, would have been expelled through the fumaroles or hot springs at the surface of the hydrothermal vent. In the course of their dispersal through the soils, rivers, and oceans, certain favorable types of these hydrothermally produced organic particles could have undergone a second, low temperature stage of evolution towards the organism.

SUMMARY

It was observed that quartz crystals from veins within two diabase dikes of precambrian age from the Klein Karas area of S. W. Africa contain particles of a brown bitumen, which closely resemble, in detailed morphology, coacervate droplets, proteinoid microspheres or fossil and recent micro-organisms. The microphotospectrographs of these minute particles revealed a strong absorbtion peak at 4000 $\overset{\circ}{A}$, which is indicative of lipids. Particles of brown bitumen type substances of relatively less elaborate biomorphic structures have been observed from hydrothermal minerals of several other localities. The theoretical significance of these organic particles is discussed with reference to problems of origin of life.

ACKNOWLEDGMENTS

I would like to thank Dr. J. R. McIver of the Department of Geology, University of Witwatersrand, S. Africa, for drawing my attention to the deposits in S. W. Africa and for exchanging and loaning some of the specimens from his museum, in the course of my visit during 1965.

I would like to express my appreciation to Dr. Sidney W. Fox for many inspiring conversations on the subject and for constructive criticisms.

From the year 1948 onwards the work was supported by many individuals and organizations, as indicated in previous publications and in publications at present in the press or in various stages of preparation. Some of the figures are from reference 30, with permission from Macmillan Journals Ltd.

The work was supported from 1966 onwards by National Aeronautics and Space Administration grants NGR 10-007-054.

REFERENCES

1. Schopf, J. W., and Barghorn, E. S., Science 152, 758 (1967).
2. Barghoorn, E. S., and Tyler, S. A., Science 147, 563 (1965).
3. Cloud, P. E., Jr., and Hagen, H., Proc. Natl. Acad. Sci. U.S.A. 54, 1 (1965).
4. Mueller, G., Geochim. Cosmochim. Acta 4, 1 (1953).
5. Fox, S. W., Harada, K., Hare, P. E., Hinsch, G., and Mueller, G., Science 167, 767 (1970).
6. Hare, P. E., Harada, K., and Fox, S. W., Geochim. Cosmochim. Acta, Supplement 1, Volume 2, 1799 (1970).
7. Bateman, A. M., "The Formation of Mineral Deposits," p. 273. John Wiley and Sons, New York, and Chapman and Hall Ltd., London, 1951.
8. Martin, H., "The Precambrian Geology of South West Africa," p. 133. Publ. Pre-Cambrian Research Unit, University of Cape Town, 1966.
9. Kvenvolden, K. A., and Roedder, E., Geochim. Cosmochim. Acta 35, 1209 (1071).
10. Stroher, G. K., and Wolken, J. J., Science 130, 1084 (1959).
11. Oparin, A. I., "The Chemical Origin of Life," p. 42. Charles C. Thomas, Springfield, Illinois, 1964.
12. Oparin, A. I., in "The Origin of Prebiological Systems" (S. W. Fox, ed.), p. 331. Academic Press, New York, 1965.
13. Fox, S. W., Harada, K., and Kendrick, J., Science 129, 1221 (1959).
14. Miquel, J., Brooke, S., and Fox, S. W., Currents in Modern Biology 3, 299 (1971).
15. Young, R. S., in "The Origin of Prebiological Systems" (S. W. Fox, ed.), p. 347. Academic Press, New York, 1955.
16. Dana, E. S., and Ford, W. E., "A Textbook of Mineralogy," p. 775. John Wiley and Sons, New York, London, Sydney, 4th Ed., 1966.
17. Ellsworth, H. V., Amer. Min. 13, 419 (1928).
18. Spence, H. S., Amer. Min. 15, 430 (1930).
19. Mueller, G., Compt. Rend. 19th Cong. Int. Geol. Sec. 12, Fas. 12, p. 279, (1954).
20. Mueller, G., in "Advances in Organic Geochemistry" (G. D. Hobson, ed.), p. 443, Pergamon Press, Oxford, 1970.
21. Mueller, G., Compt. Rend. 19-e Cong. Geol. Int., (Alger), Sec. 13, Fas. 15, p. 523 (1954).
22. Baxter, J. W., Potter, P. E., and Doyle, F. L., Ill. State Geol. Survey, Circular 342, p. 1, 1963.

23. Roedder, E., Sci. American 207, 38 (1962).
24. Wright, F. E., and Allen, E. T., Amer. Min. 15, 169 (1926).
25. Bailey, E. H., Bull. Geol. Soc. America 70, 661 (1959).
26. Calvin, M., "Chemical Evolution," p. 26. Oxford University Press, New York, Oxford, 1969.
27. Fox, S. W., Harada, K., Krampitz, G., and Mueller, G., Chem. Eng. News 48, 80 (1970).
28. Dunham, K. C., Amer. Min. 22, 468 (1937).
29. Mueller, G., Bol. Soc. Biol. Concepcion 40, 161 (1965-66).
30. Mueller, G., Nature 235, 90 (1972).

Ed. Note: As for all papers in this volume, this paper was reviewed by critics enlisted by the editors. Although the Institute for Molecular and Cellular Evolution has its own internal mechanism for review of papers, that was bypassed in this case for obvious reasons, but with official permission.

Received 27 July, 1971

STUDIES ON AN EXTREME THERMOPHILE

Flavobacterium thermophilum HB 8

Tairo Oshima

Mitsubishi Kasei Institute of Life Sciences
Machida, Tokyo, Japan

INTRODUCTION

The growth of microorganisms under such extreme environmental conditions as high salt content, low or high temperatures, and very low pH has been known for many years. Thermophiles, microorganisms that have the ability to grow at high temperature (at or above 50°C), were first described by P. Miquel in 1888 (1 - 3). Since then, thermophiles isolated from compost, sewage, cultivated soil, mud, waters, and even in ocean bottom mud and freshly fallen snow, have been objects of biological interest. Information accumulated on the physiology of growth at high temperature has been summarized in many review papers (1 - 11).

The ability of thermophiles to grow at temperatures that denature most of the macromolecular elements of mesophiles has been intensively studied in the past several years. Three theories have been proposed to explain thermophily (3). The first is the dynamic theory that suggests replenishment of proteins, which are as labile as mesophile proteins, by rapid regeneration. Second, heat stability may be conferred by the presence of protective factors or the absence of labilizing factors. A third explanation suggests that the cell constituents of thermophiles are inherently more stable to heat than those of mesophiles, presumably as the results of mutations during the adaptation to high temperature. The third theory is now generally accepted. A number of investigations of enzymes and ribosomes of thermophilic forms have shown that thermophily is not merely the result of a rapid resynthesis of heat-labile components, but that cell constituents of thermophiles are intrinsically more thermostable without specific protectors.   Most of these studies have, however, been done using spore-forming bacteria belonging to the genus Bacillus, especially B. stearothermophilus which is a moderate thermophile having a maximum growth temperature of 70 to 75°C.

Investigations of the molecular basis of thermophily are of particular importance to researchers in molecular evolution.   Comparative studies on the structural characteristics of the thermophile proteins and nucleic acids may be one of the most fruitful ways to study macromolecular evolution, because it may be anticipated that developments of these structural characteristics should be forced mainly by a single environmental factor, heat, during the evolutionary process toward a thermophile.   These investigations may also reveal the limit of biological extremes.   The information resulting from these studies could contribute to the problems of the origin and early evolution of life on our planet and of searching for extraterresterial life in the universe. Another interesting factor is the application of heat stable enzymes in chemical industry. Thermostable enzymes will be good sources for making water-insoluble enzymes. The heat-labile enzymes from mesophiles might be chemically modified to thermostable one, imitating the evolutional alterations in thermophile enzyme proteins. For these reasons, investigations on thermophiles were begun a few years ago when extreme thermophiles were isolated from a thermal spa in Japan. This article presents some aspects of the nature of cell constituents of Flavobacterium thermophilum HB 8.

ISOLATION OF EXTREME THERMOPHILES

Attempts have been made to isolate thermophilic bacteria from thermal inhabitats at Mine Hot Spring, Shizuoka, Japan.   A typical thermal water well is illustrated in Fig. 1. The boiling water (100°C) enters the system through a wooden, square pipe in the center of the derrick and is collected in small wooden tanks at the top and the bottom of the derrick.   Hot spring water is piped through bamboo tubes to nearby hotels and houses for bathing.   The temperature of the water collected in the bottom tank, pool B, was over 90°C near the source and 80°C or higher along the border.   The pH of the spring water was approximately 6.3. The thermal water at Mine Hot Spring did not contain high concentration of salts or sulfur and can be considered similar to fresh water except for temperature.   The thermal environment may be the only major factor affecting the evolutionary processes of the inhabitants of pool B.

Fig. 1. Photograph of a thermal water derrick at Mine Hot
Spring, Shizuoka, Japan.

Several thermophilic bacteria were isolated from the pool, and one strain has been investigated in some detail. Good growth of the organism was observed in a medium consisting of 0.8% polypeptone (or casamino acid), 0.4% yeast extract, and 0.2% sodium chloride, pH 7.0 (at room temperature). The cells are yellow pigmented, gram-negative and nonsporulating rods. A micrograph is shown in Fig. 2 (12, 13). The cells occur singly, in pairs, and sometimes in chains. The yellow pigments are not soluble in the culture medium, but can be extracted with 90% acetone from the intact cell paste. The absorption spectrum of the extract exhibited three absorption maxima at 430, 453, and 473 nm, suggesting the presence of carotenoid compounds. The organism is aerobic and strictly thermophilic. Growth was observed only on the surface of soft agar medium. The optimum growth temperature was 65 to 72°C. When cultivated at these temperatures, the cells had a generation time of 20 min. Cell populations of 7 to 8 X 10^8 per cm^3 were obtained. The growth was slow when incubation was above 80°C or below 50°C. The maximal growth temperature was 85°C. No growth was observed at 40°C, which indicates the organism is obligately thermophilic.

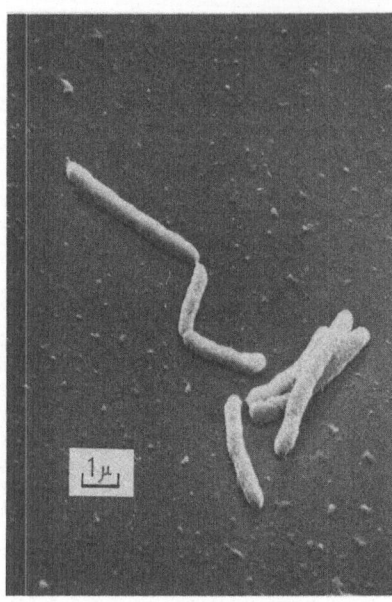

Fig. 2. Electronmicrograph of Flavobacterium thermophilum
HB 8. The picture was taken at Application Laboratory, Naka
Works, Hitachi Ltd., using a Hitachi HSM-2 scanning electron
microscope at a magnification of 10,000.

Very poor growth was observed in a synthetic medium containing glucose and ammonium salt as carbon and nitrogen sources in the presence
of mixture of vitamins, minerals, and other growth factors. Acid but no
gas was produced from glucose, galactose, maltose, and lactose. No
gas was produced from sucrose and manitol. The organism is proteolytic, since casein or albumin could serve as a good source of carbon and
nitrogen.

High susceptibility of the thermophile to antibiotics was observed.
The following antimicrobial agents inhibited the cell growth completely at
the concentration of 8 - 10 μ g per ml of culture medium: Gramicidin J,
Penicillin G, Ampicillin, Novobiocin, Methicillin, Chloramphenicol, and
Streptomycin. Slow growth was observed in the presence of 10 μ g/ml
tetracycline and no growth at the concentration of 100 μ g/ml. Actinomycin D was also a potent inhibitor for the microorganism, since no more
than 1 μg/ml was necessary to suppress bacterial multiplication. Although the isolate is a Gram-negative bacterium, it is highly sensitive to
actinomycin, novobiocin, and methicillin.

The thermophile was identified to be a new species belonging to the genus Flavobacterium (14). It was named F. thermophilum HB 8.

Organisms that are similar to the strain described above were isolated from other thermal spas at Mine Hot Spring, Shimokamo Hot Spring, and Atagawa Hot Spring in Shizuoka, thereby indicating wide distribution of F. thermophilum or closely related organisms in thermal waters in this vicinity. Although the biological entities in thermal springs in Japan have been known for many years (15, 16), the extreme thermophilic bacterium described here has not been reported. Brock and his co-worker recently reported the isolation of extreme thermophiles from hot springs and hot tap water in the United States (17). The microorganism was named Thermus aquaticus. T. aquaticus YT-1, isolated from Mushroom Spring at Yellowstone National Park, resembles F. thermophilum. The former thermophile is gram-negative, yellow pigmented, and is a nonsporulating rod. The maximal growth temperature (79°) and the nutritional requirement of the former, however, differ from those of F. thermophilum. Taking into an account that the original habitats of these two thermophilic organisms are geographically quite distant, the remarkable similarities in morphology between these two bacteria will be of interest from the viewpoint of evolutional analogies. It will be of great value to compare molecular structures of biopolymers from these thermophiles in order to elucidate the mechanisms of evolution in aquatic thermal environment.

NUCLEIC ACIDS AND NUCLEOPROTEINS

Deoxyribonucleic acid. The melting temperature (T_m) of deoxyribonucleic acid extracted and purified from F. thermophilum was $97.5^{\circ}C$ in 0.15 $\underline{M}$ sodium chloride solution containing 0.015 $\underline{M}$ sodium citrate, pH 7.0 (18). Guanine plus cytosine (G + C) content, calculated from the T_m value, was 68.8%: The density of the DNA molecule was found to be 1.727 by an ultracentrifugal analysis in cesium chloride solution, from which the G + C content was estimated to be 68.4%. These values agree with the results obtained by chemical analysis as shown in Table I. The consistency between chemical analysis, melting temperature, and density suggests that the thermophile DNA contains no minor bases such as 5-hydroxymethylcytosine, deoxyuridylic acid, etc., which are often present in phage DNAs.

The G + C content of F. thermophilum HB 8 DNA is close to those reported for Flavobacterium (19), supporting the bacteriological identifica-

Table I.   Properties of F. thermophilum HB 8 DNA

		Estimated (G + C) content
Base analysis after hydrolysis	A = 15.5% C = 36.9 G = 32.1 T = 15.4	69.0%
Melting temperature in SSC[a]	97.5°	68.8
Melting temperature in 1/10 SSC	81.3°	66.9
Density in CsC1	1.727	68.4
ε (P)	6,700	---

[a]Standard sodium citrate.

tion of the present thermophile. The DNA base composition also resembles to that of T. aquaticus YT-1, 67.4% G + C (17).

The G + C content of F. thermophilum DNA is one of the highest among the bacterial DNA. Higher values were reported for DNAs from Micrococcus lysodeikticus (20), Nocardia rubra (21), Sarcina lutea (22), and some others. It is known that many amino acids are selected by more than one codon, and that the third letter of the genetic code are often equivalence of four bases. The third nucleotide can be fixed to cytosine or guanine without any change in proteins to be coded with a few exceptions. The base composition of such DNA molecules can be expected to be about 67% G + C.

Sueoka reported the several slight, but significant, correlations of amino acid composition of proteins with G + C content of DNA (23). Alanine, proline and arginine (and some others) were positively correlated; whereas, aspartic acid and isoleucine (and some others) were negatively correlated with G + C content of DNA.

The amino acid composition of F. thermophilum was compared to that of Escherichia coli (Table II). Some enrichment of alanine, proline, and arginine was evident; however, most of individual amino acids were found in such molar proportions as those of the E. coli protein. It is proposed that the present thermophile utilizes a code book which consisted of

Table II.   Amino Acid Composition of Bulk Protein

Amino acid residues	Organism	
	F. thermophilum HB8	E. coli B
Aspartic acid	7.5 % [a]	10.2 % [a]
Threonine	5.1	5.3
Serine	3.8	4.5
Glutamic acid	12.2	11.5
Proline	6.7	4.4
Glycine	10.1	10.0
Alanine	11.5	10.4
Cystein [b]	0.2	0.8
Valine	7.8	8.4
Methionine	0.6	0.8
Isoleucine	3.8	5.8
Leucine	10.4	9.0
Tyrosine	3.1	2.5
Phenylalanine	3.5	3.4
Lysine	5.4	6.3
Histidine	1.5	2.2
Arginine	6.1	5.2

[a] The values are given in molar content (%) of amino acid per total amount of amino acid detected by analyzer.

[b] Carboxymethylated before acid hydrolysis of bulk protein.

nearly 32 codons of 64 triplets, whose third letters are mostly fixed into cytosine or guanine.

The circular dichroism (CD) of the thermophile DNA in solution differed from those of E. coli and calf thymus DNAs, as shown in Fig. 3 (18). The CD spectrum of the thermophile DNA also differs from that of DNA-RNA hybrids whose conformation has been known to be the A-form (24, 25). It has been shown that the structure of DNA in solution is generally the B-form, and that the CD spectrum of DNA in neutral solution has two characteristic features; the magnitudes of the positive peak and the negative peak (at 245 nm) are almost equal and the positive peak is rather broad with the maximum at 272 nm, whereas double stranded RNA (the A-form in solution) manifested a large, sharp positive peak at a little shorter position (260-265 nm), as is seen in Fig. 3 (26, 52). Since CD spectra of polynucleotides reflect their conformations--that is, individual forms of double stranded molecules show characteristic CD spectra (26)--

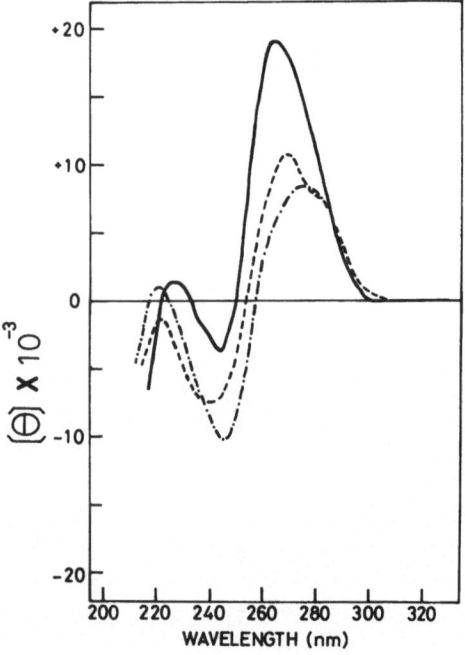

Fig. 3. Circular dichroic spectra of F. thermophilum DNA (--------), E. coli DNA (- — - — - —) and DNA-RNA hybrid (————). The spectra of E. coli DNA and DNA-RNA hybrid were kindly provided by Imahori and Watanabe (52).

the results shown in Fig. 3 suggest that the conformation of the thermophile DNA in neutral solution may not be in the usual B-form, but a mixture of A- and B-forms or a quite different form from the three (A, B or C) so far known.   X-ray analysis of the DNA is now in progress.   At present, it is not known whether the conformational difference between F. thermophilum DNA and mesophile DNA relates to the molecular mechanism of thermophily.

The biological activity of the thermophile and E. coli DNA was compared. The priming activities of these DNA for E. coli RNA polymerase (E.C. 2.7.7.6) were examined, and the results are illustrated in Fig. 4. Since F. thermophilum DNA contains a lesser amount of adenine, one would expect less uridine incorporation into the RNA synthesized by the reaction primed by the thermophile DNA.   No significant difference was

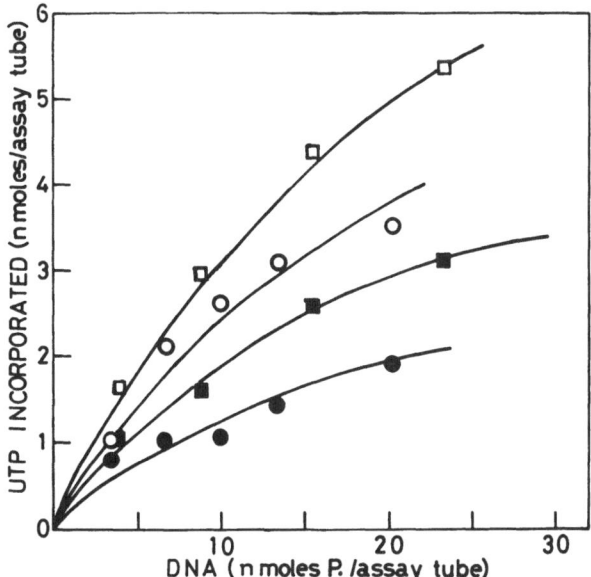

Fig. 4. Priming activity of F. thermophilum DNA. Incorporation of H^3-UTP catalyzed by E. coli RNA polymerase was plotted against the amount of DNA used as a primer. -O—O- ; F. thermophilum DNA, 30 min. reaction. -●—●- ; F. thermophilum DNA, 10 min. reaction. -□—□- ; E. coli DNA, 30 min. reaction. -■—■- ; E. coli DNA, 10 min. reaction.

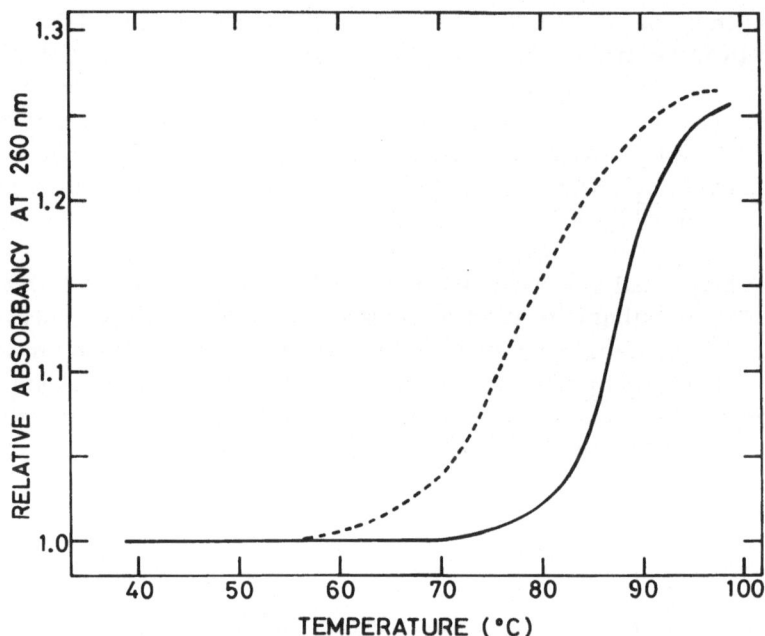

Fig. 5.   Temperature-absorbancy (at 260 nm)
profiles of F. thermophilum t-RNA (————)
and E. coli t-RNA (----).

observed in the priming activities of these DNA preparations, although
differences in the base conformation were suggested by CD spectra. Re-
versely, the thermophile DNA could be replaced by E. coli DNA in the
enzymatic reaction of thermostable DNA-dependent RNA polymerase (E.C.
2.7.7.6) extracted and partially purified from the thermophile.

   Transfer ribonucleic acids. The preparation of t-RNA purified from
F. thermophilum HB 8 showed a homogeneous peak with a 4S sedimentation coefficient by ultracentrifugal analysis.   In the presence of Mg^{2+},
the melting temperature of the thermophile t-RNA was 87.5°C, which is
9°C higher than that of E. coli t-RNA (Fig. 5).   In the absence of Mg^{2+},
F. thermophilum t-RNA had a Tm of 71°C, whereas the mesophile t-RNA
had a Tm of 62°C. In contrast to the present results, it has been reported
that the thermal stability of t-RNA of the moderate thermophile B. stearothermophilus appears to be nearly identical to that of E. coli t-RNA (27-
29). Recently, however, it has been observed that t-RNA of T. aquaticus
had a higher Tm value than that of E. coli t-RNA (86°C in the presence of

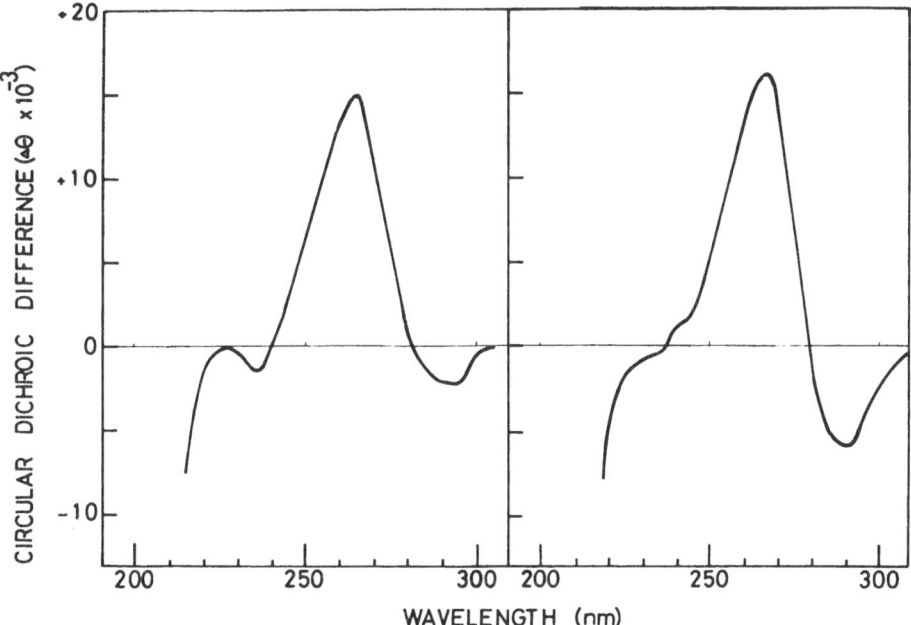

Fig. 6. Circular dichroic difference spectra of F. thermophilum
t-RNA (left, 22°C minus 91°C) and E. coli t-RNA$_f^{met}$ (right, 31°C
minus 92°C) (32).

Mg^{2+}) (30). Similarities in thermal stabilities of t-RNAs of these extreme
thermophiles may suggest an evolutional analogy in the molecule.

The difference CD spectrum of t-RNA between the native form at
room temperature and the denatured form at high temperature is fairly
sensitive to guanine-cytosine base pair content of paired regions of the
t-RNA (31-32).   The difference CD spectrum (Fig. 6) of thermophile
t-RNA (at 22°C minus at 91°C) closely resembles that of t-RNA$_f^{met}$ from
E. coli (at 31°C minus at 92°C) in the location and amplitudes of their
main peaks, as seen in Fig. 6 (32).   It can be concluded that guanine-
cytosine pair content in the stem region of the thermophile t-RNA is similar to that of E. coli t-RNA$_f^{met}$, whose G-C pair content is estimated to
be 87% (33) based on the cloverleaf model. The value is the highest among
the G-C pair contents of the fractionated t-RNA molecules of which base
sequences have been reported so far.   These findings indicate that the
thermostability of thermophile t-RNA is the result of the replacement of
most of A-U pairs with G-C pairs in the paired region of the mesophile
ancestor t-RNA molecules during the evolutionary process.

In our preliminary experiments, it was observed that both t-RNA and amino acid-sRNA synthetases were exchangeable between E. coli and F. thermophilum. Methionyl-t-RNA could be synthesized by a reaction catalyzed by E. coli methionyl-sRNA synthetase (E.C. 6.1.1.10) in the presence of F. thermophilum t-RNA instead of E. coli t-RNA (34). It is conceivable that the site of the t-RNA molecule that interacts with the enzyme has been conserved during evolution, whereas other parts, particularly the base-paired region in the clover-leaf model, have been changed to make the nucleic acid heat-stable. Comparative study of the thermophile t-RNA will be a new approach for elucidating molecular requirements for synthetase-t-RNA interaction.

Ribosomes. F. thermophilum ribosomes were also found to be thermostable. From the thermal denaturation profile, the Tm of the thermophile ribosomes was determined to be 86°C in the presence of 0.01 M mg^{2+}. This value is nearly identical to that of T. aquaticus reported by Brock and his co-workers (30), suggesting resemblance in the evolutionary alterations in ribosomes of the extreme thermophiles. Pace and Campbell have reported a positive correlation between the maximal growth temperature and Tm of ribosomes of the bacterium (35). The correlations between ribosome thermal stability and maximal growth temperatures from several studies are summarized in Table III. It is worthy of note that thermal denaturation temperatures of ribosomes and t-RNA of F. thermophilum are close to the upper limit of the growth temperature. Possibly, the nucleic acids and/or the nucleo-proteins of the organism may contribute as a limiting factor in thermal death.

Ultracentrifugal analysis revealed that the ribosomes of F. thermophilum consist of four particles in the presence of 0.01 M Mg^{2+}. The sedimentation coefficients were 30S, 50S, 70S, and 100S, respectively. In the absence of Mg^{2+}, the ribosomes were dissociated into two subunits sedimenting at 30S and 50S, suggesting that the subunit structure of the thermophile ribosomes are similar to that of mesophile procaryote ribosomes. The fact that the thermophile is sensitive to the actions of chloramphenicol and tetracycline also suggests protein-synthesizing machinery using 70S ribosomes in F. thermophilum HB 8. In a preliminary experiment, polyuridylic acid-directed incorporation of phenylalanine was observed using cell free extracts of the extreme thermophile. The detailed study on the protein-synthesizing machinery of the organism is now under way. The mechanism of thermal stability of the thermophile ribosomes remains to be solved.

TABLE III. Correlation of Thermal Stabilities of Ribosomes with Maximal Growth Temperatures of Microorganisms.

Microorganism		Maximal growth temperature ($^{\circ}$C)	Tm of ribosomes ($^{\circ}$C)
Extreme thermophile	F. thermophilum HB8	85	86[a]
	T. aquaticus YT-1	79	86[b]
Moderate thermophile	B. stearothermophilus 10	73	79[c]
	B. coagulans 43P	60	74[c]
Mesophile	B. subtilus SW-25	50	74[c]
	E. coli Q-13	45	72[c]
	D. desulfuricans	40	71[c]
Psychrophile	V. marinus 15381	18	69[c]

[a]From Oshima and Imahori (12).

[b]From Zeikus, Taylor, and Brock (30).

[c]From Pace and Campbell (35).

GLYCOLYTIC ENZYMES

A number of enzymes from the cell-free extract of F. thermophilum HB 8 have been examined. The catalytic activity thus far tested was thermostable without exception. An anzyme was designated thermostable if it had greater activity at 70°C than at room temperature. In this section, some findings on glycolytic enzymes of the microorganism are described briefly.

Thermal stability. A typical example of thermal stability of enzymatic activity in the crude extract is given in Fig. 7, which illustrates the inactivation profile of partially purified fructose 1,6-diphosphatase (E.C. 3.1.3.11). The enzyme was inactivated by heat treatment at or over 90°C; the kinetics of inactivation were first order. The enzyme preparation may have been contaminated by proteolytic enzymes, since thermostable protease activity was detected in the sonic extract, of which

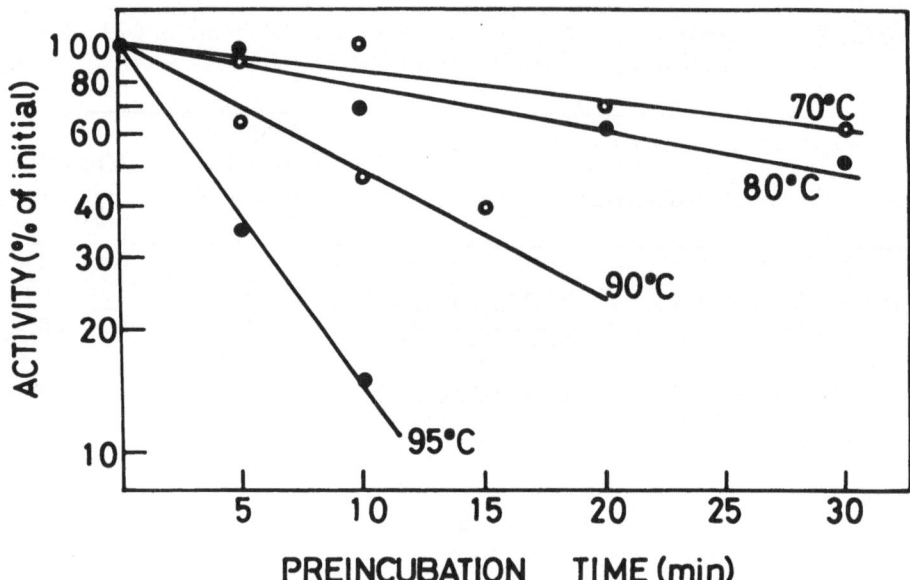

Fig. 7.   Thermal stability of fructose 1,6-diphosphatase from
F. thermophilum.   The enzyme preparation was obtained by
ammonium sulfate fractionation of the 75,000 x g supernatant of
the sonic extract.   After the enzyme solution (2 mg/ml, pH 7.5
at 25°C) was incubated at the indicated temperature, the remaining activity was assayed at 30°C.

optimal temperature was observed to be about 80°C.   Taking into account
that the curve in the figure shows inactivation by proteolysis as well as
heat inactivation, it appears that the enzyme activity is fairly resistant
toward heat treatment at or below 80°C.

The partially purified preparation of phosphofructokinase (E.C. 2.7.
1.11) demonstrated the highest thermo-stability so far observed (Fig. 8).
Only a small fraction of the initial activity was lost after heat treatment
at 70°C for 30 hours (36), and the enzyme was slowly inactivated (five hour
half life) by incubation at 90°C.   The molecular weight of the thermophile
phosphofructokinase was estimated by a gel filtration experiment to be
around 140,000.   The observation suggests a subunit structure similar to
that of the enzyme obtained from E. coli, i.e., four peptide chains per
molecule of the enzyme protein.   The thermostable interaction between
subunits would be an interesting subject in protein chemistry.

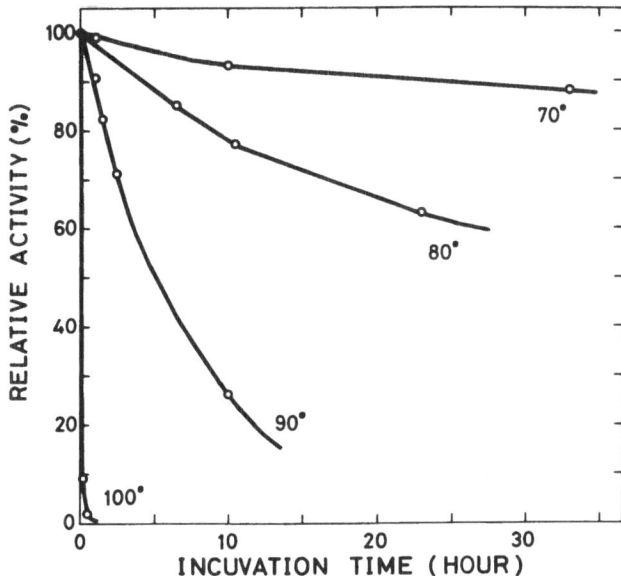

Fig. 8. Thermal stability of phosphofructokinase from F. thermophilum. The enzyme used was prepared by DEAE cellulose
column chromatography of the ammonium sulfate precipitate,
followed by a second ammonium sulfate fractionation. After the
enzyme preparation (0.5 mg/ml, pH 8.3 at 75°C) was heated,
the remaining activity was determined at 30°C.

Fig. 9 illustrates the thermal stability of crystalline glyceraldehyde-
3-phosphate dehydrogenases (E.C. 1.2.1.12). F. thermophilum HB 8
enzyme is intrinsically more stable to heat without protection. The enzyme from the extreme thermophile is more heat-resistant than its counterpart from the moderate thermophile, which suggests a positive correlation between thermostabilities of the enzyme and the maximal growth
temperatures of organisms. Similarly, Brock and his co-worker reported
that aldolase (E.C. 4.1.2.13) from an extreme thermophile was more
heat-stable than that from moderately thermophilic organism (38). There
is no significant difference between thermostabilities of the crystalline
preparation and the partial purified preparation of glyceraldehyde-3-phos-
phate dehydrogenase from the thermophile, indicating the absence of
specific protector for the enzyme protein in the cell extract.

A linear relationship was observed in Arrehnius plots of thermophile
enzymes such as phosphoglucomutase (E.C. 2.7.5.1), fructose diphos-

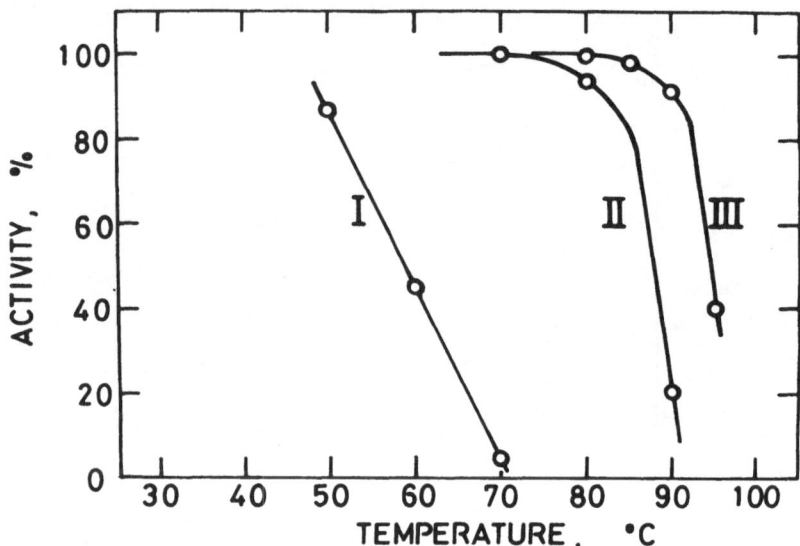

Fig. 9.   Thermostabilities of glyceraldehyde-3-phosphate dehydrogenases from mesophilic source (I, pig muscle), moderately thermophilic source (II, B. stearothermophilus) and extremely thermophilic source (III, F. thermophilum).   After a solution (2 mg/ml, pH 8.5 at 25°C) of the crystalline enzyme from
F. thermophilum was preincubated at the indicated temperature for 5 min., the activity was assayed at 30°C. The enzyme
solutions from other sources were 3 mg/ml, pH 7.0. The auth-
or is indebted to Dr. Koichi Suzuki for his kind permission to
quote the data for the pig enzyme and B. stearothermophilus enzyme (53).

phatase, phosphofructokinase, etc.   A typical example of the activity-
temperature profile is shown in Fig. 10 (39).   It has been reported that
aldolase activity of T. aquaticus increased greatly with increasing temperature above 60°C (38).   When the reported data were applied to Arrhenius
plots, the apparent discontinuity of the slope (that is, the change in activation energy) was observed at 60°C. This suggests transconformation of
aldolase into a more active form over 60°C.   The T. aquaticus adolase
might be called an "obligately thermophilic enzyme." On the contrary, a
linear line is seen in Fig. 10 for phosphoglucomutase in the range of 25°
to 60°C, indicating no transconformation to another form having a different
activation energy (above 70°C the bend is due to partial inactivation of the
enzyme by heat) (39).   The activation energy of the phosphoglucomutase

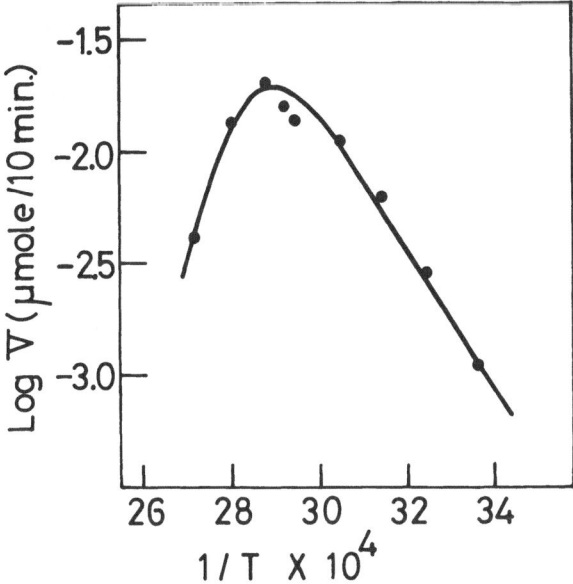

Fig. 10. Arrhenius plots for phosphoglucomutase action from F. thermophilum. For experimental details, see ref.
39.

action was estimated to be 15,000 cal/mole from Fig. 10. The F. thermophilum phosphoglucomutase might be classified as a "facultatively thermophilic enzyme" having the optimal temperature of 75°C under the assay
condition employed. Enzymes from F. thermophilum so far investigated
manifested no discontinuity of the slope in Arrhenius plots.

As might be anticipated, these heat-stable enzymes from the thermophile are resistant to high concentrations of urea or ethanol. For instance, glyceraldehyde-3-phosphate dehydrogenase and phosphoglucomutase retained their catalytic activities in the presence of 7 M urea (Fig. 11)
(37, 43). The circular dichroic spectra of the partially purified phosphoglucomutase with or without urea are shown in Fig. 12. The enzyme
retained both catalytic activity and secondary structure in the presence of
4 M urea (43). The content of α -helix in the enzyme protein can be roughly
estimated to be 20% in the presence or absence of urea. The phosphoglucomutase from F. thermophilum HB 8 was resistant to ethanol as well;
no significant change in catalytic activity was observed when up to 15%
ethanol was added into the assay mixture. Glyceraldehyde-3-phosphate

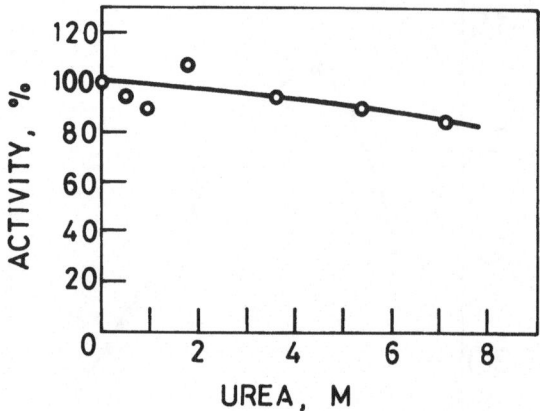

Fig. 11. Effect of urea on the glyceraldehyde-3-phosphate dehydrogenase activity of F. thermophilum.   The enzyme was preincubated for 10 min. and then assayed in the presence of the indicated concentration of urea, at 30°C, pH 8.5. The assay reaction was started by the addition of the substrate to the preincubated mixture.

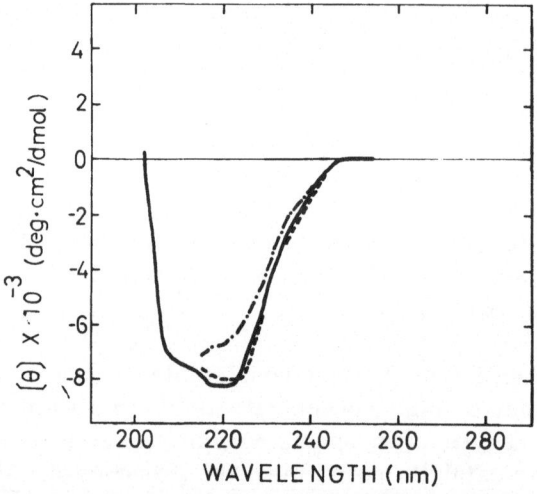

Fig. 12. Circular dichroic spectra of phosphoglucomutase from F. thermophilum.   The purification of the preparation used in this experiment was carried out by ammonium sulfate fractionation, Sephadex G-200 gel filtration, DEAE sephadex column chromatography, and elution from calcium phosphate gel.   Circular dichroism was recorded in the absence of urea (————), in the presence of 4 M urea (--------) or 8 M urea (- — - — - —). About 90% of the initial activity was observed in the presence of 4 M urea and 30% in the presence of 8 M urea.

dehydrogenase from the organism was remarkably activated by the addition of ethanol into the assay mixture (about 3 times the initial activity in the presence of 30% ethanol). The mechanism of this unusual effect of ethanol on the enzyme is now being studied in detail.

The stability of these enzymes to heat, urea, and ethanol suggests that they can be used as catalysts in the chemical industry and for preparing the water insoluble enzyme derivatives.

Allosteric nature and metabolic regulation. Phosphofructokinase and fructose 1, 6-diphosphatase are among the key enzymes in the regulation of glycolytic path (40). Phosphofructokinase, which catalyzes the essentially irreversible (under in vivo conditions) conversion of fructose-6-phosphate to fructose-1, 6-diphosphate, is specifically inhibited by ATP in an allosteric manner (41). This inhibition is competitively overcome by AMP. In contrast, fructose 1, 6-diphosphatase, which catalyzes the irreversible hydrolysis of fructose-1, 6-diphosphate to fructose-6-phosphate, is inhibited by AMP (41). AMP binds to the enzyme in a co-operative manner. The inhibitory effect of AMP can be reversed by the addition of ATP or abolished by desensitization. Of special significance is that opposite responses of these nucleotides are observed with these enzymes. The metabolic path is thus governed by the allosteric interactions of ATP and AMP with these enzymes in the cell, namely gluconeogenesis will be favored when ATP molecules are accumulated in the cell, and vice versa.

The heat stable phosphofructokinase from F. thermophilum HB 8 exhibits Michaelis-Menten kinetics in the presence of excess ATP at room temperature or at high temperature (36). Phosphoenolpyruvate was a potent inhibitor of the enzyme in the presence of low concentration of the substrate, fructose 6-phosphate (Fig. 13). In the presence of phosphoenolpyruvate, the enzyme kinetics are sigmoidal, independent of ATP concentration. ADP relieves the inhibition, changing the sigmoidal kinetics to the normal curve as seen in Fig. 13. In contrast, fructose diphosphatase from the organism is inhibited by the addition of AMP, stressing the substrate inhibition of the enzyme. The enzyme is activated by the addition of phosphoenolpyruvate, relieving the substrate inhibition (42). Both phosphofructokinase and fructose diphosphatase from the extreme thermophile are characteristic in their great thermal stabilities and their allosteric interactions with phosphoenolpyruvate, but not with ATP. These enzymes may be one of the control points in the regulation of the glycolytic path of F. thermophilum HB 8. A skillful means of regulating these enzymes in a reciprocal fashion is provided by the concentration ratio of

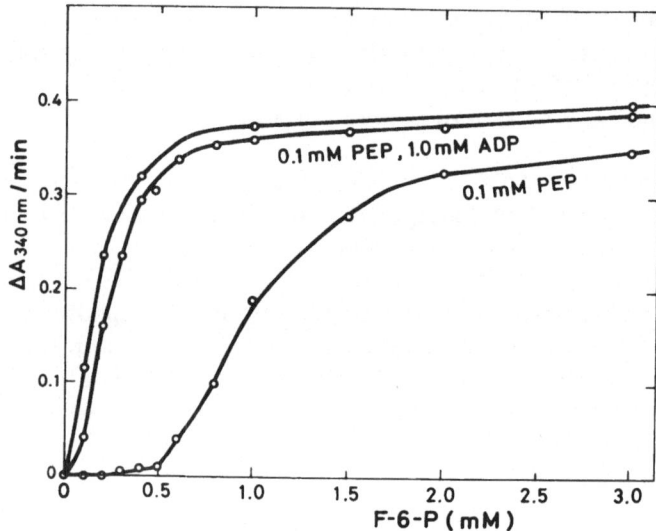

Fig. 13. Allosteric inhibition of phosphoenolpyruvate on F. thermophilum phosphofructokinase. The reaction was carried out at 75°C, pH 8.3 in the presence of 1.0 mM ATP and 1.0 mM Mg^{2+}. Formation of fructose 1,6-diphosphate was followed by measuring the corresponding NADP oxidation catalyzed by joint action of aldolase, triosephosphate isomerase and α-glycerophosphate dehydrogenase (36).

the phosphoenolpyruvate-AMP/ADP system instead of the usual ATP-AMP system. Of interest is the observation that the enzymes from the extreme thermophile demonstrate an allosteric nature at such high temperature. At these temperatures, many allosteric enzymes from mesophilic sources are desensitized or denatured. F. thermophilum phosphofructokinase and fructose diphosphatase will be interesting enzymes for studying structure-function relationship with special reference to allosterism at high temperature.

Comparison of catalytic properties. Flavobacterium thermophilum phosphoglucomutase resembles E. coli phosphoglucomutase in cofactor requirements, in inhibition by -SH reagents, in its pH optimum, and in the catalytic kinetics (39). Both magnesium ion and glucose 1,6-diphosphate are required for activity. Cysteine or dithiothreitol is necessary for optimal activity of phosphoglucomutases from the thermophile and the

TABLE IV. Comparison of Catalytic Properties of F. thermophilum and E. coli Phosphoglucomutases.

	Organism	
	F. thermophilum	E. coli
Cofactors required	Mg^{2+}, 10^{-2} M	Mg^{2+}, 10^{-3} M
	Dithiothreitol, 10^{-2} M (or cysteine)	Cystein, 10^{-2} M (or histidine, imidazole)
	Glucose 1, 6-diphosphate	Glucose 1, 6-diphosphate
Km for glucose 1-phosphate	2×10^{-4} M at 30°C	5×10^{-5} M
Km for glucose 1, 6-diphosphate	1.3×10^{-5} M at 30°C	8×10^{-8} M
pH optimum	8.8	8.8
Inhibition by pCMB	+	+
Effect of high conc. of anion	Inhibitive	Inhibitive
Reaction kinetics	"ping-pong"	"ping-pong"
References	(39)	(46, 47)

mesophile. The thermophile enzyme is partially inhibited by the addition of 10^{-2} M pCMB to the assay mixture. The enzyme exhibits "ping-pong" kinetics (44). Although the thermophile enzyme is characteristic in its extreme thermostability, it resembles phosphoglucomutases obtained from mesophiles (45, 46) as summarized in Table IV.

With the exception of thermal stability, glyceraldehyde-3-phosphate dehydrogenase from F. thermophilum was similar to that obtained from mesophiles in terms of iodoacetate inhibition, NAD requirement, and pH optimum. Inhibition studies using monoiodoacetate and monoiodoacetamide indicated the participation of cysteine residue(s) in the catalytic site of F. thermophilum glyceraldehyde-3-phosphate dehydrogenase and suggested the similarity of the chemical structure around the active center of the thermophile enzyme to the mesophile enzyme (48). NAD was required as a hydrogen acceptor in the dehydrogenase activity and could not be replaced with NADP. The dehydrogenase from the thermophilic organism was similar in pH optimum (around 8.0) to the enzyme from mesophilic organisms.

The comparative studies described above suggest that the structure around the active site of the thermophile enzyme might be similar to that of the mesophile counterpart. During the evolutionary process from the mesophile ancestor to the present extremely thermophilic bacterium, the structure around the active site of an enzyme might be preserved while other parts of the molecule are changed to make the enzyme protein thermostable. More elaborated studies on the comparison of catalytic properties and chemical structures of thermophile enzymes with mesophile counterparts, including X-ray crystallographic analysis, would, however, be required for decisive conclusions. Such work may help in the elucidation of the molecular basis of biological adaptation to high temperature.

LIPIDS

A correlation between the sensitivity to heat and the melting points of lipids of organism has been suggested in the early literature (49, 50). It has been also proposed that the integrity of the membrane might be the limiting factor in thermal death of a bacterium (6).

M. Oshima (51) compared lipid composition of F. thermophilum HB 8 with those of a mesophile and a moderate thermophile (B. stearothermophilus). It was found that cells of F. thermophilum contain unusual glycolipids and phospholipids which have not been reported elsewhere; however, there is no significant anomaly in fatty acid composition of the organism. The content of unidentified lipids from F. thermophilum is more than two-thirds of total lipid content of the cells. The glycolipid contains roughly equimolar amount of glycerol, glucose, galactose, and N-acetylglucosamine. The phospholipid contains one mole of lysine. It is hoped that structure formula of these unusual lipids will be determined in the very near future. It is quite conceivable that these unique lipids closely correlate to the molecular basis of the thermophily of the present organism.

REFERENCES

1. Miquel, P., Ann. Micrographie, 1, 3 (1888); Robertson, A. H., Tech. Bull. N. Y. State Agr. Expt. Sta., No. 130 (1927).
2. Gaughran, E. R. L., Bact. Rev. 11, 189 (1947).
3. Allen, M. B., Bact. Rev. 17, 125 (1953).
4. Keffler, H., Bact. Rev. 21, 227 (1957).
5. Kempner, E. S., Science 142, 1318 (1963).
6. Brock, T. D., Science 158, 1012 (1967).
7. Campbell, L. L., and Pace, B., J. Appl. Bact. 31, 24 (1968).
8. Gillespy, T. G., and Thorpe, R. H., J. Appl. Bact. 31, 59 (1968).
9. Friedman, S. M., Bact. Rev. 32, 27 (1968).
10. Brock, T. D., Ann. Rev. Ecology and Systematics, 1, 191 (1970); "Microbial Growth" (P. M. Meadow and S. J. Pirt, eds.), p. 15. Cambridge University Press, London, 1969.
11. Farrell, J., and Rose, A., Ann. Rev. Microbiol. 21, 101 (1967); Loginova, L. G., Golovacheva, R. S., and Egorova, L. A., "Microbial Life at High Temperatures," Nauka, Moscow, 1966.
12. Oshima, T., and Imahor, K., J. Gen. Appl. Microbiol. 17, 513 (1971).
13. Oshima, T., and Imahori, K., in preparation.
14. Breed, R. S., Murray, E. G. D., and Smith, N. R., "Bergey's Manual of Determinative Bacteriology" Seventh Edition, Williams and Wilkins, Baltimore, 1957.
15. Emoto, Y., Onsenkogaku Zasshi 6, 29 (1968).
16. Goto, E., Imai, H., and Ito, Y., Onsen Kagaku 16, 144 (1966).
17. Brock, T. D., and Freeze, H., J. Bacteriol. 98, 289 (1969).
18. Oshima, T., and Imahori, K., in preparation.
19. Sueoka, N., in "The Bacteria" (I. C. Gunsalus and R. Y. Stanier, eds), Vol. V, p. 419. Academic Press, New York and London, 1964).
20. Scaletti, J. V., and Naylor, H. B., J. Bacteriol. 78, 422 (1959).
21. Zimmer, C., and Venner, H., Hoppe-Seyler's Z. Physiol. Chem. 333, 20 (1963).
22. Dutta, S. K., Jones, A. S., and Stacey, M., J. Gen. Microbiol. 14, 160 (1956).
23. Sueoka, N., Proc. Natl. Acad. Sci. U. S. 47, 1141 (1961).
24. Milman, G., Langridge, R., and Chamberlin, M. J., Proc. Natl. Acad. Sci. U. S. 57, 1804 (1967).

25. Tsuboi, M., and Higuchi, S., Proteins, Nucleic Acids, Enzymes (Tokyo) 13, 533 (1968).
26. Tunis-Schneider, M. J. B., and Maestre, M. F., J. Mol. Biol. 52, 521 (1970).
27. Mangiantini, M. T., Tecce, G., Toschi, G., and Trentalance, A., Biochim. Biophs. Acta 103, 252 (1965).
28. Saunders, G. F., and Campbell, L. L., J. Bacteriol. 91, 332 (1966).
29. Friedman, S., and Weinstein, I., Biochim. Biophys. Acta, 114 593 (1966).
30. Zeikus, J. G., Taylor, M. W., and Brock, T. D., Biochim. Biophys. Acta, 204 512 (1970).
31. Gannis, R. B., and Cantor, C. R., Biochemistry 9, 4714 (1970).
32. Watanabe, K., Seno, T., Nishimura, S., Oshima, T., and Imahori, K., Polymer J. submitted.
33. Dube, S. K., Marcker, K. A., Clark, B. F. C., and Cory, S., Nature 218, 232 (1968).
34. Takasaki, Y., Oshima, T., and Imahori, K., to be published.
35. Pace, B., and Campbell, L. L., Proc. Natl. Acad. Sci. U. S. 57, 1110 (1967).
36. Yoshida, M., Oshima, T., and Imahori, K., Biochem. Biophys. Res. Commun. 43, 36 (1971).
37. Fujita, S., Oshima, T., and Imahori, K., Seikagaku 43, 505 (1971).
38. Freeze, H., and Brock, T. D., J. Bacteriol. 101, 541 (1970).
39. Yoshizaki, F., Oshima, T., and Imahori, K., J. Biochem. 69, 1083 (1971).
40. Scrutton, M. C., and Utter, M. F., Ann. Rev. Biochem. 37, 249 (1968).
41. Stadtman, E. R., Advan. Enzymol. 28, 41 (1966).
42. Yoshida, M., and Oshima, T., Biochem. Biophys. Res. Commun., 45, 495 (1971).
43. Yoshizaki, F., Oshima, T., and Imahori, K., Seikagaku 43, 582 (1971).
44. Cleland, W. W., Biochim. Biophys. Acta 67, 104 (1963).
45. Najjar, V. A., in "The Enzymes" (P. D. Boyer, H. Lardy and K. Myrback, eds.), 2nd ed., Vol. 6, p. 161. Academic Press, New York and London, 1962.
46. Joshi, J. G., and Handler, P., J. Biol. Chem. 239, 2741 (1964).
47. Hanabusa, K., Dougherty, H. W., Del Rio, C., Hashimoto, T., and Handler, P., J. Biol. Chem. 241, 3930 (1966).
48. Harris, J. I., in "Pyridine Nucleotide-Dependent Dehydrogenases" (H. Sund Ed.) p. 57. Springer, Berlin, 1970.

49. Heilbrunn, L. V., Am. J. Physiol. 69, 190 (1924).
50. Belehradek, J., Protoplasma 12, 406 (1931).
51. Oshima, M., Seikagaku 42, 432 (1970).
52. Imahori, K., and Watanabe, K., Polymer Symposia 30, 633 (1970).
53. Suzuki, K., and Harris, J. I., FEBS Letters 13, 217 (1971).

NOTES ADDED IN PROOF

1. Recently it was found that 30S and 50S ribosome subunits of the thermophile were exchangeable with those of E. coli in an E. coli cell-free protein biosynthesis. When R-17 phage RNA was used as a messenger in the thermophile cell free protein synthesizing system, the patterns of protein formed were gel-electrophoretically similar to those produced by the E. coli system--that is, the main product was R-17 phage coat protein. The results indicate that the mechanisms of ribosome subunits interaction, the initiation, elongation, and termination mechanisms of polypeptide synthesis are similar to those of E. coli protein synthesizing machinery. The essential sites of ribosomes for their biological functions would be preserved during evolution to the present thermophile, although heat stabilities of the subunits were greatly increased. (Y. Ohno, T. Oshima and K. Imahori, Ann. Meeting Jap. Biochem. Soc., 1972).

2. The chemical structure of the major component of glycolipids of F. thermophilum HB8 was recently determined except for a sequence between glucose and glucosamine residues. The lipid contains a glucosamine residue linked with a C_{17}-iso acid through an amide bond. This unique lipid has, to my knowledge, not been found in living world. The proposed structure is galactofuranosyl (1-2)galactopyranosyl(1-2)glucopyranosyl(1-6)N-(15 methylhexadecanoyl)-glucosaminyldiglyceride, or gal-gal-glcN(15 methyhexadecanoyl)-glc-diglyceride. (M. Oshima, private communication)

3. The A-form conformation in the thermophile DNA, which was suggested by CD measurements, would contribute to reduce the mutation rate at physiological temperature, since it is difficult to form thymine dimer in A-form DNA. Such protective mechanisms would be necessary for an extreme thermophile DNA, because otherwise the mutation rate is greater at higher temperature.

Received 24 August, 1971

EVOLUTIONARY SIGNIFICANCE OF AMOEBA-

FLAGELLATE TRANSFORMATION

Shuhei Yuyama

Department of Zoology, University of Toronto
Toronto, Ontario, Canada

Amoebo-flagellates include taxonomically diverse amoebae which can transform into flagellates under certain environmental conditions (1-7). Studies on the transformation of amoebo-flagellates appear to have implications in two separate aspects of evolutionary biology. One is the selective advantages of acquiring secondary motile organs on an organism which already possesses one--the amoeboid locomotion. An understanding of the precise environmental requirements for the transformation process would enable us to surmise more accurately the behaviour of these organisms in nature. The other aspect pertains to the evolution of motile organelles in eukaryotes. In the past, studies on the life cycle of an organism have frequently contributed to our understanding of the evolution of a certain trait and the phylogenic position of the organism. It is anticipated, therefore, that studies of the development of the flagellar apparatus in amoebo-flagellates, one of the most primitive organisms, will enhance our understanding of the evolution of motility in eukaryotes.

There is tremendous variation in the phenomenon of amoeba-flagellate transformation among different amoebo-flagellates. In some amoebo-flagellates such as Tetramitus rostratus, the transformation process involves complex changes in cell structures and functions, indicating that the formation of flagella is only part of the cytodifferentiation of the whole cell (8, 9). Transformation of Myxomycetes amoebae, however, appear to involve the formation of flagella only (10). Another amoebo-flagellate, Naegleria gruberi, has been extensively studied in the laboratory, and

425

in this case, complexity of transformation can be considered to lie between the two extremes already mentioned (11).

Formation of the flagellar apparatus is quite ubiquitous in nature and it is not limited to the amoebo-flagellates. Many different cells in unicellular and multicellular eukaryotes can or do form the flagellar apparatus under certain conditions or during the course of normal development.

In the present paper, the phenomenon of amoeba-flagellate transformation will be discussed primarily in its ecological and evolutionary context. The discussion will be mainly focused on one species, Naegleria gruberi, since most of our information on the transformation phenomenon comes from this organism. Readers who are interested in using Naegleria as a model system for studying cell differentiation should refer to two recent reviews (7, 12).

BIOLOGY OF NAEGLERIA GRUBERI:
AMOEBA AND FLAGELLATE

The following account is a brief summary of observations on Naegleria reported by many investigators (5, 11, 13-21).

Naegleria gruberi is a small phagotrophic amoeba (10 - 20 μ in diameter) with a nucleus containing a prominent nucleolus. It is known to assume three morphologically distinct forms: amoeba, flagellate, and cyst. The amoeba is the only form known to feed and divide. Naegleria amoebae can be cultured for many generations without showing any trace of morphogenesis under favorable nutritional conditions. Since synchonous amoeba-flagellate transformation can be induced in these amoebae independent of their growth experiences, the genetic material responsible for transformation must be transmitted from generation to generation without need for expression. Clones isolated from a population of either amoebae, flagellates, or cysts can exhibit any of the three morphogenetic phenomena, thus the genotype of Naegleria is stable and morphogenesis involves only phenotypic changes (12, 20). During mitosis, the nucleolus and the nuclear membrane are preserved. The Feulgen positive material seemingly divides equally in a manner typical of mitoses in higher eukaryotes, although individual chromosomes are too small to be distinguished (16). Amoebae possess numerous food vacuoles and one or two contractile vacuoles. They ordinarily assume a typical limax shape and move actively (50-100 μ/min.) with blunt pseudopodia when they are placed in a non-nutritive environment at room temperature (5). In the presence of food bacteria, they actively phagocytose, but their movement is much slower (12). Recent electron microscopical observations indicate that there are no centrioles, centriole-like structures, or apparent centriole-precursors in the amoeboid form (18, 19, 22). During mitosis, there is no centriole at the poles, although numberous parallel and longitudinal microtubules have been observed inside the nuclear membrane.

The cytological characteristics of Naegleria flagellates are essentially identical to those of Naegleria amoebae except for the presence of a flagellar apparatus which is typical of those found in other eukaryotes (18, 19). It is generally agreed that the entire ensemble of the flagellar apparatus must be assembled during the transformation process. The numerous food vacuoles in the amoebae diminish in the flagellates (19).

Naegleria has been found in soil samples and fresh water samples all over the world (7). Evidently, Naegleria can survive under various climatic conditions.

Amoeba-flagellate Transformation

Structural changes during transformation are relatively simple. The amoebae become rounded and about 10 minutes after this occurs, the flagella start to emerge and grow (ca. 0.5μ/min.) (7) until they reach a size of 15-25 μ. Electron microscopic studies reveal that the basal bodies first appear deep inside the cytoplasm and then migrate to the cell periphery where the flagella emerge from this position (19). Two flagella of equal length usually appear, but the number can vary with the strain, ploidy, and culture conditions (7). Flagella number can also be experimentally increased by thermal treatment of the amoebae (23).

Naegleria amoebae can be grown with food bacteria (polyxenically or monoxenically) on an agar surface or in liquid (24-26) or axenically with or without particulate in liquid (7). On the agar surface, amoebae encyst when the food bacteria depleted. The cysts of Naegleria are not stable and lose their viability within a few months (7, 27).

Transformation can be obtained by harvesting and suspending amoebae in distilled water or in a dilute buffer solution (24, 25).

Conditions for obtaining synchronous transformation in Naegleria have been described (20, 28). Briefly, the following procedures are employed. Logarithmically growing cells are harvested and washed three times with a low pH buffer (pH 3.8) to remove bacteria. Transformation is not initiated at the low pH. When the cells are suspended at time zero in a neutral buffer at room temperature (23°C), flagellates start to appear at 70 minutes and reach the 90% level within the next 20-30 minutes (Fig. 1). A neutral buffer can be substituted for the low pH buffer at a lower temperature (8°C-10°C) with the same results (27).

Transformation is temperature dependent and amoebae do not transform below 10°C or above 39°C. Between 15°C and 39°C, the Arrhenius plot of the interval between the initiation (pH change) and the time of transformation in 50% of the population (T_{50}) is linear (20, 21). T_{50} is constant at a pH between 5.2 and 9.0 (21, 29).

Ionic concentration effects transformation (24). At a concentration 25 mM or above of Na^+, K^+, Ca^{++}, or Mg^{++}, transformation is inhibited, although lowering of ionic concentration causes amoebae to transform synchronously (27). Since transformation occurs either in distilled water or in any dilute solutions of Na^+, K^+, Ca^{++}, Mg^{++}, PO_4^{--}, CO_3^{--}, and tris, no specific ion is required for transformation.

Proper oxygen tension is necessary for transformation. Transformation is blocked in a buffer solution which has been bubbled with N_2 overnight. In a heavy cell suspension (2×10^6 cell/ml or higher), transformation is delayed unless aeration or shaking is employed (21, 29). When the cell concentration is below 10^6 cells/ml, the cell density has no effect on T_{50} or synchrony. There is no interaction between cells during transformation (29).

Neither the age of the cell culture (logarithmically growing phase or stationary phase) nor the presence of cysts influences the time course of transformation (21).

Transformation does not occur in a non-aqueous environment. Occasionally, some flagellates (1-2%) were found to be present among amoebae grown on an agar surface, but the presence of flagellates was invariably associated with the presence of water droplets on the agar surface (27).

Thus, the transformation of Naegleria amoebae occurs in a ubiquitous aqueous environment; the conditions required for transformation,

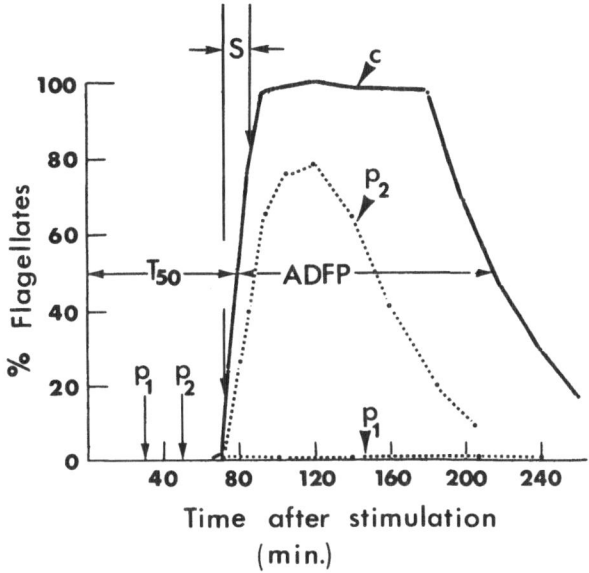

Fig. 1. The typical time course of amoeba-flagellate transformation in Naegleria gruberi and the effects of puromycin
on the transformation. The Naegleria amoebae were suspended
in a pH 3.8 buffer and stimulated to transform by changing the
pH to 7.0. The solid line (C) indicates the transformation and
reversion of control cells at $25^{\circ}C$. Two broken lines (P_1 and
P_2) indicate the two different effects of puromycin on transformation. When amoebae were treated continuously with $5 \times 10^{-4}M$
puromycin initiated at 30 minutes after the stimulation, transformation is blocked. When amoebae were treated at about 50
minutes, the duration of the flagellated stage is shortened.
T_{50} = the interval between the time of stimulation and the time
of 50% transformation in the population; ADFP = average duration of the flagellated period; S = the interval between the time
of 17% and 83% transformation indicating the degree of synchrony. The time of puromycin treatment is indicated by the arrows
(P_1 and P_2).

i.e., proper pH, oxygen tension, temperature, ionic concentration, and
absence of nutritive substances, are commonly found in rivers, lakes,
and swamps. It is quite probable that Naegleria amoebae living in soil
transform into flagellates when rain creates water puddles.

Stimulus for Transformation

Since the total population of Naegleria amoebae grown on an agar surface converts into flagellates after washing and resuspending the amoebae in a favorable environment (as previously described), the amoebae must have mechanisms which enable them to detect the environment favorable for transformation.

Since the T_{50} is a well reproducible value under a certain set of conditions, it seems reasonable to assume that a certain environmental change associated with the manipulation of cells is detected by amoebae and triggers a series of biochemical reactions leading to transformation some 70 minutes later. The following possibilities were conceivable: changes in osmotic pressure, ionic concentration, or ionic composition; changes in pH; changes in O_2, CO_2, or N_2 tension; changes in temperature; removal of bacteria or bacterial products; changes accompanying transfer of amoebae from an agar surface to a liquid environment; mechanical agitation; removal of amoebae-specific factors bound to the cell surface; any combination of the above. In addition, there was a possibility that many environmental changes stimulate a common trigger mechanism. These factors were carefully examined (12, 21), but no definite environmental change was positively identified as the stimulus for transformation. It has been proposed that the change in the Gibbs-Donnan equilibrium is related to the stimulus (30), but such a contention is not compatible with the fact that transformation occurs with amoebae cultured in a liquid environment without changing the ionic composition (31).

The main difficulty in identifying the stimulus resides in the fact that the putative triggering reaction is unknown and thus, the assay is totally dependent upon the emergence of flagella. There is evidence indicating that amoebae which have been stimulated are not totally committed until later times (ca. 40 minutes after stimulation) in the latent period (27). Thus, stimulated amoebae do not necessarily transform when they are resuspended or plated in an environment favorable for growth. It is obvious from these observations that the non-emergence of flagella does not guarantee that the amoebae have not been stimulated. Fulton (32), using the Naegleria strain NEG which had been cultured axenically in a medium containing glucose, phosphate, methionine, yeast extract and dialyzed fetal calf serum, reported that the removal of yeast extract enable the amoebae to transform into flagellates, while the removal of the other components does not. However, the problem is still complicated since the removal of yeast extract reduces the osmotic pressure and the ionic balance of the environment. In addition, the removal of yeast extract may cause immediate cessation of growth which may be triggering the transformation. In fact, earlier studies by Fulton and Gurerrini (31) suggested that starvation is the probable cause of transformation. The addition or removal of yeast extract has no effect on the transformation of Naegleria 1518/1s strain cultured axenically (27).

Thus, it would seem that our understanding of the initial events which convert amoebae from the growth-division phase to the transformation phase at the molecular level holds the key to the solution of the problems of the stimulus.

So far, most of our information on the biochemical events during the latent period of transformation comes from the use of metabolic inhibitors (12, 28, 29). Inhibition of transformation with actinomycin D, cycloheximide and puromycin (Fig. 1) suggests that RNA and protein synthesis are required for transformation. Walsh (33) recently reported that the pattern of ribosomal RNA and heterogeneous RNA synthesis is markedly different in cells during growth and cells during transformation.

Concerning the relationship between the cell cycle and transformation, it is established that amoebae are both capable and ready to transform into flagellates at any time during the cell cycle, since synchronous transformation can be obtained from both the randomly growing cells and from the stationary cells. Since the degree of synchrony expressed by the interval between 17% to 83% transformation (see Fig. 1S) is as short as 15 minutes in some experiments, while the average generation time of amoebae is 3-4 hours, all cells at time of harvest are ready for transformation. The G_1, G_2 and S periods of Naegleria amoebae have not yet been determined because of difficulties in labelling Naegleria DNA and because of the lack of mitotic inhibitors; e.g., colchicine and vinblastin have no effect on Naegleria (27, 29).

Reversion of Flagellates

Reversion of flagellates to amoebae occurs spontaneously in the laboratory. Although the duration of the flagellated period can be controlled, it is quite limited, and eventual reversion seems to be an inevitable event since the Naegleria flagellates do not feed. During reversion, flagellates lose their typical pear shape (become amoeboid), and the flagella are absorbed into the amoeba by the formation of successive swellings in the flagella (16). The flagellates are more stable as the

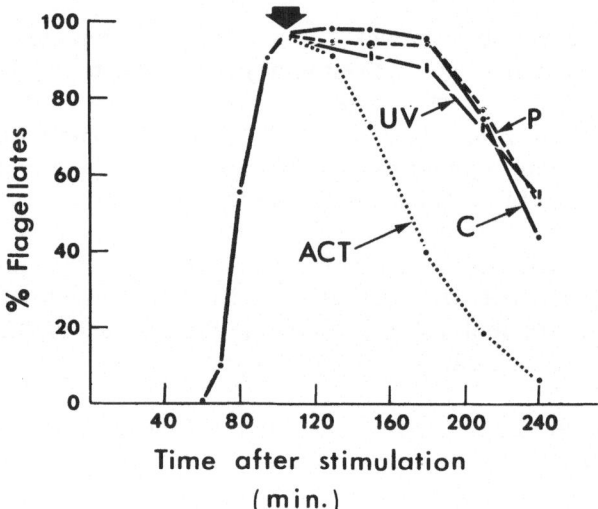

Fig. 2. Effects of three anti-transformation agents on the reversion of flagellates into amoebae. The large arrow indicates
the time of the initiation of the treatments with actinomycin D
(150 μg/ml), puromycin (5 x 10^{-4}M) or U. V. irradiation (ca.
1,000 ergs/mm^2). ACT = actinomycin treated; P = puromycin
treated; UV = U. V. treated; C = control.

temperature is lowered to about 10°C where they become unstable. Since
reversion and retransformation occur repeatedly until the amoebae begin
to encyst, spontaneous reversion is not due to loss of conditions favorable for the transformation (7, 31). However, flagellates can be readily
converted to amoebae by violent mechanical agitation, high viscosity, or
high temperature (7, 12).   Treatment by actinomycin D quickly reverts
the flagellates (Fig. 2), although treatment of amoebae by actinomycin D
just before the emergence of flagellates does not shorten the flagellated
period.

     The actinomycin D sensitivity and the low termperature stability of
flagellates suggest that synthesis of RNA is involved in the reversion
process.   Microtubules in other organisms are less stable at low temperatures (34). Treatments of flagellates with cycloheximide, puromycin, or ultraviolet irradiation have no effect on their stability (Fig. 2).
However, puromycin treatment of flagellates at the late stage during the
latent period (Fig. 1) shortens the flagellated period (39), suggesting
that stability of the flagellates is programmed during the latent period
by protein synthesis.

Although the flagellated period is brief, flagellates are capable of regenerating flagella which have been mechanically removed.

TRANSFORMATION IN OTHER ORGANISMS

Transformation in Tetramitus rostratus, an amoeba-flagellate which appears to be closely related to Naegleria (7) was found to occur more frequently under crowded conditions (35).   Subsequent investigations by Fulton (7, 36) revealed that the lowering of oxygen tension in the environment holds the key to the transformation process.   Apparently, Tetramitus transformation is more sensitive than Naegleria, and many conditions must be met to induce synchronous transformation in the total population; such conditions involve proper growth conditions of amoebae, proper washings to remove bacteria, proper ionic environment, and lower oxygen tension (7, 36).   Transformation takes much longer than with Naegleria, and flagellates start to appear about 3 hours after the initiating manipulation.   In nature, environmental requirements may be met at the bottom of stagnant ponds, and transformation may be the adaptive form to seek an oxygen rich environment.   Fulton (7) also indicates that flagellates appear to divide as the consequence of transformation, suggesting the possible meiotic event.   Flagellates of Tetramitus are stable in sharp contrast to Naegleria and they grow, feed, and divide.

Transformation in Myxomycetes amoebae also appears to be dependent upon the availability of an aqueous environment, and amoebae form flagella almost immediately upon exposure to water (37-39).   During growth, amoebae undergo mitosis with the typical mitotic apparatus including centrioles (also dissolution of the nuclear membrane), and flagellum outgrowth initiates from centrioles which migrate to the cell periphery and act as basal bodies (10, 40).   Flagellated cells tend to remain amoeboid.

Many marine invertebrates develop flagellar apparatus at the blastula stage during the embryonic development. Mechanisms of the flagellar development have not been explored.   In developing sea urchin eggs, RNA synthesis is not required after fertilization, since the enflagellation of blastula occurs in the presence of actinomycin D that blocks uridine incorporation (41, 42).   There is evidence that the signal for enflagellation is related to the increasing ratio of nucleus to cytoplasmic volume, since eggs which have been reduced in cytoplasmic volume by microsurgery prior to fertilization develop the flagellar apparatus before the controls; the number of cells comprising the blastula are reduced in

proportion to the reduction in cytoplasmic volume (43).   If we consider
fertilization (or activation) as the initiating stimulus, the enflagellation
of blastula can be considered analogous to the transformation in Naegleria
or Tetramitus.

   Many parasitic protozoans also develop the flagellar apparatus in
their life cycle. Leishmania donovani grows inside the reticuloepithelial
cells of mammals as a non-flagellated form.   This form has been found
to grow only inside another living cell.   However, the flagellated form
of the organism (leptomonad) can grow extracellularly in nutrient broth,
suggesting that some of the synthetic processes were activated to enable
the flagellate to grow outside the cell with a simpler nutritional environment at lower temperatures (44, 45).

ECOLOGICAL SIGNIFICANCE

   There are two contrasting views on whether or not the phenomenon
of amoeba-flagellate transformation has a selective advantage; one is
that transformation has a definite selective advantage and the other is
that it is an evolutionary relic and does not serve any purpose.

   Naegleria flagellates move approximately 20 times faster than amoebae (27). Although nothing is known about the tropisms of the flagellates,
it is apparent that flagellates disperse much faster than the amoebae in
an aqueous environment.   Three different possibilities exist to explain
the selective advantages of acquiring a faster dispersion system.   The
first is that rapid dispersal obviously lowers the probability of predation, if there is any aqueous predator for Naegleria. At least this argument is quite convincing with larvae of marine invertebrates.   It might
also be argued that rapid dispersion of the population lowers the probability of extinction, provided that there is no selective advantage of con-
gregation, e.g., crossfeeding.   The second is that rapid dispersion increases the probability of finding a new habitat better suited to the organism. The third argument is that the flagellates are gametes and that
simultaneous transformation of mating types after the rain undoubtedly
would increase the probability of mating.   Enflagellation of one mating
type equipped with the proper tropism is essential for finding the non-
motile mate. With Naegleria or Tetramitus, these arguments are attractive but remain as pure hypothesis, since no mating types have been
found with these organisms.

In general, in the amoeba-flagellate transformation, the acquisition of the flagellar apparatus is accompanied by various structural and physiological changes of the organism. In the case of <u>Naegleria</u> or <u>Tetramitus</u>, there is evidence that RNA and protein synthesis is required for transformation, indicating the involvement of gene activation (28). Thus, transformation can be designated as one type of cytodifferentiation. Acquisition of flagella may be an evolutionary relic, but it seems necessary to consider whether or not the physiological changes occuring simultaneously with transformation may have some selective advantage. Enflagellation may play an important role during the activation of a series of genes, while the acquisition of the flagella itself may have little selective advantage.

Phylogeny and taxonomy of amoeba-flagellates have been reviewed by Fulton (7), who concludes that they are not well understood. Further elucidation of the transformation process at the molecular level, as well as discoveries of additional life cycles in these organisms, may help to enhance our understanding of their phylogenic designations.

EVOLUTION OF THE FLAGELLAR APPARATUS

Symbiotic association of spirochaeta-like organisms with the primitive eukaryotes is viewed by some as the origin of the flagellar apparatus (e.g., 46). The strength of the argument resides in the observations that (a), symbiotic association of spirochaeta and protozoans does exist (47) and (b), that basal bodies are presumably self-replicating entities (48) containing DNA (49-51), and (c), no locomotory organ of intermediate structure between the typical flagellar apparatus of eukaryotes (9 + 2 arrangement) and bacterial flagella is known. The last observation assumes that the origin of the 9 + 2 structure is the same in all eukaryotes.

Although the arguments are attractive, each evidence needs critical evaluation. First, although spirochaetes contain bundles of microtubule-like fibrous proteins, they do not have the 9 + 2 structures, and it is not known if there is any similarity between the basic unit of proteins in these fibrous proteins and flagella. Secondly, whether or not centrioles (basal bodies) are self-replicating entities is not settled. Recent evidence suggests that centrioles do not replicate but tend to assemble <u>de novo</u> close to the old centrioles (52-58). There are instances where centrioles emerge from cells in which no centrioles were found previously. For example, numerous centrioles were found in parthoge-

netically activated sea urchin eggs (59), although no centrioles were
found in unfertilized eggs. It can be argued, however, that centrioles
may still exist inside the egg, since the enormous size of the egg makes
a search for centrioles difficult and since the egg apparently undergoes
meiotic divisions with centrioles prior to maturation. However, with
Naegleria amoebae, in which such obstacles do not exist, an extensive
and systematic effort to discover centrioles has failed (22). In addition,
the number of basal bodies can be increased by experimental manipulation (23). With Naegleria, it seems reasonable to conclude that basal
bodies and thus the 9 + 2 structures are formed de novo during transformation (22). The presence of DNA in basal bodies of ciliates has
been reported (49-51, 60, 61), but the apparent technical difficulties
severely limit the validity of these observations (62). Finally the ubiquitous and unique 9 + 2 structure of the basal bodies (and the flagella)
does not necessarily reflect the identical origin. It appears rather that
no other arrangement is possible at the molecular level for an efficient
motile apparatus. If not, there should be thousands of motile apparati
with different structures even if the origin is identical.

Recent studies have revealed that the motility of cells is primarily
based on two different fibrous proteins, microtubules and microfila-
ments (63). An alternative hypothesis of the origin of cell motility is
simply to assume the evolution of these fibrous proteins (64). Existence
of the equilibrium of microtubule proteins between an assembled state
(microtubules) and unassembled state suggests the feasibility of the evolution of proteins with such morphopoietic properties (34).

Another problem associated with the formation of the flagellar apparatus is the formation of the membrane protrusions at the site of the
basal body attachment. The mitotic apparatus appears to play a role in
determining the position of the cleavage furrow (65-67). The cytoaster-
centriole complex appears to influence the cell surface in such a way that
the cleavage furrows are formed (68). The identical structure shared between the basal body and the centriole has been well documented (62, 65).
It is not difficult to imagine that basal bodies have evolved into a cell
organelle to perform a dual role, i.e., assembling the microtubules from
unit proteins (69) and causing the membrane protrusion during the formation of the flagella. Which have evolved first, basal bodies or centrioles?
While all flagella with the conventional 9 + 2 arrangement or its variables
have basal bodies, centrioles are not always associated with mitosis.
Hence, it appears that the participation of centrioles in mitosis came after
the evolution of basal bodies.

CONCLUDING REMARKS

Amoebo-flagellates are organisms widely distributed throughout the world. Studies on the environmental requirements of amoeba-flagellate transformation indicate that rainfall and the formation of a liquid environment are sufficient conditions in nature to enable the amoebae to transform into flagellates. Such conditions would convert the total population of amoebae into flagellates, since transformation can be induced at any stage of the cell cycle.

The acquisition of the flagellar apparatus endows amoebae with faster motility and thus faster dispersion of the cell population in the aqueous environment. This would certainly increase the probability of survival of the organism.

Analysis of the stimulating factors of transformation during the environmental changes have led to the conclusion that amoebae have mechisms which can detect a favorable environment for transformation, although they are not totally committed and can reverse their course before the emergence of flagella when the environment becomes favorable for growth but not for rapid dispersion of the population. Such a feed-back control mechanism would give organisms an additional advantage.

Studies on the formation of the flagellar apparatus in Naegleria suggest that the theory of the symbiotic origin of the flagellar apparatus may not be sound. De novo origin of basal bodies suggests that proteins with the unique morphopoietic features have evolved. The striking conservation of the structure of basal bodies reflects the fact that no alternative arrangements have been possible within the mutable variation of such proteins.

ACKNOWLEDGEMENT

The present research is supported by the National Research Council of Canada.

438 S. YUYAMA

REFERENCES

1. Wenyon, C. M., "Protozoology," Woods and Co., New York, 1926.
2. Hyman, L. H., "Invertebrates," McGraw-Hill, New York 1940.
3. Corliss, J. O., Systematic Zool. 8, 169 (1959).
4. Klein, R. M., and Cronquist, A., Quart. Rev. Biol. 42, 105 (1967).
5. Page, F. C., J. Protozool. 14, 499 (1967).
6. Kudo, R. R., "Protozoology," 5th Ed., Thomas, Springfield, Ill. 1966.
7. Fulton, C., in "Methods in Cell Physiology," (D. M. Prescott, ed.), Vol. IV, p. 341, 1970.
8. Bunting, M., Proc. Natl. Acad. Sci. U. S. 8, 294 (1922).
9. Bunting, M., J. Morph. 42, 23 (1926).
10. Cadman, E. J., Trans. Roy. Soc. Edinburgh 57, 93 (1931).
11. Schardinger, F., Sitzungsb. Akad. Wiss. (Wien) Math. Nat. Cl. Abt. I 108, 713 (1899).
12. Yuyama, S., in "The Developmental Aspects of Cell Cycle" (I. L. Cameron, G. M. Padilla, and A. M. Zimmerman, eds.), Academic Press, New York, 1971.
13. Calkins, G. N., Trans. 15th Intern. Congr. Hygene Dem. Washington 2, 287 (1913).
14. Wilson, C. N., Univ. Calif. Pub. Zool. 16, 241 (1916).
15. Pietschmann, K., Arch. Protistenk. 65, 379 (1929).
16. Rafalko, J. S., J. Morph. 81, 1 (1947).
17. Singh, B. N., Phil. Trans. Roy. Soc. London B 236, 405 (1952).
18. Schuster, F. L., J. Protozool. 10, 297 (1963).
19. Dingle, A. D., and Fulton, C., J. Cell. Biol. 31, 43 (1966).
20. Fulton, C., and Dingle, A. D., Develop. Biol. 15, 165 (1967).
21. Yuyama, S., Ph. D. Thesis, Case-Western Reserve University, Cleveland, Ohio (1967).
22. Fulton, C., and Dingle, A. D., J. Cell. Biol. in press.
23. Dingle, A. D., J. Cell. Sci. 7, 463 (1970).
24. Willmer, E. N., J. Exp. Biol. 33, 583 (1956).
25. Willmer, E. N., J. Embryol. Exp. Morph. 6, 187 (1958).
26. Chang, S. L., J. Gen. Microbiol. 18, 579 (1958).
27. Yuyama, S., unpublished observation.
28. Yuyama, S., Biophys. J. (Suppl.) 6, 110 (1966).
29. Yuyama, S., J. Protozool. 18, 337 (1971).
30. Perkins, D. L., and Jahn, T. J., J. Protozool. 17, 168 (1970).
31. Fulton, C., and Guerrini, A. M., Exptl. Cell Res. 56, 194 (1969).
32. Fulton, C., J. Cell. Biol. 47, 67a (1970).

33. Walsh, C., J. Cell. Biol. 47, 219a (1970).
34. Inoue, S., Rev. Mod. Phys. 31, 402 (1959).
35. Outka, D. E., J. Protozool. 12, 85 (1965).
36. Fulton, C., Science 167, 1269 (1970).
37. Kerr, N. S., J. Protozool. 7, 103 (1960).
38. Kerr, N. S., J. Protozool. 12, 85 (1965).
39. Haskins, E. F., Can. J. Microbiol. 14, 1309 (1968).
40. Schuster, F. L., Protistologia 1, 49 (1965).
41. Gross, P. R., J. Exp. Zool. 157, 21 (1964).
42. Raff, R. A., Greenhouse, G., Gross, K. W., and Gross, P. R., J. Cell. Biol. 50, 516 (1971).
43. Rustad, R. C., and Yuyama, S., in preparation.
44. Adler, S., and Theodor, O., Proc. Roy. Soc. London B 108, 453 (1931).
45. Trager, W., J. Exp. Medicine 97, 177 (1953).
46. Sagan, L., J. Theor. Biol. 14, 225 (1967).
47. Copeland, H. F., "The Classification of the Lower Organisms," Pacific Books, Palo Alto, Calif. 1956.
48. Lwoff, A., "Problems of Morphogenesis in Ciliates," Wiley and Sons, New York, 1950.
49. Randall, J. T., and Disbrey, C., Proc. Roy. Soc. London B 162, 473 (1965).
50. Smith-Sonneborn, J., and Plaut, W., J. Cell Sci. 2, 225 (1967).
51. Smith-Sonneborn, J., and Plaut, W., J. Cell Sci. 5, 365 (1969).
52. Gall, J. G., J. Biophys. Biochem. Cytol. 10, 163 (1961).
53. Andre, J., J. Microscopie 3, 23 (1964).
54. Sorokin, S., J. Cell. Sci. 3, 207 (1968).
55. Robins, E., Jentzsch, G., and Micali, A., J. Cell. Biol. 36, 329 (1968).
56. Dirksen, E. P., J. Cell Biol. 51, 286 (1971).
57. Kalnins, V. I., and Porter, K. R., Z. Zellforsch. 100, 1 (1969).
58. Dippel, R. V., Proc. Natl. Acad. Sci. U. S. 61, 461 (1968).
59. Dirksen, E. R., J. Biophys. Biochem. Cytol. 11, 244 (1961).
60. Sukhanova, K. M., and Nilova, V. K., Tsitologiya 7, 431 (1965).
61. McDonald, B. R., and Weijer, J., Canad. J. Genet. Cytol. 8, 42 (1966).
62. Fulton, C., in "Results and Problems in Cell Differentiation" (J. Reinert and H. Ursprung, eds.), Vol. 2, p. 170, Springer-Verlag Berlin, 1971.
63. Wessells, N. K., Spooner, B. S., Ash, J. F., Bradley, M. O., Luduena, M. A., Taylor, E. L., Wrenn, J. T., and Yamada, K. M., Science 171, 135 (1971).

64. Satir, P., and Satir, B., J. Theoret. Biol. 7, 123 (1964).
65. Mazia, D., in "The Cell" (J. Brachet and A. E. Mirsky, eds.),
     Vol. 3, p. 77, Academic Press, New York, 1961.
66. Rappaport, R., J. Theor. Biol. 9, 51 (1965).
67. Yuyama, S., Biol. Bull. 140, 339 (1971).
68. Wolpert, L., Int. Rev. Cytol. 10, 164 (1960).
69. Stephens, R. E., in "Biological Macromolecules" (S. N. Timasheff
     and G. D. Fasman, eds.), Vol. 5, Marcel Dekker, New York, 1971.

Received 26 October, 1971

ACADEMIC ASPECTS

THE HETEROTROPH HYPOTHESIS AND

HIGH SCHOOL BIOLOGY

Claude A. Welch

Department of Biology, Macalester College
Saint Paul, Minnesota 55101

Theory is finally coming of age in biology. It was to the physical sciences that philosophers invariably turned when they discussed the role of theory as a source of fruitful and generative ideas leading to new experiments and revolutionary concepts. But the biologists have broken loose from their routine role of description and classification. They are caught up in the exciting interaction of facts and ideas and this relatively new affection "has changed their complexion"--to paraphrase an old song. Broad explanatory theories and conceptual schemes have finally crept into biology, but it has been only slightly more than a century that the three major concepts--cell theory, evolution theory, and gene theory-- have attained operational respectability. This "operational respectability" however, has not come easily in respect to evolution theory, particularly in the American secondary schools. Evolution, especially Darwinian natural-selection, is still a bete noire in a sizable portion of American high schools. But the corner has been turned, and new directions for high school biology are in process.

PROCESS AND PRODUCT

The new directions in biology teaching came when biologists and educators decided that they were going to let the young students in on the real nature of science. Science teaching had become one long rhetoric of conclusions. Most of the "conclusions" of science had been stripped of the exciting detective work which had created them and these conclusions,

instead, were recited as cold statistics.   The product of science was
sold without any indication of the process which produced it.

Most of the high school curriculum studies, such as the Biological
Sciences Curriculum Study, accepted the nature of scientific inquiry as
a major goal to be taught.   The process of science was to be at least as
important as the product.   The detective work and the nature of the detective were to become as important as the products and conclusions of
the detections.

This switch in emphasis had a tremendous impact on the nature of
the narrative which now had to be told. If the detective work was now to
be respectable, then scientific "hunches," hypotheses, theories, explanations, predictions, experimental tests and the like must be seen as a
major thread and pattern in the scientific fabric.   The stage was set to
put life back into biology. And, fortunately, biology had become a lively
science thanks to the theoretical framework which had gained momentum
from the imaginative minds of Darwin and Mendel.

HYPOTHESES AND CREATIVITY

Although there are books on experimental design and even on "the
scientific method," few scientists have dared to suggest that there is a
technique on how to come up with bright hypotheses.   To confront a student with an unsolved problem for which he is asked to "dream up" an
explanatory hypothesis often results in a frustrating experience for both
the teacher and the student. But if scientific process is to be understood,
witnessed and "felt," the frustration of "blind alleys," "blank walls,"
and even the "dead calm of zero insight" must be experienced.   There is
always the hope that one little original spark of an idea for one little problem will introduce a student to the creativity and art within our science.
The experience of going through the "dead calm of zero insight" is usually
enough to convince even the most confident student that bright hypotheses
do not leap forward as automatically as a high powered car from the touch
of the accelerator.   The calm, zero insight cannot be a steady diet, of
course.   Once or twice is enough of a humbling experience so that one
can soon recognize the power and vitality of a fresh, fruitful hypothesis.

The importance of a fruitful hypothesis was emphasized by James
Conant in his development of the case history approach to science. The
tactics and strategies of science are brilliantly developed in Conant's

case histories and they still serve as a standard by which the process approach of science teaching can be judged. If there is any limitation to them for the high school student it is that the case histories are, after all, histories. The Phlogiston Hypothesis is a beautiful example of a fruitful, imaginative concept which served its adherents well. After all, Priestley made his mark without ever feeling the necessity to forsake phlogiston. But to the "now" generation of high school students, at least to some portion of them, outworn concepts in their historical setting do not carry the message of scientific inquiry. Phlogiston, as pointed out by Thomas Kuhn, is certainly no less scientific just because it became outmoded. To many students, however, the only path to an understanding of the fragile nature of our concepts is the path of the present. The Heterotroph Hypothesis provides just such a path in the present and it can serve as an excellent vehicle for understanding the tactics and strategies of science as well as an exial thread for the content of biology.

HETEROTROPHS AND SPONTANEOUS GENERATION

Rarely do we find a mind-expanding problem which can match the question of the origin of life on Earth. The concept of Special Creation provided an easy answer to this question, but it was also an ignava ratio, or lazy argument. The revolutionary idea of an evolving universe certainly does not negate an act or continuing acts of creation. But our current understanding of fossils, geological history, and molecular biology provides a framework which demands a careful, exacting set of postulates to serve as an explanatory system. The awakening awareness of the tenth grade biology student is a suitable medium for posing the problem of the origin of life. The problem is suitable not only because it is exciting but, just as important, it is a problem for which only a partial answer has been found. The Heterotroph Hypothesis is an exciting, speculative working hypothesis and it needs plenty of refinement. It is an idea that the students can understand, but they can also see its difficulties. They can also see and read how the assumptions are being continually reexamined and modified. They can witness the current arguments pro and con concerning the nature of the Earth's primitive atmosphere. They are not reading history. They are living during the development of an important hypothesis about the origin and evolution of our planet and its living systems. They can sense the simple riddle of the "only L" amino acids and feel the dead calm of zero insight as they try to dream up a solution to this problem. But they can also feel a sense of unity with the great scientists in that they see that a great scientist is not necessarily all that is needed for the solution of a great problem. Even though the tenth-grader is certainly unable to appreciate the sophistication of complex organic chemistry, he can usually recognize a poor answer to a good question. He can also appreciate the opportunity to try to answer some of these questions himself. He knows that many of the truly creative scientific ideas have come to very young minds, and it is encouraging to the students that they are being introduced to new ideas and as-yet unsolved problems rather than just the opposite. So one of the truly great advantages of incorporating the Heterotroph Hypothesis into the high school biology curriculum is that it serves as an excellent vehicle for illustrating in an on-the-spot manner how the process of science works. The nature of the postulational system, the fruitfulness of the ideas, the elegant predictions, the nature of the experimental tests, the corrective feedback from experiments, the competitive nature of the experimentalists, the use of ad hoc postulates, the imposition of parsimony, the resurrection of an old idea (spontaneous generation)--all these aspects of process are profusely illustrated as the Heterotroph Hypothesis grows and develops in contemporary science.

A second, yet very important, function of the Heterotroph Hypothesis for high school as well as advanced biology, is that it provides a beautiful sequence for discussing the content of cell biology. One almost gets the same feeling Darwin expressed when, after reading Malthus, he claims that now, at last, he had a theory by which to work. The Heterotroph Hypothesis provides at least an organizational beginning to place cell topics in some logical order. The ideas concerning the evolution of metabolism gives good reason to introduce fermentation just before photosynthesis, and photosynthesis, in turn, just before oxidative respiration. Of course, one can "cover" catabolism all in one swoop by following glucose metabolism to CO_2, H_2O and ATP. But his approach ignores the perennial and important question that biologists should always ask: how did things get the way they are? Biologists do want to know "what is" and then how the "what is" works. But they, perhaps uniquely, also must always ask: how did the "what is" get the way it is? This is what the great idea of evolution has done to the biologist and we are stuck with it (gladly) until something better comes along. The Heterotroph Hypothesis does for cell biology what evolution did for multicell biology: it helps us organize and explain great, seemingly independent classes of observations. At least some high school students see this advantage, and it makes a world of difference.

The introduction of ideas relative to the origin of life has added even greater excitement to the already explosive developments in molecular as well as classical biology.   The high school biology curriculum has been ready, willing, and able to respond to new ideas.   The origin of life is one of the oldest of all "happenings," but it is a relatively new idea in the teaching of biology.

Received 11 November, 1971

THE PLACE OF THE ORIGIN OF LIFE IN THE

UNDERGRADUATE CURRICULUM

Allen Vegotsky

Associate Professor of Biology
Wells College, Aurora, New York

The subject of the prebiological or abiotic synthesis of living matter has a great appeal to all questioning minds. At the same time, this topic is a central theme of biology, essential for comprehension of the basic nature of life and its early evolution. Thus, it is not surprising that students find life's origin an absorbing subject and that authors of textbooks in many areas of biology frequently include at least some consideration of this matter.

Some thought on how to integrate the origin of life into the curriculum is needed to be sure that students do get some exposure to this subject without undue repetition. A brief survey of a random selection of some of the frequently used undergraduate textbooks was made in an attempt to evaluate the degree of treatment of prebiological evolution in texts (Table I). Six undergraduate courses were considered. This topic is seen to be almost as pervasive in the undergraduate curriculum as the Krebs cycle. However, with the exception of the microbiology texts which emphasize spontaneous generation and Pasteur's classic experiments, the manner of coverage of biological origins is not very different in courses at different levels. Repetition of a theme of this significance is not objectionable per se, but the lack of novel development of this theme in different courses is likely to have a negative effect on students. At Wells College, we cover biochemical origins at two levels, first for beginning students and later in an advanced course. The methods of treatment will constitute the remainder of this brief essay.

TABLE I. The Origin of Life in College Textbooks

Type of Textbook	Number of Texts Surveyed	Number of Texts Including Subject	Average Pages of Coverage[a]
General Biology	8	8	10
Genetics	6	2	7
Microbiology	6	6	7
Cell Biology	4	2	3
Physiology	5	1	9
Biochemistry	5	1	23

[a]Based on column 3.

For a number of reasons, I chose to incorporate one or two lectures on the origin of life into the first course in biology. Principally, I believe that the matter of biochemical origins is sufficiently urgent to place it in a course that most science-oriented students will take. Secondly, it provides a logical entree into another major theme of biology, evolution. My lectures have included a short historical introduction to man's changing ideas on biogenesis and a more lengthy discussion of some of the key modern experiments in this field. These lectures are based largely on source material from the research of Sidney Fox and coworkers, partly because I am fortunate to have been one of those coworkers and partly because of the satisfaction derived by students and myself from an experimental approach to the synthesis of proteins, enzymes, nucleic acids, cells, etc. based on a simple simulated model of the primordial world. More specifically, the topics include a brief historical review of earlier ideas on the origin of life (i. e. spontaneous generation, panspermia) and a major emphasis on the studies of Oparin, Calvin, Miller, Fox, among others. For supplementary reading, the students might be referred to Wald (1) or Fox and Vegotsky (2), and for a treatment in greater depth to Fox, et al (3). The instructors may also find these articles useful as well as several books on the subject (4–8). One important fringe benefit derivable from inclusion of the origin of biochemical substances in the introductory course is the framework provided for the students to observe the relevance of chemistry to life. Some students seem resistant to including chemistry in a course designated as biology; relating chemistry to the origin of life, however, makes the chemistry more palatable. Indeed, at least one recent introductory biology textbook considers abiogenesis in an early chapter in which chemical principles are first treated rather than in later chapters on evolution (9).

There is an obvious limitation to what material on prebiological evolution can be conveyed in a beginning course and this is the lack of appropriate background in the students. In my experience, a senior level course in biochemistry or molecular biology provides a logical opportunity for a more sophisticated treatment of abiogenesis. Presumably, the student taking such a course will have had a thorough grounding in the physical and biological sciences and will appreciate the challenge and complexities of attempts to synthesize life. It came as a surprise to me that the five sizable biochemistry books reviewed for Table 1 had no mention of the origin of life with the single exception of Lehninger's excellent treatment (10). Nevertheless, I think there is room for this topic in such a course. My molecular biology course approaches the origin of life in three ways: lectures, seminar and laboratory. One or two lectures reinforce the topic covered in the freshman course at the suitable level. The students are allowed to do more reading on their own for the seminar; the subject matter is sufficiently interesting and controversial to make for stimulating discussions. The proceedings of the Wakulla Springs conference on this subject (6) is an excellent source for student reading and contains not only research papers but some animated discussions of these papers by specialists.

In this course, the more unique feature of the treatment of biochemical origins is not the lectures or seminar but rather the integration of these components with a laboratory. There is surprisingly much that can be accomplished in an undergraduate laboratory based on simulated prebiological origins. These experiments which will be described shortly not only are exciting for the students to do but they are interelated; they convey a sense of research rarely experienced by undergraduates in their curriculum. Students too often complete their undergraduate biology courses with a feeling that research composes three hour slots of independent disconnected experiments. In our course, a series of laboratories are devoted to synthesis of proteinoids (11), dialytic purification, hydrolysis, and two dimensional paper chromatography. In the process of doing these experiments, the students develop some important biochemical techniques as well as a sense of continuity in experimentation that is so much a part of research.

SYNTHESIS OF PROTEINOID

The series of experiments begins with a laboratory synthesis of proteinoids essentially utilizing the procedures of Fox and Harada (11). A
mixture of 18 amino acids is used in the proportions given in Table II.
These proportions were derived from a consideration of comparative biochemistry (12) and represent the approximate composition of an average
modern protein with one critical exception; the ratio of aspartic acid to
the other amino acids is increased by a factor of ten. In the absence of
sufficient aspartic acid, very little high molecular weight product is obtained (11, 13, 14). The standard mixture of amino acids in these experiments is designated 10A (i.e. ten times the amount of aspartic acid
in a calculated average protein). It has been my practice to prepare a
large batch of the amino acid mixture in advance of the laboratory period
and grind these 18 amino acids together in a mortar for at least 30 minutes.
Such a stock mixture is usable for several years if kept dry. The students then weigh out 1.00 g. from the stock mixture into labelled pyrex
test tubes. To encourage the feeling of individual research, I have allowed the students to consider the effect of specific variables on the polymerization of amino acids. Thus, while several students follow the
standard instructions for a 10A mixture, others may wish to consider the
effect of temperature (in the range of 150-180°C.) or of duration of heating period (in the range of 3-6 hours) or of addition of ATP (up to 0.01 g.)
or addition of phosphoric or polyphosphoric acid (up to 0.1 g.). Within
the ranges indicated, the variables will affect the yield and nature of
polymers obtained (11, 13, 14).

TABLE II. Weight Proportions of Amino Acids in Reaction Mixture.

Amino Acid	Proportion	Amino Acid	Proportion
Alanine	0.60	Lysine	0.85
Arginine	0.85	Methionine	0.19
Aspartic Acid	9.80	Phenylalanine	0.51
Cystine	0.24	Proline	0.53
Glutamic Acid	1.29	Serine	0.58
Glycine	0.62	Threonine	0.47
Histidine	0.34	Tryptophan	0.18
Isoleucine	0.60	Tyrosine	0.44
Leucine	0.60	Valine	0.54

The experimental design is deliberately simple in keeping with the restrictions of the prebiological world. The standard conditions employed were a temperature of 170 $\pm$ 2°C., a six hour reaction period, and an atmosphere low in oxygen. For temperature control, we have partially immersed the test tubes in a preheated oil bath maintaining approximately 170°C. with a hot plate. The reaction should be performed in a hood. During the long heating period, the students should take turns observing the reaction, keeping records of the temperature in the oil bath, and making temperature adjustments when necessary. In the course of the reaction, foaming should be observed due mainly to the release of steam in the dehydration synthesis of peptide bonds, and the reaction mixture should darken gradually to a tan color.

A reaction period of six hours duration preceded by perhaps an hour of organization may come as a shock to students oriented to think that all experiments take three hours, but I believe this awakening to be part of a liberal education since long hours of work are so characteristic of actual research. It has been my practice to schedule this laboratory very early in the semester when the students are less committed to outside activities. During part of the long reaction period, I have lectured on the theory behind the thermal synthesis (3) and the experiments to follow and discussed what is thought about primordial world conditions. In this context, it should be pointed out that many specialists in this field prefer atmospheres of the reducing type (e.g., methane, ammonia, hydrogen). However, to obtain proteinoids one needs only to avoid excess oxygen in the reaction mixture, thus limiting oxidative decomposition. This can be accomplished by heating the amino acids in sealed test tubes in an atmosphere of nitrogen. Alternatively, and more simply, the test tubes may be capped with one-hole rubber stoppers arranged with a rubber hose outlet partially clamped shut. During the reaction, pressure builds up inside the tube, steam evolves, and little oxygen can enter.

After the reaction, the test tubes are removed from the oil bath and allowed to cool for about ten minutes after which 5 ml. of distilled water is added. The product is then stirred with water and refrigerated. The brief trituration makes the reaction product easier to handle in later stages. The total time for the procedure thus far is about 7-8 hours, if the amino acid mixtures are prepared in advance and if the heating bath is adjusted to 170°C. prior to the laboratory. It is, of course, not necessary for all of the students to be present for all of the experiment.

PREPARATION OF MICROSPHERES

One of the unique features of the proteinoids is their propensity to form microspheres (spherules) with morphology and other properties similar to cells (6, 15, 16). In the prebiological context, we might visualize polycondensation of amino acids on a volcanic surface yielding proteinoids; if this process is followed by rain or by contact of the reaction products with the ocean, then microspheres could form and eventually evolve to living cells. It is relatively easy to perform a laboratory experiment simulating the production of microspheres in the primordial world. In our classes, the students prepare microspheres essentially by the method of Fox, Harada and Kendrick (15). They added about 1 ml. of hot 2% NaCl solution to unpurified reaction product of the thermal synthesis. The suspension is brought to a boil in a test tube and allowed to cool. As the suspension cools, it should become turbid due to the vast numbers of microspheres produced. These microspheres are in the size range of a few microns and can be seen on a microscope slide at a magnification of about 400 times or more. All of this can be done in much less than an hour. Variations of the reaction conditions leads to microspheres of different morphology and properties (17, 18), and students may wish to explore this possibility. It is desirable to have on hand photographs of microspheres taken through a phase contrast and electron microscope for this laboratory session (3, 6, 18).

DIALYTIC PURIFICATION OF THERMAL POLYMER

The thermal reaction product is a mixture of small compounds (i.e., unreacted amino acids and degradation products) and higher molecular weight polymers. These may be conveniently separated by dialysis, a technique of value in itself to potential molecular biologists or biochemists. In a typical procedure, students transferred their triturated reaction products into ten inch lengths of dialysis tubing tied to seal both ends. Each tube was then immersed in a one liter beaker of tap water in a refrigerator. The water surrounding the dialysis sacs was replaced three times daily with tap water for the first three days and with distilled water on the fourth day. Aliquots of a few mls. were saved from each of the first four changes and refrigerated. These aliquots were later tested for amino acids by cospotting on filter paper one drop of aliquot and one drop of ninhydrin (0.2% in water-saturated butanol) and drying. The students should find a strong positive color test (purple) for amino acids in the first aliquot and progressively weaker tests thereafter, demonstrating the significance of dialysis.* In our class, the dialysis is generally performed between two weekly labs to save time. The students were allowed to collaborate in replacing the dialysis water so that the time demand on each student was not excessive. This continuity of experiments conveys a greater feeling for research activity and keeps the goals of the experiment constantly in the minds of the students.

DRYING THE PROTEINOID

Prior to characterization by paper chromatography, it is necessary to dry the dialyzed thermal polymer. This can be done by any method that does not require excessive heating. Freeze drying is one of the gentlest ways to dehydrate the product. Alternatively, the products may be concentrated down to <u>near dryness</u> in a beaker or watch glass under an infrared lamp so long as overheating is avoided. The last few mls. may be removed by vacuum desiccation. The thoroughly dried pulverized polymer should be tan colored. Careful exclusion of oxygen during the polymerization process and drying by lyophilization would lead to a product closer to white in color. Under the conditions described, a weight yield of 15% may be anticipated for a 10A mixture. Addition of small quantities of ATP or other phosphates should lead to a significant increase in yield (13, 19), while a decreased reaction temperature or duration should lead to a lowered yield (11, 14).

---

*It should be noted that there is much room for variation in dialytic procedures. It would be preferable to perform the dialysis in a cold room; ideally the water surrounding the dialysis sac should be agitated, for example, with a magnetic stirrer. The rationale for the use of cold temperature in biochemical investigation (to prevent denaturation and bacterial contamination) should be pointed out to the students. Another consideration for discussion is the fact that the effectiveness of dialysis of insoluble materials is questionable. It can be argued that the nondialyzable fraction may include low molecular weight compounds that did not escape from the dialysis sac simply because they were insoluble. In actual research on proteinoids, it is generally the practice to dialyze the proteinoids in the form of a sodium salt (11).

HYDROLYSIS AND PAPER CHROMATOGRAPHY

For paper chromatography of amino acids, about 10 mgs. of hydrolyzed thermal polymer is ample.   The polymers may be hydrolyzed in
sealed test tubes containing 2 mls. of 6 N HCl (105°C., 20 hours) or by
refluxing with 6 N HCl for 20 hours.   A very simple but adequate procedure we have used in some student experiments was to hydrolyze the
polymers in closed weighing bottles (ground glass lids) containing 10 mls.
of HCl (in an oven set at 105°C., 12 hours); under these conditions, very
little HCl evaporated out of the weighing bottles.   Prior to paper chromatography, it is necessary to remove most of the HCl and water. This
may be done in a vacuum desiccator if NaOH is used to trap the acid
fumes.   Removal of the bulk of the acid will take several days and frequent changes of sodium hydroxide and desiccant.

The dried hydrolysates were dissolved in 0.5 ml. of water after the
students applied 5 μl on chromatography paper.   It may be necessary
to impress students with the need to keep the spots small.   For separation of the amino acids, two dimensional paper chromatography was employed.   In the first dimension, phenol:water (80:20) was used.   After
drying in the hood for about a day to remove the phenol fumes, the chromatography paper was rotated 90° and rechromatographed in butanol:
acetic acid: water (4:1:1) and then allowed to dry in the hood.   To make
the amino acid positions visible, the paper was sprayed with ninhydrin
(0.2% in water-saturated butanol). The spots were allowed to develop at
room temperature for several hours. If immediate observations are required, the chromatogram may be heated in an oven at 100-110°C. for a
few minutes.   The students may need to be forewarned to avoid touching
their chromatograms and leaving confusing fingerprints.

Identification of the amino acids in the hydrolysate may be accomplished by comparing the positions of unknown with known R_f values for
amino acids (19).   It is, however, preferable for the students to make
their own "maps" based on R_f values obtained with amino acid standards
chromatographed in one dimension in each of the solvent systems used.
(The same apparatus should be used for the standards as was used for
the hydrolysate.)   Each standard spot should contain 2-5 μg. of each
amino acid in the reaction mixture except for cystine, histidine, methionine, phenylalanine, proline, tryptophan, and tyrosine; for the latter
group, the spots should contain twice as much of each amino acid. I have
permitted students to collaborate on the chromatography of the standards.
From the R_f value obtained, the unknown amino acids may be tentatively

classified. With good technique, about 18 spots should be visible. The novice should be pleased with 10 or more distinct amino acid spots.

The above experiments constitute the equivalent of approximately four laboratory periods. Together, and coordinated with appropriate lectures and seminars, they provide a learning experience that is not only conceptual but also practical and memorable. The student develops techniques in the context of a research-type project. If desirable, and if appropriate facilities are available, these experiments may be extended in many ways such as quantitative analysis of amino acids in the hydrolysate (21), electrophoresis (11, 14), infrared spectrum analysis (11), and enzymatic assays (6, 22).

REFERENCES

1. Wald, G., Scientific Am. 192, 44 (1954).
2. Fox, S. W., and Vegotsky, A., in "Collier's 1965 Year Book," p. 28. The Crowell-Collier Publishing Co., New York, 1965.
3. Fox, S. W., Harada, K., Krampitz, G., and Mueller, G., C. and E. News 48, 80 (1970); Fox, S. W., ibid. 49, 46 (1971).
4. Bernal, J. D., "The Origin of Life," The World Publishing Co., Cleveland, 1967.
5. Calvin, M., "Chemical Evolution," Oxford University Press, New York, 1969.
6. Fox, S. W., ed., "The Origin of Prebiological Systems and of their Molecular Matrices," Academic Press, New York, 1965.
7. Kenyon, D. H., and Steinman, G., "Biochemical Predestination," McGraw Hill Book Co., New York, 1969.
8. Keosian, J., "The Origin of Life," 2nd ed., Reinhold Publishing Corp., New York, 1968.
9. Brown, R. A., "General Biology," McGraw Hill Book Co., New York, 1970.
10. Lehninger, A. L., "Biochemistry," p. 769. Worth, New York, 1970.
11. Fox, S. W., and Harada, K., J. Am. Chem. Soc. 82, 3745 (1960).
12. Vegotsky, A., and Fox, S. W., in "Comparative Biochemistry" (M. Florkin and H. S. Mason, eds.), Vol. IV, p. 185. Academic Press, New York, 1962.
13. Vegotsky, A., and Fox, S. W., Federation Proc. 18, 343 (1959).
14. Vegotsky, A., Ph.D. Dissertation, Florida State University, 1961.
15. Fox, S. W., Harada, K., and Kendrick, J., Science 129, 1221 (1959).

16. Fox, S. W., McCauley, R., Joseph, D., Windsor, C. R., and Yuyama, S., in "Life Sciences and Space Research" Vol. IV, (A. H. Brown and M. Florkin, eds.), Spartan, Washington, D. C., 1966.

17. Fox, S. W., in "Encyclopedia of Polymer Science" Vol. IX, (H. F. Mark, N. G. Gaylord, and N. M. Bikales, eds.), p. 284. Inter-science, New York, 1969.

18. Miquel, J., Brooke, S., and Fox, S. W., Currents in Modern Biology 3, 299 (1971).

19. Fox, S. W., and Harada, K., Arch. Biochem. Biophys. 86, 281 (1960).

20. Lederer, E., and Lederer, M., "Chromatography: A Review of Principles and Applications," Elsevier Publishing Co., New York, 1957.

21. Fox, S. W., Harada, K., Woods, K. R., and Windsor, C. R., Arch. Biochem. Biophys. 102, 439 (1963).

22. Rohlfing, D. L., and Fox, S. W., Arch. Biochem. Biophys. 118, 122, 127 (1967).

Received 1 October, 1971

MAN AND EVOLUTION

SOME SOCIAL AND PHILOSOPHICAL IMPLICATIONS OF PROGRESS ON THE ORIGIN AND SYNTHESIS OF LIFE

Charles C. Price

Benjamin Franklin Professor of Chemistry
University of Pennsylvania, Philadelphia, Pa. 19104

There has been remarkable progress in synthesizing more complex essential components of living organisms and of obtaining ever greater understanding of the processes by which life may have originated on this planet. These developments will have significant practical consequences, but perhaps as important will be some of the implications they provide for the philosophical view of man--what he is, where he came from, where he may be headed.

One facet of this is the role of order in evolution. From the big bang onward, the major steps leading to human society here on Earth have had some common features. They have involved an increase in local order at the expense of energy. The increase in order has created capabilities in the ordered system not evident in the less-ordered arrangement. And the new capabilities so created have been utilized to build the next step in the process.

In every step, the enhanced capability due to increase of order and organization has led to the next. In the early steps, such as formation of atomic nuclei in stars and of simple chemical compounds from these elements on primordial Earth, the steps were very largely guided by the fundamental properties of the elementary particles of nature and to a somewhat lesser degree by the circumstances. As these molecules elaborated more and more complex and ordered structures, the basis of the chemistry of life was established. Out of this welter of proliferating chemical systems, eventually a living cell was organized. By mutation and adaptation this has led to the amazing variety of living beings. Now the human brain, with its remarkable capacity to originate, organize, and communicate ideas, has led to the development of human society.

As evolution produced more and more complex molecules and then life, the element of choice between alternate pathways became a more and more significant aspect of the steps actually taken. Choices made in earlier steps in turn imposed limits on the choices available for later steps. Despite this aspect of chemical and biological evolution, a recent book on the subject by Kenyon and Steinman has the startling title of "Biochemical Predestination." The concept embodied in the title itself poses a philosophical challenge to traditional views of creation!

As new knowledge develops, it has increasingly provided natural explanations for facts and phenomena formerly ascribed to the supernatural. Perhaps new understanding of chemical evolution and biological function can develop a philosophical view of man more unified, less divisive, less of a major breeding ground for man's inhumanity to man than the many religious dogmas now so much used to inflame feelings of hatred, suspicion and prejudice in human society.

In addition to altered concepts of man's origin, the new views of evolution raise fascinating speculation on the course of the future, when further increases in order, organization, and cooperation may provide enhanced capabilities for the quality of life on this planet. Are these developments "predestined" and if so in what sense? Obviously we have many choices before us and the various choices will have differing consequences. To a very large degree our choices are limited by biological and societal choices made in the past and in that sense they are "predestined." On the other hand, they will to some degree also be influenced by chance, in the sense that fortuitous (or calamitous) combinations of circumstances and choices may occur. This has unquestionably been a factor throughout chemical and biological evolution. The enormous variety of living species is a testimony to the successes, the even larger variety of extinct species to the failures!

This concept of evolution toward increasing order and organization of structure and information presents an interesting paradox. The increased demands of society for organization, order, and cooperation would, at first, appear to diminish individual freedom and initiative. Actually, the more complex society which demands the increasing order also provides a far wider choice for the individual. The exponential increase in specialization, together with liberation from menial tasks necessary for survival, have enormously increased man's opportunities for service, for recreation and for creative endeavour.

In addition to the philosophical challenges to traditional values from greater understanding of the natural origin and purpose of life, and from its expression in the evolution of human society, there is another major challenge to society presented in the progress of "synthetic biology"--the enhanced potential being developed to alter existing living systems. The developing techniques for the synthesis of genetic information and for inserting it into existing living systems raises the question of whether and how such an awesome possibility should be utilized to "improve" the biological species. A basic philosophical question is whether society should control such a remarkable power, whether it should be exercised by free individual choice, or whether it should be "outlawed."

I am of the view that it is impractical, if not impossible, to impose a world-wide ban on experimental inquiry. If that be so, then "outlawing" a search for knowledge, apart from the question of whether it would ever be desirable, is not a realistic solution. Putting the final power to make personal decisions for all individuals on the basis of central governmental planning and authority is conceivable, but seems to me repugnant and unlikely to provide a flexible basis for the development of society. Therefore, of the three possible alternatives suggested, I would strongly favor leaving the basic decisions of how or whether to utilize such knowledge as may be developed with the individual. The society may feel compelled to provide some regulations, but should, in my opinion, leave the basic decisions to the informed choice of individuals, so long as these individual choices do not seriously infringe on the rights of others. This does indeed still leave much room for differences, so that this question is one on which there will certainly be strong and emotional opinions of great diversity.

Received 20 January, 1972